THE DEATH OF ADAM

THE DEATH OF ADAM

*Evolution and Its Impact
on Western Thought*

JOHN C. GREENE

The Iowa State University Press, *Ames,* Iowa, U.S.A.

Second ISUP paperback printing, 1973

Third ISUP paperback printing, 1974

Fourth ISUP paperback printing, 1977

Fifth ISUP paperback printing, 1981

Sixth ISUP paperback printing, 1983

JOHN C. GREENE is professor of history at the University of Connecticut.

For

Professor Josey and Dean Akeley
My father and mother
And Ellen

PREFACE

THIS IS a book about the rise of evolutionary views of nature and the decline of static creationism in the two centuries separating Isaac Newton and Charles Darwin. The title suggests the latter side of the story. I write as a historian, not as a biologist, theologian, or philosopher. My purpose has been to describe analytically and synthetically the tremendous revolution in human thought which took place in the interval between John Ray's *The Wisdom of God Manifested in the Works of the Creation* (1691) and Charles Darwin's *Origin of Species* (1859) and *Descent of Man* (1871). I do not pretend to have covered all of the relevant developments in this period, but I have done my best to trace the leading ideas which entered into Darwin's great synthesis. I have allowed the men who accomplished this intellectual revolution to speak for themselves as much as possible. They have been portrayed, not as "pure scientists" (there are no such beings), but as flesh and blood individuals influenced by the general ideas abroad in their own age, yet working, wittingly or unwittingly, to transform them.

Since science is an indivisible whole, I have proceeded on the assumption that the breakdown of the traditional view of nature and the rise of evolutionary concepts could not be understood by concentrating on biology alone but must be studied across the whole range of the sciences, from astronomy and geology on the inorganic level to biology and anthropology on the organic level. I have argued that facts in themselves, whether fossil bones, or new stars, or specimens of anthropoid apes, were not sufficient to shake the hold of the static view of nature. They acquired revolutionary import only when interpreted in the light of an incipient counterview, the concept of nature as a law-bound system of matter in motion.

Finally, in describing the intellectual unrest and spiritual anxiety generated by the gradual dissolution of the traditional view of nature, I have called attention to the religious aspect of scientific thought. As faith in the stability and wise design of the structures of nature declined, there was a compensating effort to find in the idea of progress a new world view which would give meaning to science and direction to human history.

I am deeply grateful to Professors Crane Brinton and Arthur M. Schlesinger, Sr., of Harvard University for their loyal and understanding support during the long work of preparing this book. Others, notably Professors Charles Mowat, Erwin Ackerknecht, Donald Bucklin, Andrew Clark, Hunter Dupree, Richard Pohl, and Joseph O'Mara, have given me the benefit of their criticisms of one or more chapters. To the staff of the Iowa State University Press I owe a debt of gratitude for the interest and understanding they have shown in all the problems connected with the publication of this book. Mrs. Rowena James has been of invaluable assistance in helping to fashion the narrative into a coherent and readable whole.

This book would have been impossible without the assistance afforded by the system of interlibrary loans of rare books. The librarians at the Chicago, Wisconsin, and Iowa State Universities have been most helpful in procuring needed materials.

My appreciation goes also to Mr. Louis Facto, photographer of the Agricultural Experiment Station, Iowa State University, who, with much care and skill, prepared the photographs for this book. My sincere thanks also to Professor Robert Haupt who assisted in this work and drew the map showing Buffon's idea of the geographic distribution of the quadrupeds of the world.

JOHN C. GREENE

October, 1959

TABLE OF CONTENTS

THE DEATH OF ADAM

...by the Works of Creation...I mean the Works created by God at first, and by Him conserved to this Day in the same State and Condition in which they were first made....

JOHN RAY, 1701

Give me extension and movement, and I will remake the world.

RENÉ DESCARTES, 1596–1650

Setting for Conflict

I N NOVEMBER, 1690, as the Glorious Revolution against James II and the divine right of kings settled into respectability, the Reverend John Ray took up his pen to begin work on a book he had wanted to write for a long time. At sixty-three years of age he was a writer of great reputation, but not in the field of divinity to which he had consecrated his life some thirty years earlier. A strange turn of events had diverted the intellectual talents intended for the service of God in the ministry to the study of natural history, thus producing one of the great biologists of all time.

The son of a village blacksmith, Ray had entered Cambridge University on scholarship in 1644, at a time when the University was a center of military operations in the warfare between the Puritan party and the supporters of King Charles I. Among those supporters was the King's physician, William Harvey, already famous for his treatise on the circulation of the blood. The year 1644 was to be memorable in the history of science, too, for in that year appeared Descartes' *Principles of Philosophy* and the first Latin edition of his stirring *Discourse on Method*.

Taking his bachelor's degree in 1648, Ray had continued at Cambridge as a Fellow of Trinity College, teaching Greek and Latin, studying divinity, and devoting his leisure hours to natural history and comparative anatomy. He was ordained to the ministry in 1660, but found his chosen career closed to him soon after by his refusal to make the anti-Puritan affidavit required of the English clergy by the new king, Charles II. Deprived of his Fellowship and his right to preach by this act of conscience, Ray had turned wandering natural historian, setting off with a former student, Francis Willughby, on a three-year Continental tour that was to prepare him, though he did not realize it, for a great scientific career (1).

And what a glorious time for a young man with scientific interests to visit the Continent! The Dutch cities, united in their newly won freedom from Spanish oppression, teemed with intellectual as well as commercial activity. Nowhere else in Europe was a man so free to think his own thoughts and publish them to the world. There in 1638 the condemned Galileo had published his *Observations and Mathematical Demonstrations on Two New Branches of Science.* There Descartes had promulgated his revolutionary ideas unmolested. There Spinoza lived and worked; there, too, John Locke would find refuge to write his *Essay on Toleration* and meditate his great treatise on the human understanding. Holland had her own sons, too: Christiaan Huygens, the link of genius connecting Galileo with Newton; Anthony van Leeuwenhoek and Jan Swammerdam, explorers of the wonderful new world opened up by the invention of the microscope; Franciscus Sylvius, a teacher whose fame drew medical students to Leyden from the whole of Europe; Jakob de Bondt, George Marcgrave, and Jan de Laet, naturalist voyagers to the exotic lands Holland had wrested from Portugal in the Far East and South America. In the Dutch museums Ray and his companions saw some of the natural curiosities which trade of empire had brought to Europe.

The Germanies, which the travellers next visited, had been prostrated by thirty years of warfare between Catholics and Protestants, but the young Leibniz, laden with scientific laurels won at Paris, would soon rally the German literati to form the Berlin Academy.

The glory of Italian science, dimmed though it had been by the humiliation of Galileo thirty years earlier, was by no means totally eclipsed. In Padua the English travellers witnessed dissections at the university where Vesalius had restored anatomy to the rank of an observational science a century before. At Bologna they hoped for an opportunity to see and hear the great Marcello Malpighi, founder of microscopic anatomy and physiology, but he was on leave of absence. In Florence they found the flame kindled by Galileo kept alive by the members of the Academy of Experiments: Vincenzo Viviani, Galileo's biographer and editor; Evangelista Torricelli, inventor of the barometer; Giovanni Borelli, a physicist who applied mechanical principles to the study of anatomy; Francesco Redi, whose

2

experiments on putrefied meat discredited the idea of spontaneous generation; Domenico Cassini, soon to be called to France to direct the newly established Paris Observatory. At Naples, too, they found a group of virtuosi who were active in scientific pursuits and well acquainted with the work of the Royal Society of London. Sicily and Malta were next, then back through Italy to Switzerland for a summer of botanizing in the Alps.

Thence to Montpellier, in southern France, where they found many of their countrymen engaged in medical studies, among them Martin Lister, a physician and naturalist who was to become one of Ray's closest friends. There, too, Ray met the great Danish geologist Nicolaus Steno and the French botanist Pierre Magnol, with whom he doubtless discussed the specimens he had been collecting.

Last of all came Paris, her salons astir with the doctrines of René Descartes, whose brilliant career had ended at mid-century. Soon her coteries of virtuosi would draw together under the patronage of the Grand Monarch, Louis XIV, to form the Royal Academy of Sciences, a brilliant company recruited from the whole of Europe— Christiaan Huygens from Holland; G. D. Cassini from Italy, Olaus Roemer from Denmark; Picard, Auzot, Mariotte, and Perrault from France. At the Jardin des Plantes, Ray met the professors of botany, among them Léon Marchand, "the best herbarist I met with in France."

On his return to England in the spring of 1666, Ray took his place among the remarkable group of scientific men who had banded together four years earlier to form the Royal Society of London: Robert Boyle, Robert Hooke, Sir Christopher Wren, and others. Soon they were joined by a young Fellow of Ray's old college, Isaac Newton. Offered the secretaryship of the new society in 1667, Ray declined, preferring to devote himself to the study of God's works at his leisure, scraping together a livelihood as best he could. Working at first with Willughby on their Continental and British collections, then retired with his growing family on a tiny income in a village near his birthplace; afflicted with leg sores which plagued him constantly and resisted all efforts at treatment, he poured forth a series of books which laid the foundations of systematic natural history, among them his *History of Plants, History of Fishes, Synopsis of Quadruped Animals and of Serpents,* and *Synopsis of British Plants.*

Amid the acclaim these works brought him from his scientific colleagues, Ray never forgot his original calling, never ceased longing to serve God more immediately and directly. In remaining true to his conscience he had sacrificed the opportunity to preach the divine Word from the pulpit. True, he had made amends as best he could by studying the works of God diligently and by taking every opportunity to remind his readers of their duty to do likewise. Thus, he had written in the "Preface" to his *Synopsis of British Plants*:

> I am full of gratitude to God that it was His will for me to be born in this last age when the empty sophistry that usurped the title of philosophy [science] and within my memory dominated the schools has fallen into contempt, and in its place has arisen a philosophy solidly built upon a foundation of experiment: against it elderly professors protest and struggle in vain....It is an age of noble discovery, the weight and elasticity of air, the telescope and microscope, the ceaseless circulation of the blood through veins and arteries, the lacteal glands and the bile duct, the structure of the organs of generation, and of many others — too many to mention. ...
>
> There are those who condemn the study of Experimental Philosophy as a mere inquisitiveness and denounce the passion for knowledge as a pursuit unpleasing to God, and so quench the zeal of the philosopher. As if Almighty God were jealous of the knowledge of men. ... As if He were unwilling that man should employ the intelligence which He had bestowed on the objects of which He had made it capable and which He had provided for its investigation. Those who scorn and decry knowledge should remember that it is knowledge that makes us men, superior to the animals and lower than the angels, that makes us capable of virtue and of happiness such as animals and the irrational cannot attain (2).

This was all very well, but something more was needed, a book which, instead of remarking occasionally on the proofs of divine contrivance in the great theater of nature, would be devoted from beginning to end to demonstrating the wisdom and benevolence of God. The nucleus of the book was already at hand in a series of lectures Ray had delivered at Trinity College in 1659 and 1660. Now, thirty years later, he took up his pen again and within a year produced a fervent tribute to the Creator entitled *The Wisdom of God Manifested in the Works of the Creation*. This was no ordinary piece of natural theology, for its author was an eminent naturalist, fully attuned to the scientific revolution which had reached its climax in 1687 with the publication of Isaac Newton's *Mathematical Principles of Natural Philosophy*. The concept of nature set forth in Ray's pages was to dominate the study of natural history for nearly two hundred years to come. Profoundly nonevolutionary in character, it was to constitute the chief obstacle to the rise of evolutionary views. For the ideas expressed by Ray were not mere intellectual tools, to be adopted or rejected as scientific problems required. Rather, they were presuppositions of thought, habitual modes of apprehending the world. They defined the human situation, the purpose of science, the means of knowledge, the basis of social obligation, the relations between nature, man, and God. It is important, therefore, to grasp Ray's conception of nature and, at the same time, to note a lurking anxiety in his praise of the Creator, indicating the direction from which the threat to the static world view was to come.

The works of the creation, said Ray, were "the Works created by God at first, and by Him conserved to this Day in the same State and Condition in which they were first made" (3). These words expressed the prevailing conviction of the stability of the fundamental structures of nature — stars, mountains, oceans, species, and the like. In Ray's mind there was no *scientific* problem concerning the origin of these structures. They were created by God in the beginning. If one wished to know why they had the particular shapes, patterns, and properties which they exhibited, the answer was to be sought, not in the daily operations of nature, but in the purposes which God had intended them to serve when he created them. It was precisely this adaptation of structure to function in all the productions of nature that illustrated God's wisdom and benevolence: "The admirable Contrivance of all and each of them, the adapting all the Parts of Animals to their several Uses: The Provision that is made for their Sustenance, which is often taken notice of in Scripture . . ." The sense of the permanence of the basic structures of nature carried with it a conviction of their wise design. This, in turn, implied the perfect equilibrium of nature and the subservience of the lower forms of existence to the higher. Matter existed to provide a theater for life, plants and animals to serve the uses of intelligent, moral beings whether on this planet or on others, and the whole to manifest the Creator's wisdom, power, and benevolence to his creatures, ministering to their intellectual and spiritual as well as to their physical needs.

FIG. 1.1 — John Ray (1627–1705), clergyman and naturalist, one of the founders of systematic natural history.

But what of chance and change? Were they not as real as order and stability? Ray did not deny their reality, but he assigned them a subordinate role in the economy of nature. They served to add variety to nature's panorama, but they could not alter its fundamental aspect. Change might consist in cyclical movement (as in the motion of the planets) or in random variation (as in deviations from the norm of the species) or in decline from original perfection (as in the degeneration of man), but it could never alter the blueprint of creation, either by destroying the old or by creating the new.

Such were the ideas and attitudes implicit or explicit in every chapter of Ray's great tribute to the Creator. Speaking of the "fixed stars" (the very name suggested immutability), Ray conjectured that each of them was surrounded by planets inhabited by intelligent beings. Such an arrangement was in keeping with "the Divine Greatness and Magnificence"; it exhibited the material world as a theater for the activities of intelligent beings. As for the earth, since it existed to sustain life, it "ought to be firm, and stable, and solid, and as much as is possible secured from all Ruins and Concussions." And so it was, as Copernicus had shown by revealing a system of things suited to the convenience of man and other animals. The uses of mountains had been pointed out by no less a scientist than Edmond Halley, discoverer of Halley's comet. In Halley's view, the mountain ranges had been so placed that their ridges "might serve as it were Alembicks, to distil fresh Water for the use of Man and Beast; and their heights to give a descent to those streams to run gently like so many Veins of the Macrocosm, to be the more beneficial to the Creation." To which might be added, said Ray, that these ranges were "so disposed and situated...in the mid-land parts, and in continued Chains running *East* and *West*, as to render all the Earth habitable, a great part whereof otherwise would not have been so: but the *Torrid Zone* must indeed have been such a place as the Ancients fancied it, unhabitable for heat" (4).

In the realm of animate nature the proofs of divine wisdom and contrivance were even more apparent. Observing the constancy in

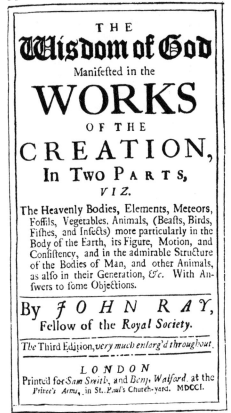

FIG. 1.2 — Title page of the third edition of Ray's famous treatise on natural theology, a classic expression of the static version of the Christian doctrine of the creation.

THE

𝕿𝖑𝖎𝖘𝖉𝖔𝖒 𝖔𝖋 𝕲𝖔𝖉

Manifested in the

WORKS

OF THE

CREATION,

In Two Parts,

V I Z.

The Heavenly Bodies, Elements, Meteors, Fossils, Vegetables. Animals, (Beasts, Birds, Fishes, and Infects) more particularly in the Body of the Earth, its Figure, Motion, and Confistency, and in the admirable Structure of the Bodies of Man, and other Animals, as also in their Generation, &c. With Answers to some Objections.

By *J O H N   R A Y*, Fellow of the *Royal Society.*

The Third Edition, *very much enlarg'd throughout.*

L O N D O N

Printed for *Sam Smith*, and *Benj. Walford.* at the *Prince's Arms*, in St. Paul's Church-yard. MDCCI.

form, figure, and situation of the essential parts of living organisms and the variability of their less essential parts, Ray was led to contrast the perfection of nature with the imperfection of man's works. "Man," he declared, "is always mending and altering his Works: But Nature observes the same tenor, because her works are so perfect, that there is no place for amendments; nothing that can be reprehended ...no blot or error in this great Volume of the World, as if any thing had been an imperfect Essay at the first...This could not have been, had Man's Body been the work of Chance, and not Counsel and Providence. Why should there constantly be the same Parts?... Nothing so contrary as Constancy and Chance" (5).

Ray's philosophy of nature was, then, a reaffirmation of the Christian doctrine of the creation, infused with the hopefulness generated by the scientific revolution of his day. But although the dominant tone of *The Wisdom of God* was confident and optimistic, there was an undertone of anxiety, an uneasy awareness that the interpretation of nature so joyfully proclaimed was not the only possible interpretation, nor that shared by all thinking men. There were, Ray acknowledged, two rival interpretations which had been advanced by "Atheistical persons." The first of these, propounded originally by Aristotle, asserted the eternity of the world in more or less its present form. The threat from this quarter seemed not to worry Ray greatly, for he dismissed it by referring his readers to Archbishop Tillotson's refutation in his *Principles of Natural Religion*. Ray's main concern was with the second heresy, namely, the atomic hypothesis of Democritus and Epicurus, according to which the universe and all of its productions had resulted from chance collisions of atoms moving at random in empty space. In its ancient, atheistic form this doctrine had been amply refuted many times, said Ray, but of late a theistic version of the same hypothesis had been advanced by Descartes and his followers on the Continent. These writers explicitly rejected the idea that mankind could understand the final causes, or purposes, for which things had been made and undertook "to solve all the *Phoenomena* of Nature, and to give an account of the Production and Efformation of the Universe, and all the corporeal Beings therein, both celestial and terrestrial, as well animate as inanimate, not excluding Animals themselves, by a slight *Hypothesis* of Matter so and so divided and mov'd" (6). But, whereas the ancient atomists had conceded that such adaptation of structure to function and such order as appeared in the universe were the outcome of a long process of trial and error, struggle for existence, and survival of the fittest, the modern "mechanick theists" held that God "had no more to do than to create the Matter, divide it into parts, and put it into motion according to some few Laws, and that would of it self produce the World and all Creatures therein." The design, if any, in this scheme of things was in the original ordaining of matter and its laws.

In refuting these "mechanick theists" Ray was embarrassed by the fact that he himself had been converted to a kind of atomism, as had his illustrious countrymen Robert Boyle and Isaac Newton. Aristotle's four elements — air, fire, earth, and water — would no longer suffice to explain the composition and qualities of inanimate bodies. The explanation must rather be sought in "the divers Figures of the minute Particles of which they are made up":

> And the reason why there is a set and constant number of them [inanimate bodies] in the World, none destroy'd, nor any new ones produc'd, I take to be, because the sum of the Figures of those minute Bodies into which Matter was at first divided, is determinate and fix'd. Because those minute parts are indivisible, not absolutely, but by any natural force; so that there neither is nor can be more or fewer of them: For were they divisible into small and diversely-figur'd parts by Fire or any other natural Agent, the Species of Nature must be confounded, some might be lost and destroy'd, but new ones would certainly be produc'd; unless we could suppose these new diminutive Particles should again assemble and marshal themselves into corpuscles of such figures as they compounded before ... (7).

But although Ray accepted the existence of ultimate particles which formed the productions of nature by their various combinations, he would not for a moment concede that these combinations could take place without the supervision of an intelligent directing agent. Nature was full of phenomena which could not be understood apart from final causes and "some vital Principle": "As for example, that of Gravity or the Tendency of Bodies downward, the Motion of the Diaphragm in Respiration, the Systole and Diastole of the Heart, which is nothing but a Muscular Constriction and Relaxation, and therefore not mechanical, but vital." The explanation of organic

9

phenomena, in particular, required the admission either of something corresponding to Aristotle's "vegetative soul" or of some kind of "Plastick Nature." "For my part," said Ray, "I should make no scruple to attribute the Formation of *Plants,* their growth and nutrition to the vegetative Soul in them; and likewise the formation of *Animals* to the vegetative power of their Souls; but that the Segments and Cuttings of some Plants, nay, the very Chips and smallest Fragments of their Body, Branches, or Roots, will grow and become perfect Plants themselves; and so the vegetative Soul, if that were the Architect, would be divisible, and consequently no spiritual or intelligent Being; which the *Plastick Principle* must be...I therefore incline to Dr. *Cudworth's* opinion, that God uses for these Effects the subordinate Ministry of some inferiour *Plastick Nature;* as in his Works of Providence he doth of Angels" (8).

For his part, the Honorable Robert Boyle, founding member of the Royal Society of London and author of a widely read attack on Aristotelian doctrines entitled *The Sceptical Chymist,* was ready to dispense with vegetative souls, substantial forms, and plastic nature and to regard all natural phenomena as products of a system of matter in motion. "What we would prove," he declared in his *Origin of Forms and Qualities,* published in 1666,

> is...that the matter of all natural bodies is a substance extended and impenetrable. That all bodies thus agreeing in the same common matter, their distinction is to be taken from these accidents which diversify it. That motion, not being essential to matter, and not originally producible by other accidents, as they are from it, may be look'd upon as the first and chief mode or affection of matter. That motion variously determined, naturally divides the matter, it belongs to, into actual fragments; and this division obvious experience manifests to have been made into parts exceedingly minute, and very often too minute to be singly perceivable by our senses. Whence it necessarily follows, that each of these minute parts, or *minima naturalia* (as well as every particular body, made up by the coalition of any number of them) must have its determinate size, and shape; and that these three, bulk, figure, and either motion or rest, are the primary and most universal modes of the insensible parts of matter, consider'd each of them apart...We would further show, that there being men in the world, whose organs of sense are so differently contrived, that one is fitted to receive

impressions from some, and another from other sorts of external objects; the perceptions of these impressions are differently express'd by words, as heat, colour, sound, odour; and are commonly imagined to proceed from certain distinct and peculiar qualities in the external object, bearing a resemblance to the ideas which their action upon the senses excited in the mind; tho' all these sensible qualities, and the rest to be met with in the bodies without us, are but the effects or consequents of the primary affections of matter, whose operations are diversified according to the nature of the organs or other bodies they affect.

That when a portion of matter, either by the access or recess of corpuscles, or by the transposition of those it consisted of before, or by any two or all of these ways, obtains a concurrence of all the qualities men commonly agree to be necessary and sufficient to denominate the body possess'd of them, a metal, a stone, or the like, and to rank it in any peculiar and determinate species of bodies; then a body of that denomination is said to be generated. And this convention of essential accidents being taken together for the specific difference that constitutes the body and discriminates it from all other forms, is by one name, because consider'd as a collective thing, called its form; which, consequently, is but a certain character, or a peculiar state of matter, or an essential modification (9).

Here was atomism with a vengeance, but Boyle was no less anxious than his colleague John Ray to avoid the charge of atheism. He was, as a matter of fact, a devout Christian and governor of a society for propagating the Gospel in His Majesty's colonies in America. But, whereas Ray sought to evade the deterministic implications of the atomic hypothesis by postulating the existence of formative forces in nature, Boyle preferred to make room for God and his purposes in the original formation of the system of matter in motion. "By embracing the corpuscular, or mechanical philosophy," he explained, "I am far from supposing, with the *Epicureans,* that, atoms accidentally meeting in an infinite vacuum, were able, of themselves, to produce a world, and all its phenomena: nor do I suppose, when God had put into the whole mass of matter, an invariable quantity of motion, he needed do no more to make the universe; the material parts being able, by their own unguided motions, to throw themselves into a regular system. The philosophy I plead for, reaches but to things purely corporeal; and distinguishing between the first origin of things and the subsequent course of nature, teaches, that God, indeed, gave motion to matter; but that, in the beginning, he so guided the various motions of the parts of it, as to contrive them into the world he design'd they should compose; and established those rules of motion, and that order amongst things corporeal, which we call the laws of nature. Thus, the universe being once form'd by God, and the laws of motion settled, and all upheld by his perpetual concourse, the general providence; the same philosophy teaches, that the phenomena of the world, are physically produced by the mechanical properties of the parts of matter; and, that they operate upon one another according to mechanical laws" (10).

Newton was of like opinion, and for similar reasons. Convinced that "God in the Beginning form'd Matter in solid, massy, hard, impenetrable, moveable Particles of such Sizes and Figures, and with such other Properties, and such Proportion to Space, as most conduced to the End for which he form'd them," he was equally certain that the same God who created the ultimate particles arranged that they should interact to form a stable habitation for the creatures who were to occupy the universe.

> For it became who created them to set them in order. And if he did so, it's unphilosophical to seek for any other Origin of the World, or to pretend that it might arise out of a Chaos by the mere laws of Nature; though being once form'd, it may continue by those laws for many Ages. For while Comets move in very excentrick Orbs in all manner of Positions, blind Fate could never make all the planets move one and the same way in Orbs concentrick some inconsiderable Irregularities excepted, which may have arisen from the mutual Actions of Comets and Planets upon one another, and which will be apt to increase, till this System wants a Reformation. Such a wonderful Uniformity in the Planetary System must be allowed the Effect of Choice (11).

So Newton, like Boyle and Ray, attempted to weld into a single philosophy of nature two not entirely compatible conceptions: one, the idea of nature as a law-bound system of matter and motion, and two, the idea of nature as a habitation created for the use and edification of intelligent beings by an omnipotent, omniscient, and benevolent God. These two conceptions dwelt side by side in men's minds in a state of tension. The doctrine of the creation postulated a stable framework of structures "created by God at first, and by Him conserved to this Day in the same State and Condition in which they were first made." But the idea of a self-contained system of matter in motion seemed to imply the transiency of all particular structures produced by the combinations and permutations of the system. It opened the door, as Newton uneasily noted, to philosophers like Descartes, "feigning Hypotheses for explaining all things mechanically." These "mechanick theists" would end by eliminating God from His works and overthrowing the chief argument for His existence: namely, the wise adaptation of the present frame of nature to the needs of living creatures, especially man.

In vain did Newton and his colleagues seek to restrict science to the study of the existing order of nature, its beauty, regularity, and wise contrivance. As the principles of the new mechanical philosophy were applied to the study of the earth's crust and its productions, man himself not excepted, there was a growing realization of the mutability of the seemingly permanent structures of nature. Mutability once assumed, a vastly extended time perspective and a sense of the relativity of human conceptions followed as attempts were made to explain nature's phenomena in terms of her daily operations. The habit of regarding the physical environment as subservient to the sentient creation then lost its hold, and the adaptation of living organisms was seen to be a matter of dire necessity rather than of wise contrivance. Chance and struggle, the antitheses of pre-established harmony and providence, claimed a share in the process of creation as biologists sought to understand the production of new varieties and species. Throughout this entire process of discovery the search for workable scientific explanations was colored and often confused by a sustained effort to save the great doctrines of revelation and creation which had oriented Western man to his universe for centuries and had infused science itself with meaning and purpose. To describe the successive stages of the great adventure of the human intellect which led from Boyle's *Origin of Forms and Qualities* to Darwin's *Origin of Species* and from John Ray's *Wisdom of God Manifested in the Works of the Creation* to Herbert Spencer's *First Principles* is the task of the ensuing chapters.

1 3

How? cry'd the Countess, *can Suns be put out?*.

Yes, without doubt, said I, *for People some thousand years ago saw fix'd Stars in the Sky, which are now no more to be seen; these were suns which have lost their Light, & certainly there must be a strange desolation in their Vortexes...*

You make me tremble, reply'd the Countess...

Oh Madam, said I, *there is a great deal of time required to ruine a World.*

Grant it, said she, *yet 'tis but time that is required.*

I confess it, said I; *all this immense mass of Matter that composes the Universe, is in perpetual motion, no part of it excepted, and since every part is moved, you may be sure that changes must happen sooner or later; but still in times proportioned to the Effect. The Ancients were pleasant Gentlemen, to imagine that the celestial Bodies were in their nature unchangeable, because they observed no change in them; but they did not live long enough to confirm their opinion by their own experience; they were Boys in comparison of us.*

Truly, said the Countess, *I find the Worlds are... like the Roses themselves, which blow one day, and die the next: For now I understand, that if old Stars disappear, new ones will come in their room, because every species must preserve itself.*

No species, Madam, said I, *can totally perish;... why may not that matter which is proper to make a Sun, be dispers'd here & there, and gather it self again at long run, into one certain place, and lay the foundation of a New World? I am very much inclin'd to believe such new productions, because they suit with that glorious and admirable idea which I have of the works of Nature.*

FONTENELLE, Plurality of Worlds, 1686

...for the sun and stars are born not, neither do they decay, but are eternal and divine.

ARISTOTLE, 384–322 B.C.

2.

The Inconstant Heavens

FROM THE TIME of Plato and Aristotle the heavens had been a symbol of perfection and stability: "for the sun and stars are born not, neither do they decay, but are eternal and divine." For nearly two thousand years this symbol held unchallenged sway in men's minds. Then, beginning in the sixteenth century a swift succession of new discoveries weakened its power. First, Copernicus made the earth a planet. Then Tycho Brahe saw with astonishment a new star in Cassiopeia's Chair, shining brighter than Venus. Soon after this, Galileo used his "optic reed" to shatter forever the illusion of celestial incorruptibility and immutability. Were there not craters on the moon and spots on the face of the sun? Did not those spots appear and disappear in a most surprising fashion? If so, was not change as natural to heavenly as to earthly bodies? So thought Galileo. To him it seemed evident "that none of the conditions, whereby Aristotle distinguisheth the Coelestial Bodies from Elementary, hath other foundation than what he deduceth from the diversity of the local motion of those and these; insomuch that it being denied, that the circular motion is peculiar to Coelestial Bodies, and affirmed, that it is agreeable to all Bodies naturally moveable, it is behooful upon necessary consequence to say, either that the attributes of generable, or ingenerable, alterable, or inalterable, partable, or impartable, &c, equally and commonly agree with all worldly bodies, namely, as well to the Coelestial as to the Elementary; or that Aristotle hath badly and erroneously deduced those from the circular motion, which he hath assigned to Coelestial Bodies." "I cannot," he added, "hear it to be at-

15

tributed to natural bodies, for a great honour and perfection that they are impassible, immutable, inalterable, &c....It is my opinion that the Earth is very noble and admirable, by reason of so many and so different alterations, mutations, generations, &c. which are incessantly made therein; and if without being subject to any alteration, it had been all one vast heap of sand, or a masse of Jasper, . . . wherein nothing had ever grown, altered, or changed, I should have esteemed it a lump of no benefit to the World, full of idlenesse, and in a word superfluous, and as if it had never been in nature; and should make the same difference in it, as between a living and a dead creature: The like I say of the Moon, Jupiter, and all other Globes of the World" (1).

Galileo's argument was to have profound consequences for the Western world view. Simplicius, spokesman for Aristotle in Galileo's *Dialogues on the Two World Systems,* might well protest it tended to "the subversion of all Natural Philosophy, and to the disorder and subversion of Heaven and Earth, and the whole Universe." For it implied that the present frame of nature might have a long and turbulent history and that the key to that history was to be sought in the laws governing the motions of matter — the very laws which Galileo had set out to discover by studying the motions of projectiles and falling bodies. Galileo had concluded from these studies that motion was as natural to bodies as rest; in his view, forces acted to alter rather than to maintain the motions of bodies. Intoxicated by this new concept, he had gone on to suggest that God, in the beginning, had created each planet at a certain distance from the orbit in which it was destined to revolve, so that in falling through that distance it should acquire the speed at which it was henceforth to revolve.

But who was to say where "the beginning" was? Must God begin with the sun and planets already created? Would it not suffice for Him to create matter, endow it with a fixed quantity of motion, and subject it to regular laws, leaving it to the operation of those laws to produce the present world? "Give me extension and movement, and I will remake the world," wrote Descartes, and proceeded to do so in his theory of celestial vortexes, published in 1644, two years after Galileo's death. A new way had been opened. From the theory of vortexes it was but a step to Bernard Fontenelle's *Conversations on the Plurality of Worlds,* to the Reverend Dr. Burnet's *Sacred Theory of the Earth,* and a thousand other speculations concerning the history of the earth and the cosmos. "Mechanick theism" was loose in the world. Newton would make short work of Descartes' theory of vortexes in his exposition of the laws of motion, but the general idea that worlds might be generated by the operations of the system of matter in motion was not to be banished. In another century and a half it would triumph over Newtonian creationism.

► **MOSES AND NEWTON**

By the time of the second edition of the *Principia* (1713) Newton sensed the danger that his own laws of motion might be used to spin theories of the origin of the earth and the solar system, as his protest against "feigning hypotheses" showed. His apprehension was well founded. Within a decade of its publication the axioms of the *Principia* were being used in exactly this way. William Whiston, the learned young divine who was to succeed Newton in the Lucasian chair at Cambridge, dedicated to him in 1696 *A New Theory of the Earth, from Its Original, to the Consummation of All Things, Wherein the Creation of the World in Six Days, the Universal Deluge, and the General Conflagration, As Laid Down in the Holy Scriptures, Are Shewn to Be Perfectly Agreeable to Reason and Philosophy* (2). This work purported to

William Whiston,
1667 • 1752

prove: first, that the earth had been formed from a nebulous comet; second, that Noah's Flood had resulted from the near approach of a comet; and third, that the terrestrial world would eventually be reconstituted in a conflagration ignited by still another comet. He would show how the events narrated in Scripture might have been produced by the operation of natural laws; then the testimony of sacred and profane history would be cited to prove the hypothetical account a true one. Science and Scripture would serve to confirm each other, the one providing principles of explanation, the other historical data.

The idea that a comet might have caused the Deluge was not original with Whiston. In December, 1694, a year and a half before he published his *Theory,* the noted astronomer Edmond Halley had propounded a similar hypothesis before the Royal Society but had refrained from publishing it "lest by some unguarded Expression he might incur the Censure of the Sacred Order" (3). Whiston rushed in where Halley feared to tread, but he was careful to preface his book with an explanation of the proper mode of interpreting the Bible. Those who interpreted the Bible literally, he declared, had discredited Scripture by construing it always in the vulgar sense, no matter how inconsistent the result with science and common observation. At the other extreme, equal damage had been done by theorists like the Reverend Thomas Burnet, who, finding their theories at variance with the plain language of Genesis, resorted to allegorical interpretations. True wisdom would dictate a middle course: "That we never forsake the plain, obvious, easie and natural sense, unless where the nature of the thing itself, parallel places, or evident reason, afford a solid and sufficient ground for so doing" (4). In most cases, said Whiston, careful study of all the relevant texts and circumstances would reveal an interpretation consistent with *both* reason and Scripture. But if in some cases it should prove impossible to reconcile the sacred text with reason and observation without forcing or twisting its meaning, there was nothing to do but wait for more light, trusting that no ultimate conflict could exist between the truth manifested in God's works and that revealed in His Word.

Applying these canons of interpretation to the Biblical account of the creation, Whiston found that all the difficulties and obscurities traditionally associated with it vanished once its intended scope was properly understood. "The Mosaic Creation," he declared, "is not a Nice and Philosophical account of the Origin of All Things; but an Historical and True Representation of the Formation of our single Earth out of a confused Chaos, and of the successive and visible changes thereof each day, till it became the habitation of Mankind" (5). The Hebrew words for *make, create, heavens, world,* and the like were used in various senses in Scripture and hence might be given a broad or narrow construction as the circumstances required. Moreover, Moses had no reason to present a full account of the origin of the universe; a passing reference to the creation of all things, such as that contained in the first verse of Genesis, was quite adequate for his purposes. Indeed, no one would have assumed that the narrative which followed applied beyond the earth and its atmosphere had not an erroneous natural philosophy prevailed in antiquity. The earth was then thought to be the center around which the planets and fixed stars revolved in crystalline spheres. So long as this idea persisted, it was natural to identify earth and cosmic history.

> But though such a Scheme, and such an Apprehension were passable enough in the days of our Forefathers, 'tis by no means so now... 'Tis now evident, That every one of the Planets, as well as that on which we live, must have a right in its proportion to share in the care of Heaven, and had therefore in all probability a suitable space or number of Days allow'd to its proper Formation; much what the same Separations of Parts, Digestions, and Collections, being no doubt suppos'd in the Original Formation of any other, as in that particular Planet, with which *Moses* was concern'd. And if one or two on account of their smallness, might be finish'd in less; the rest on account of their bigness, from a parity of Reason, would take up much more than that six days time which was spent in our Earth's Formation (6).

Having restricted the Mosaic narrative to the earth and its atmosphere, Whiston found little difficulty in showing how the various phenomena connected with comets could account satisfactorily for the formation of the earth, its inundation, and its eventual reconstitution by fire. Suppose, he said, that a comet were diverted from its eccentric orbit into a concentric one. The vapors in its tail would settle down around the burning core according to their specific gravities, forming successive layers of heavy fluid, water, and earth. In the event that the newly formed body had no daily rotation but only an annual revolution around the sun, a day and a year would be one and the same and a perpetual equinox would prevail, providing ideal conditions for the support of a large and flourishing population. The sun and planets would rise in the west and set in the east. But these were exactly the conditions which an unbiased reader of Scripture and of classical authors would discover to have prevailed on the earth before the Fall of Man.

The Deluge, Whiston continued, might easily have resulted from a comet's approach to the newly formed planet, swamping it with great masses of water the weight of which would expel the underground waters of the planet from their caverns, augmenting the general flood. Such a comet would probably alter the planet's orbit from a circular to a slightly eccentric one. That this had actually happened to the earth was apparent from the fact that the eccentricity of the earth's orbit equals exactly the difference between the solar and the lunar year, a coincidence scarcely to be explained except on the supposition that the difference was produced by a change in the once circular orbit of the earth. It could even be proved, said Whiston, that on the day assigned by the best chronologers for the occurrence of the Deluge the earth occupied the very place in its orbit required by the hypothesis in question.

> The very day of the Comets passing by, or of the beginning of the Deluge determin'd from the Astronomical Tables of the Conjunctions of the *Sun* and *Moon,* is exactly coincident with that before nearly determin'd by the place of the *Perihelion,* and *exactly* by the *Mosaick* History. It has been before prov'd, that seeing the *Moon* still accompanies the Earth, it must needs have been three Days past the New or Full, at the passing by of the Comet. It has also been before prov'd, that the Flood began in the Year of the Julian Period 2365, or the 2349th before the Christian *Aera.* Now it appears by the Astronomical Tables of the Conjunctions of the *Sun* and *Moon,* that the mean New Moon happen'd at the Meridian of *Babylon* just before Eleven a Clock in the Forenoon, on the 24th day of November, (in the Julian Year) and so at Eleven a Clock on the 27th of November, 'twas three days after the New. Which being the 17th day of the Second Month, from the Autumnal Equinox, is the very same pitched upon from the place of the *Perihelion,* and expressly mention'd in the Sacred History: And by so wonderfully corresponding therewith, gives the highest Attestation to our *Hypothesis* that could, for the completion and consummation of the foregoing Evidence, be reasonably desir'd (7).

These hypotheses and proofs, Whiston admitted, would seem farfetched to those accustomed to regard the present order of things as natural and permanent. But the assumption of an unvarying earth had no warrant in either Scripture or reason. If, as revelation made clear, the earth was contrived as a stage for the drama of human probation, it must have undergone changes designed to adapt it to the altered condition of human nature. Thus, the commencement of the earth's daily rotation at the Fall of Man and the reconstitution of its surface at the Deluge had destroyed the paradisiacal state of nature and substituted a habitation fit for fallen creatures. Perhaps on other

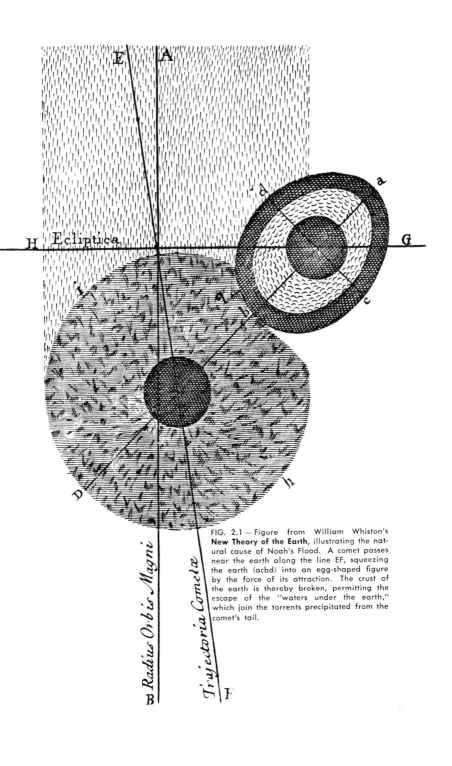

FIG. 2.1 — Figure from William Whiston's **New Theory of the Earth,** illustrating the natural cause of Noah's Flood. A comet passes near the earth along the line EF, squeezing the earth (acbd) into an egg-shaped figure by the force of its attraction. The crust of the earth is thereby broken, permitting the escape of the "waters under the earth," which join the torrents precipitated from the comet's tail.

planets, too, irregularity and eccentricity had entered into nature concomitantly with sin.

> I do by no means question but these uncertain *Eccentricities* and various Positions of the *Aphelia* of the Planets...happen'd by a particular Providence, and were all one way or other fitted to the state of each Species of Creatures Inhabiting the several Planets...But my meaning is this: That before any good or bad actions of Creatures, when every thing was just as the Wisdom of God was pleas'd to appoint; when each Creature was compleat and perfect in its kind, and so suited to the most compleat and perfect state of external Nature; 'tis highly probable that the outward World...was even, uniform, and regular, as was the temper and disposition of each Creature that was to be plac'd therein...Such a state, 'tis natural to believe, obtain'd through the Universe till succeeding changes in the Living and Rational, requir'd proportionable ones in the Inanimate and Corporeal World. 'Tis most Philosophical, as well as most Pious, to ascribe only what appears wise, regular, uniform, and harmonious, to the First Cause; (as the main *Phaenomena* of the Heavenly Bodies, their Places, and Motions, do, to the degree of wonder and surprize) but as to such things as may seem of another nature, to attribute them intirely to subsequent changes, which the mutual actions of Bodies upon one another, foreordain'd and adjusted by the Divine Providence, in various Periods, agreeably to the various exigencies of Creatures, might bring to pass (8).

One further objection remained to be considered. If the Deluge was produced by mechanical causes, how could it be regarded as a divine punishment for human wickedness? Whiston was of two minds on this point. On the one hand, he conceded that some links in the chain of events which produced the Deluge seemed more evidently to require divine interposition than others. In general, however, he preferred to think that God had prearranged the whole series of events so as to make the punishment fit the crime both judicially and chronologically. Surely, it was more consonant with God's wisdom and power to suppose that He had foreseen and synchronized the course of human and natural events than to suppose Him forced to intervene miraculously on the spur of the moment .

Whiston's book was well received and went through many editions. The great philosopher John Locke thought the author "more to be admired that he has laid down an Hypothesis whereby he has explained so many wonderful, and before inexplicable Things in the great Changes of this Globe, than that some of them should not easily go down with some Men; when the whole was intirely new to all" (9). Newton himself "well approved" of the work, if we can believe Whiston. It seems probable that we can, judging from Newton's long letter to the Reverend Thomas Burnet concerning that writer's *Sacred Theory of the Earth,* the book which inspired Whiston to attempt a theory of his own (see below, page 39).

Of our present sea, rocks, mountains, &c., I think you have given the most plausible account [wrote Newton]. And yet if one would go about to explain it otherwise, philosophically, he might say that as saltpetre dissolved in water, though the solution be uniform, crystallizes not all over the vessel alike, but here and there in long barrs of salt; so the limus of the chaos, or some substances in it, might coagulate at first, not all over the earth alike, but here and there in veins or beds of divers sorts of stones and minerals. That in other places which remained yet soft, the air which in some measure subsided out of the superior regions of the chaos, together with the earth or limus by degrees extricating itself gave liberty to the limus to shrink and subside, and leave the first coagulated places standing up like hills; which subsiding would be encreased by the draining and drying of that limus. That the veins and tracts of limus in the bowels of those mountains also drying and consequently shrinking, crackt and left many cavities, some dry, others filled with water. That after the upper crust of the earth by the heat of the sun, together with that caus'd by action of minerals had hardened and set; the earth in the lower regions still going closer together left large caverns between it, and the upper crust filled with the water, which upon subsiding by its weight, it spread out by degrees till it had done shrinking, which caverns or subterraneal seas might be the great deep of Moses, and if you will, it may be supposed one great orb of water between the upper crust or gyrus and the lower earth, though perhaps not a very regular one. That in process of time many exhalations were gather'd in those caverns which would have expanded themselves into 40 or 50 times the room they lay in, or more, had they been at liberty.... That at length somewhere forcing a breach, they by expanding themselves, forced out vast quantities of water before they could all get out themselves, which commotion caused tempests in the air, and thereby great falls of rain in spouts, and all together made the flood, and after the vapours were out, the waters retired into their former place. . . .

As to Moses [Newton went on] I do not think his description of the creation either philosophical or feigned, but that he described realities in a language artificially adapted to the sense of the vulgar. . . . To describe them distinctly as they were in themselves, would have made the narrative tedious and confused, amused the vulgar, and become a philosopher more than a prophet. . . . Consider, therefore, whether any one who understood the process of the creation, and designed to accommodate to the vulgar not an ideal or poetical, but a true description of it as succinctly and theologically as Moses has done, without omitting any thing material which the vulgar have a notion of, or describing any being further than the vulgar have a notion of it, could mend that description which Moses has given us (10).

Thus the great Newton, despite his own admonitions, could not resist "feigning hypotheses" about the history of the world, though he was careful to say that these were mere conjectures, not conclusions which he would be willing to defend.

► A COMET SIDESWIPES THE SUN

A half century later, however, Georges Louis Leclerc, Comte de Buffon, took a different view of the matter. A true son of the French Enlightenment, Buffon was named to the Royal Academy of Sciences in Paris at the age of twenty-six. He subsequently traveled to England and there, like his contemporary Voltaire, was filled with admiration for Newtonian science. On returning to France, he published a translation of Newton's treatise on the calculus and another of the Reverend Stephen Hales' *Vegetable Staticks,* a pioneer work in plant physiology. In 1739 he was appointed Intendant of the Royal Garden and Keeper of the Royal Cabinet of Natural History. It was then he set out to write a natural history which would ultimately embrace the whole range of nature's productions from minerals to man — an undertaking that was to fill thirty-six volumes before his death in 1788.

Comte de Buffon,
1707 · 1788

"Mr. Whiston," Buffon declared in the "Theory of the Earth" which opened his *Natural History, General and Particular,* "neither doubted of the truth of the deluge, nor of the authenticity of the sacred writings. But, as physics and astronomy occupied his principal attention, he mistook passages of Holy Writ for physical facts, and for results of astronomical observations; and so strangely jumbled divinity with human science, that he has given birth to the most extraordinary system that perhaps ever did or ever will appear" (11). The true naturalist, declared Buffon, will leave the interpretation of Scripture to theologians and confine himself to probable hypotheses based on accurate observation of nature.

Buffon did not stop at excluding the Bible from the domain of natural history. He went on to banish natural theology as well. He agreed with Newton that the revolution of the planets and their satellites around the sun in one and the same direction and in approximately the same plane could scarcely be accidental, but he favored seeking a natural cause of these uniformities instead of ascribing them to God. Might not the solar system have derived its origin, arrangement, and motions from a collision between a comet and the sun?

> The comet, by falling obliquely on the sun...must have forced off from his surface a quantity of matter equal to a 650th part of his body. This matter, being in a liquid state, would at first form a torrent, of which the largest and rarest parts would fly to the greatest distances; the smaller and more dense, having received only an equal impulse, would remain nearer the sun; his power of attraction of the particles of matter would cause all the detached parts to assume the form of globes, at different distances from the sun, the nearer moving with greater rapidity in their orbits than the more remote....But the obliquity of the stroke might be so great as to throw off small quantities of matter from the principal planet, which would necessarily move in the same direction. These parts, by mutual attraction, would reunite, according to their densities, at different distances from the planet, follow its course around the sun, and at the same time revolve about the body of the planet, nearly in the plane of its orbit. ...Thus the formation, position, and motion of the satellites correspond, in the most perfect manner with our theory; for they all move in the same direction, and in concentric circles round their principal planets, and in the plane of their orbits (12).

Buffon defended his theory against various scientific objections but made little attempt to forestall Scriptural critics, except for a

25

half-hearted suggestion that the shearing off of the planets from the sun must have produced the separation of light from darkness described in Genesis. This gesture proved totally inadequate. Buffon was required to retract his theory of planetary origins before he was permitted to proceed with the publication of his *Natural History*, but he was to return to it in his *Epochs of Nature* a quarter of a century later (13).

▶ WORLDS WITHOUT END

Buffon's bold hypothesis was soon surpassed by still more audacious speculations. In 1750 Thomas Wright of Durham, an eccentric genius who made his living by tutoring the sons and daughters of English noblemen in natural philosophy, published a handsomely illustrated *Original Theory or New Hypothesis of the Universe* (14), in which he suggested that the so-called fixed stars constituted a system of bodies in motion around some

Thomas Wright, 1711 · 1786

undiscovered central globe. That some stars had a proper motion had been shown, said Wright, by Halley's comparison of the present positions of Sirius and Arcturus with the positions ascribed to them by the ancients. But if these stars were in motion, then why not others — why not all the stars, including the sun?

> 'Tis true, the Sun may be said to be the Governor of all those bodies round him; but how? no otherwise than he himself may be governed by a superior Agent, or a still more active Force; and methinks it is not a little absurd to suppose he is not, since we have discovered by undoubted Observations, that the same gravitating Power is common to all; and that the Stars themselves are subject to no other Direction than that which moves the whole Machine of Nature (15).

But if the sun and other stars were in motion, Wright continued, could there be any doubt that their movements displayed a harmony comparable to that discerned by Newton in the motions of the solar system? Surely God would not make part of His creation regular and leave the rest in confusion. True, the stars seemed to be strewn across the heavens carelessly, but might not this appearance of disorder be an illusion arising from man's local and eccentric point of view in the universe? If, for example, the stars were all in motion about a central body, all moving in the same direction and in approximately the same

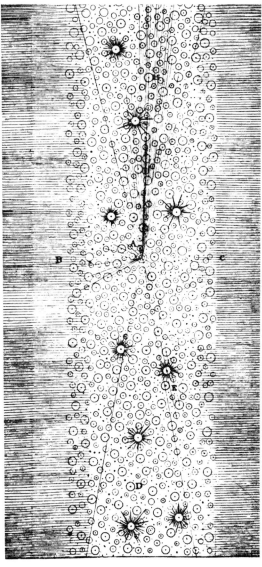

FIG. 2.2 — The first scientifically important explanation of why we see the Milky Way, from Thomas Wright's **Original Theory or New Hypothesis of the Universe.** Wright imagines all of the stars, including our sun, to be revolving in approximately the same plane about a central globe, thus forming a thin disk, or wheel, of stars. Hence, an observer at A, looking through the disk, will see innumerable stars crowding on each other (the Milky Way) but will see many fewer stars if he looks across the disk toward B.

plane (as the planets move around the sun), would not a person on a planet circling a star located somewhere near the center of this great stellar wheel see the stars stretched out in a narrow band across the vault of the heavens, with but a scattering of stars on each side of the band? And was this not exactly what appeared to observers on the earth? Was it not likely, then, that the so-called Milky Way was a vast assemblage of separate stars moving in perfect order around a central Body? And was that central globe not the logical place in the universe for the immortal soul of man to find eternal rest?

27

Here and here alone centered in the Realms of inexpress-
ible Glory, we justly may imagine that primogenial Globe or
Sphere of all perfections, subject to Extreams of neither Cold
nor Heat, of eternal Temperance and Duration. Here we may
not irrationally suppose the Vertues of the meritorious are
at last rewarded and received into the full Possession of every
Happiness, and to perfect Joy (16).

Wright was a prophet without honor in his own country, but
his ideas bore fruit on the Continent, where he was not entirely unknown, having declined the offer of a professorship at the Imperial
Academy of St. Petersburg eight years earlier. By a happy chance a
brief account of Wright's ideas in a German journal caught the eye
of Immanuel Kant, the Königsberg professor whose writings were

**Immanuel Kant,
1724 • 1804**

soon to turn the philosophical world
upside down. Kant was interested in
mathematical physics, geography, and
anthropology, as well as philosophy.
His brilliant scientific imagination transformed Wright's speculations
into a general theory of cosmic evolution, set forth in 1755 in a
work entitled *Universal Natural History and Theory of the Heavens;
or an Essay on the Constitution and Mechanical Origin of the Whole
Universe Treated According to Newton's Principles* and dedicated to
Frederick the Great of Prussia. Like Descartes, Kant viewed the present order of nature as a necessary outcome of the laws governing the
motions of matter.

> I accept the matter of the whole world at the beginning as in
> a state of general dispersion, and make of it a complete chaos.
> I see this matter forming itself in accordance with the es-
> tablished laws of attraction, and modifying its movement by
> repulsion. I enjoy the pleasure, without having recourse to
> arbitrary hypotheses, of seeing a well-ordered whole produced
> under the regulation of the established laws of motion, and
> this whole looks so like that system of the world which we
> have before our eyes, that I cannot refuse to identify it with
> it (17).

To illustrate the process of world formation, Kant undertook to
trace the beginning of the solar system. Matter being dispersed evenly
throughout space in the first instance, Kant explained, the denser
elements would attract the lighter ones and thus form a nucleus

which would increase reciprocally in size and attractive power, extending its gravitational pull through an ever wider region. But, since matter exerts repulsive as well as attractive force, some particles falling toward the nucleus would be diverted by neighboring particles and thrown into circular paths around it. These revolving particles would jostle each other in their orbits until they either fell into the central body or moved harmoniously around it in parallel circles more or less in the same plane. The particles moving in orbits at approximately the same distance from the center of attraction would then draw together to form planets, which, when formed, would set in motion a similar process of satellite formation.

This hypothesis, said Kant, accounted for the known phenomena of the solar system. It explained why all the planets rotated and revolved in the same direction and in approximately the same plane. It accounted for the highly eccentric, or irregular, orbits of the comets by assuming that the attractive force of the sun was too weak in the outer reaches of the system to draw the bodies formed there into the pattern of the planets. It explained why, in general, the denser planets were nearest the sun, the lighter particles of matter experiencing greater difficulty than the heavy in making their way toward the center of attraction. This point was confirmed almost beyond doubt by the calculations of Buffon showing that the density of the sun was approximately equal to the average density of the planets, a coincidence intelligible only on the supposition that the sun was composed of the same primary matter from which the planets were formed. Thus the characteristics of the solar system were intelligible in terms of its mode of origin, while the densities of the planets, which Newton regarded as divinely contrived to enable the planets nearest the sun to endure its fierce heat, were seen to be a product of the laws appointed by God for the government of matter.

These same laws, Kant continued, must apply throughout the universe, operating to form matter everywhere into systems of bodies similar to the solar system. Thus, the Milky Way was undoubtedly, as Thomas Wright had suggested, a vast system of stars governed in their motions by the same balance of centripetal and centrifugal forces which held the planets in their courses. Beyond the Milky Way might be still other star systems, their existence revealed to man only by those faint luminous patches, or "nebulous stars," noted by various astronomers. "Their analogy with the stellar system in which we find ourselves, their shape, which is just what it ought to be according to our theory, the feebleness of their light which demands a presupposed infinite distance: all this is in perfect harmony with the view that these elliptical figures are just universes (*Welt-inseln*) and, so to speak, Milky Ways, like those whose constitution we have just unfolded." Thus, the process of world formation extended through infinite space, requiring infinite wisdom for its regulation and infinite time for its completion. "Millions and whole myriads of millions of centuries will flow on, during which always new worlds and systems of worlds will be formed...It needs nothing less than an eternity to animate the whole boundless range of the infinite extension of space with worlds, without number and without end" (18).

Here with a vengeance was the "mechanick theism" which John Ray had feared and deplored. To Kant, however, the tendency of matter "to fashion itself by a natural evolution into a more perfect constitution" seemed the strongest possible proof of the divine contrivance of the system of nature.

> Matter, which is the primitive constituent of all things, is...bound to certain laws, and when it is freely abandoned to these laws it must necessarily bring forth beautiful combinations. It has no freedom to deviate from this perfect plan. Since it is thus subject to a supremely wise purpose, it must necessarily have been put into such harmonious relationships by a First Cause ruling over it; and *there is a God, just because nature even in chaos cannot proceed otherwise than regularly and according to order* (19).

Thus, one hundred years before the Darwinian controversy Kant elaborated a doctrine of evolutionary theism. He was careful, however, to leave room for the Christian revelation. On it he rested man's hope of immortality, the hope without which the human spirit, contemplating the creation and destruction of worlds throughout infinite space and time, might well sink into despair.

► THE WHEEL COMES FULL CIRCLE

Kant's book attracted even less attention than Wright's, owing partly to the bankruptcy of his publisher and the temporary seizure of the unsold copies. Kant and Wright had held out little hope that their ideas could be confirmed by observation, but they had failed to reckon with the ingenuity of telescope makers. Before the end of the century William Herschel, a German musician turned Englishman and astronomer, had constructed several telescopes of unrivaled power and had inaugurated a systematic investigation of the structure of the heavens. Leaving Hanover to escape military service, Herschel applied himself to a musical career, securing in 1776 the post of organist at the Octagon Chapel in Bath, England. His days were devoted to music, but his nights were taken up with scientific studies, especially mathematics, optics, and astronomy. "The great run of business," he wrote to James Hutton later in life, "far from lessening my attachment to study, increased it, so that many times after a fatiguing day of 14 to 16 hours spent in my vocation, I retired at night with the greatest avidity to unbend the mind (if it may be so called) with a few propositions in Maclaurin's *Fluxions* or other books of that sort" (20).

Sir William Herschel, 1738 . 1822

Determined to see for himself the wonders described in works on astronomy, he set to work to construct a telescope, then another and another, his sister Caroline reading to him far into the night while he performed the tedious labor of grinding and polishing the mirrors. His labor was not in vain, for the telescopes he built were far more powerful than any hitherto constructed, so powerful, indeed, that the stars appeared round through them — "round as a button," as he told the astonished Lord Cavendish. Soon the Royal Society began to hear from him, and in 1781 the world was electrified by his discovery of a new planet, named by him *Georgium Sidus* in honor of King George III but subsequently called *Uranus*. A royal pension now enabled him to devote himself to astronomy. In 1783, two years after his dramatic discovery, Herschel communicated to the Royal Society some observations "On the Proper Motion of the Sun and Solar System" (21). It was now clear beyond reasonable doubt, he declared, that Arcturus, Sirius, Aldebaran, and many other "fixed" stars were in motion. But if some stars were in motion all of them must be, since they presumably attracted each other in accordance with the law of gravitation. Hence the sun must be in motion, and the direction of its movement should be an important subject of inquiry.

31

Herschel's main interest, however, was in the "interior construction" of the heavens. The astronomer, he declared in 1784, must stop thinking of the starry sky as a concave surface and regard it instead "as a naturalist regards a rich extent of ground or chain of mountains, containing strata variously inclined and directed, as well as consisting of very different materials" (22). In support of this view, Herschel described how the whitish appearances in the hand and club of Orion and in other parts of the Milky Way had been resolved by his telescope into clusters of stars. This clustering effect, he went on, could not be purely fortuitous; the odds were all against it. But suppose that the laws of attraction were operative throughout space and that the stars had once been more evenly distributed throughout the heavens. Would not the operation of attractional force have grouped them eventually into various kinds of nebulae, depending upon the sizes of the stars and their distances from each other? And would not the relative antiquity of each nebula be indicated by the degree of accumulation and condensation obtaining among its component stars? If so, the nebula called the Milky Way, containing the earth and solar system, would appear to be still in its youth. The nebulae called "planetary," on the other hand, would represent the next to the last degree of compression or clustering; in the last stage, perhaps, the component stars would unite to form a new body "either in succession, or by one general tremendous shock." Might not the blazing new star observed briefly in Cassiopeia's Chair in 1572 have been produced by such an event? (23).

It might be objected, Herschel noted, that this hypothesis provided no guarantee against the eventual destruction of the system of the universe by the collision of stars. To this it could be answered that the Author of Nature had undoubtedly provided for its maintenance, although the exact nature of the provision remained undiscovered by man. There might, for example, be projectile forces capable of counteracting the force of gravitation in star clusters, "if not forever, at least for millions of ages."

> Besides, we ought perhaps to look upon such clusters, and the destruction of now and then a star, in some thousands of ages, as perhaps the very means by which the whole is pre-

served and renewed. These clusters may be the *Laboratories* of the universe, if I may so express myself, wherein the most salutary remedies for the decay of the whole are prepared (24).

In 1791 Herschel reported some observations which had caused him to modify his conception of stellar evolution:

> November 13, 1790. A most singular phaenomenon! A star of about the 8th magnitude, with a faint luminous atmosphere, of a circular form, and of about 3' in diameter. The star is perfectly in the center, and the atmosphere is so diluted, faint, and equal throughout, that there can be no surmise of its consisting of stars; nor can there be a doubt of the evident connection between the atmosphere and the star. Another star not much less in brightness, and in the same field with the above, was perfectly free from any such appearance (25).

From these phenomena Herschel inferred the existence of diffuse luminous matter in the heavens and surmised that stars might be formed from nebulous matter as well as nebulae from the clustering of stars. In 1811 and 1814 he supported this idea further by arranging various types of nebulous appearances in a continuous series ranging from diffuse nebulosity to globular clusters of highly compressed stars. These successive states of nebulosity, he declared, must represent stages in the evolution of stars and star systems. In that case, the Milky Way must be in the process of breaking up into globular clusters. Thus the progress of clustering provided a kind of celestial timepiece, "and although we do not know the rate of going of the mysterious chronometer, it is nevertheless certain, that since the breaking up of the parts of the Milky Way affords a proof that it cannot last forever, it equally bears witness that its past duration cannot be admitted to be infinite" (26).

Despite the novelty of his discoveries and ideas, Herschel seems never to have wavered in his sanguine conviction that the universe was divinely contrived as a theater for the activities of intelligent creatures. When, in 1802, Napoleon took exception in Herschel's presence to the boastful claim of the great French mathematician Pierre-Simon Laplace that he could dispense with God in accounting for the arrangement and stability of the solar system, Herschel recorded in his diary his own conclusion that the argument, when pushed to its conclusion, must lead "to 'Nature and nature's God.'" Not only did he suppose the moon and the planets to be inhabited, but his theory that solar heat was generated in the atmosphere of the sun, rather than in its body, led him to propose that the sun and stars might also be habitable worlds. In some star clusters, he argued, there may not be room for the planets which are presumed to encircle the stars. Hence the stars in these clusters must be habitable themselves, "unless we would make them mere useless brilliant points" (27).

To most of Herschel's contemporaries, however, his suggestion that the seemingly unchanging heavens were subject to ceaseless change and evolution was highly disturbing. In 1790 the English journal *Monthly Review* hinted that Herschel's "ebullitions" were much less welcome than his empirical discoveries and that the Royal Society would do well to discourage unfounded speculation of the kind contained in Herschel's early papers (28). Some of Herschel's friends, afraid for his reputation, urged him to make a greater display of his mathematical talents, presumably to offset the unfavorable impression created by his novel ideas. In Geneva, Marc Auguste Pictet, Director of the Geneva Observatory, openly denounced Herschel's nebular hypothesis as impious.

FIG. 2.3 — Various types of nebulosity and star clusters observed by Sir William Herschel and figured by him in the **Philosophical Transactions** of the Royal Society (1814). Fig. 8 shows the appearance observed by Herschel in 1791, which first gave him the idea that stars might be formed from diffuse luminous matter in space. Previously he had explained nebulous appearances as resulting from dense clustering of stars, as in Fig. 17.

Fig. 1.   Fig. 2.   Fig. 3.    Fig. 4.

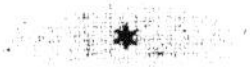

Fig. 5.   Fig. 6.   Fig. 7.   Fig. 8.

Fig. 9.    Fig. 10.    Fig. 11.

Fig. 12.   Fig. 13. 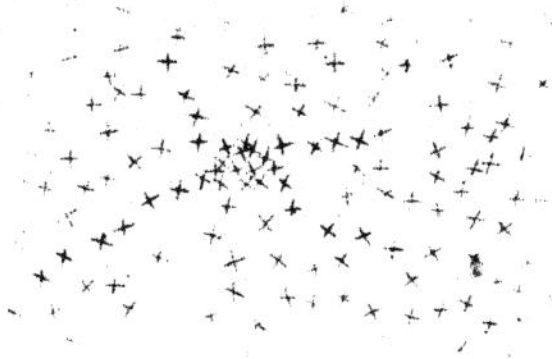   Fig. 14.

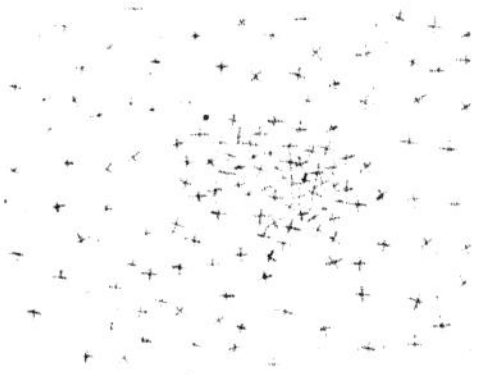

Fig. 15.    Fig. 16. 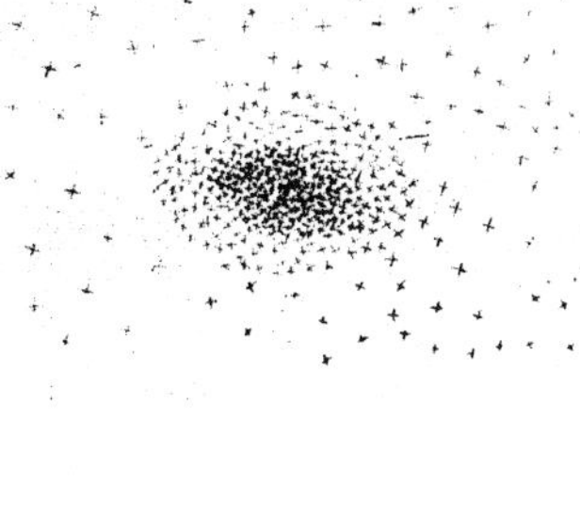   Fig. 17.

In France, however, Herschel found a staunch adherent in Pierre-Simon Laplace, the former peasant boy whose mathematical investi-

Pierre-Simon Laplace, 1749 · 1827

gations, combined with those of Joseph Louis Lagrange, were unraveling such knotty problems in celestial mechanics as the acceleration of the moon's mean motion, the libration of the moon, and the acceleration and retardation of the motions of Jupiter and Saturn with respect to each other. In his popular *Exposition of the System of the World,* first published in 1796, Laplace used Herschel's idea of nebular condensation to solve the problem of the origin of the solar system. Beginning by supposing that the sun's atmosphere had once extended far beyond the orbits of the present planets, he tried to show how known laws would operate to form this atmosphere into concentric rings and these rings into planets accompanied by satellites (29). This hypothesis, expounded very briefly and tentatively in the last chapter of the *Exposition,* attracted little attention at first. Its author was much better known for his prodigious *Celestial Mechanics,* in which a century of mathematical analysis of the solar system was summed up and extended. His demonstration of the stability of the solar system was hailed as proof positive of its divine origin, but Laplace himself drew no such conclusion. For him, as for Buffon, the uniformities observable in the arrangement and motions of the planets were a challenge to scientific explanation, not a proof of divine handiwork. In a note to the fifth edition of his *Exposition,* Laplace took issue with Newton on this subject:

> But could not this arrangement of the planets be itself an effect of the laws of motion; and could not the supreme intelligence which Newton makes to interfere, make it to depend on a more general phenomenon? such as, according to us, a nebulous matter distributed in various masses throughout the immensity of the heavens. Can one even affirm that the preservation of the planetary system entered into the views of the Author of Nature? The mutual attraction of the bodies of this system cannot alter its stability, as Newton supposes; but may there not be in the heavenly regions another fluid besides light? Its resistance, and the diminution of its emission produced in the mass of the Sun, ought at length to destroy the arrangement of the planets, so that to

maintain this, a renovation would become evidently necessary. And do not all those species of animals which are extinct, but whose existence Cuvier has ascertained with such singular sagacity, and also the organization in the numerous fossil bones which he has described, indicate a tendency to change in things, which are apparently the most permanent in their nature? The magnitude and importance of the solar system ought not to except it from this general law; for they are relative to our smallness, and this system, extensive as it appears to be, is but an insensible point in the universe. If we trace the history of the progress of the human mind, and of its errors, we shall observe final causes perpetually receding, according as the boundaries of our knowledge are extended. These causes, which Newton transported to the limits of the solar system, were, in his time, placed in the atmosphere in order to explain the cause of meteors: in the view of the philosopher, they are therefore only an expression of our ignorance of the true causes (30).

The wheel had come full circle. The sense of the permanency of the structures of nature was gone and with it the belief in final causes and the argument from design. Every state of the system of matter in motion appeared to flow irresistibly from its predecessor; none could be regarded as the first, none as the last. Every scientific law was presumed to be a particular consequence of still more general principles of explanation. God was at the end of an infinitely long succession of events and an infinitely long chain of reasoning. Mankind was adrift on a tiny globe in a sea of space and time, with no knowledge of its origin and destiny except what science could provide. A sobering prospect, Laplace conceded, yet not without its compensations. For science, though it destroys man's illusions, consoles him by ministering to his needs and banishing his superstitious terrors. "Let us," he concluded, "carefully preserve and augment the accumulation of these sublime discoveries, the delight of thinking beings" (31).

...*the* Terraqueous Globe *is to* this Day *nearly in the same* Condition *that the* Universal Deluge *left it; being also like to* continue *so till the Time of its final* Ruin *and* Dissolution, *preserved to the* same End *for which 'twas first* formed.

JOHN WOODWARD, 1695

...*we find no vestige of a beginning, — no prospect of an end.*

JAMES HUTTON, 1788

A Wreck of a World

W<small>HILE ASTRONOMERS GROPED</small> their way toward a new concept of the heavens, a similar revolution was taking place in ideas about the earth. Evidences of change in the heavens were few in number and difficult to interpret, but changes on the earth's surface — volcanic eruptions, earthquakes, floods, landslides, and the like — were fairly common and easy to observe. It was, as Franklin observed, "the wreck of a world we live on," and the wreckage seemed to invite explanation by all and sundry. The success of Newtonian physics gave a great impetus to observation and speculation in other fields of science. It stimulated the search for "useful knowledge" and held out the hope that all of the phenomena of nature could eventually be derived from the operation of a few fundamental laws. The application of mechanical principles to explain terrestrial phenomena produced the "theory of the earth," a favorite form of speculation among natural historians for more than a century. Descartes himself led the way by attempting to show how physical principles could account not only for the genesis of the solar system but also for the stratification and irregular contours of the earth's surface. Thus cosmogony and geology were twin offspring of the new physics.

► "DIRTY LITTLE PLANET"

In England, the general belief in the complete inspiration of Scripture strongly colored and conditioned geological speculation. This was notably true of the work of the Reverend Thomas Burnet, a prominent churchman under Charles II. Like Descartes, Burnet delighted "to take in Pieces this Frame of Nature, and melt it down into its first Principles; and then to observe how the Divine Wisdom wrought all these Things out of Confusion into Order, and out of Simplicity into that beautiful Composition we now see them in" (1). To this end he produced in 1681 his *Sacred Theory of the Earth: Containing an Account of Its Original Creation, and of All the General Changes Which It Hath Undergone, Or Is to Undergo, until the Consummation of All Things.* As the title implied, Burnet did not propose

Thomas Burnet, 1635? • 1715

to rely solely on reason and observation to decipher the globe's history. These would play their part, but they would be supplemented by the testimony of sacred and profane history, especially the former, since the Bible was obviously intended to provide man with clues to his past history and future destiny.

> It seems to me very reasonable to believe [Burnet explained] that besides the Precepts of Religion, which are the principal Subject and Design of the Books of Holy Scripture, there may be providentially conserved in them the Memory of Things and Times so remote, as could not be retrieved, either by History, or by the Light of Nature; and yet were of great Importance to be known, both for their own Excellency, and also to rectify the Knowledge of Men in other Things consequential to them: Such Points may be, *Our great Epocha,* or the Age of the Earth, the Origination of Mankind, the First and Paradisiacal State, the Destruction of the old World by an Universal Deluge, the Longevity of its Inhabitants, the manner of their Preservation, and of their Peopling the second Earth; and lastly, the Fate and Changes it is to undergo. These I have always look'd upon as the Seeds of great Knowledge, or Heads of Theories fix'd on Purpose to give us Aim and Direction how to pursue the rest that depend upon them (2).

Thus Burnet derived from the Bible two basic presuppositions: that the earth was created by an omnipotent, omniscient, and benevolent God as a theater for human probation and redemption, and that the main events of earth history were the creation, the Deluge, and the final conflagration. Taking these things for granted, science would confirm them and show *how* the great terrestrial changes had been or were to be accomplished.

The Deluge provided a convenient starting point in this undertaking, since it was an event recorded in the Bible. Besides, it was attested independently by a great number of observable phenomena on the surface of the earth. It was obvious, said Burnet, that the earth had undergone a great transformation since the creation. Surely God would not have permitted the "dirty little planet" which we now behold — a senseless jumble of deserts, overturned rocks, and yawning caverns — to proceed directly from His almighty hand. Both divine dignity and the laws of nature required that the earth should have originated from Chaos as a regular sphere, a fluid mass contained within a smooth, solid surface. This pristine world, unmarred by rough terrain or change of season must indeed have been a paradise, but it was doomed to early extinction. As the crust of the earth dried, great cracks opened, giving vent to the waters under the earth and precipitating a universal Deluge. Thus, by the simple operation of the laws of nature the first world was destroyed and the post-diluvian earth formed from its ruins.

For the moment Burnet seemed to have forgotten that Scripture represented the Deluge as a punishment visited upon mankind for sin. He recovered himself by explaining that the whole train of causes and effects which preceded the Deluge had been divinely synchronized with the events of human history so as to make the great crisis in earth history coincide with that in human affairs. It was, he declared, "the great Art of divine Providence, to adjust the two Worlds, human and natural, material and intellectual, as seeing thro' the Possibilities and Futuritions of each, according to the first State and Circumstances he puts them under, they should all along correspond and fit one another, and especially in their great Crises and Periods" (3). He hesitated, however, to make the pre-established harmony of human and earth history complete. He conceded grudgingly that God probably interposed special providences at crucial moments, but he stressed the difficulty of distinguishing special from ordinary providence.

Burnet had a strong sense of the variability of nature. Observing how the rivers washed the mountains into the sea, he concluded that this process must eventually level the continents, but he viewed these evidences of universal decay as proofs, not of the great antiquity of the earth, but rather of its extreme youth. He used them to refute Aristotle's theory of the eternity of the world:

41

For 'tis certain that the Mountains and higher Parts of the Earth grow lesser and lesser from Age to Age; and that from many Causes, sometimes the Roots of them are weaken'd and eaten by subterraneous Fires, and sometimes they are torn and tumbled down by Earthquakes, and fall into those Caverns that are under them; and tho' those violent Causes are not constant, or universal, yet if the Earth had stood from Eternity, there is not a Mountain would have escaped this Fate in one Age or other....But there are other causes that consume them insensibly, and make them sink by degrees; and those are chiefly the Winds, Rains, and Storms, and Heat of the Sun without; and within, the soaking of Water and Springs, with Streams and Currents in their Veins and Crannies. These two Sorts of Causes would certainly reduce all the Mountains of the Earth, in tract of Time, to Equality; or rather lay them all under Water....I do not say the Earth would be reduced to this uninhabitable Form in ten thousand Years time, tho' I believe it would: But take twenty, if you please, take an hundred thousand, take a million, 'tis all one, for you may take the one as easily as the other out of Eternity....Nor is it any matter how little you suppose the mountains to decrease, 'tis but taking more time, and the same Effect still follows (4).

How like, and yet how unlike, the conceptions of modern geology! Burnet saw clearly that everyday processes of matter in motion could radically transform the aspect of nature in the course of time, but his habit of regarding change as decline from original perfection prevented him from conceiving the processes at work on the earth's surface as a system of forces in equilibrium, some tending to the destruction of the present order, others operating to create a new order. He could picture the transformations the earth had undergone, but he looked to Scripture rather than nature to discover their pattern.

Despite his scripturalism, Burnet was roundly denounced for the liberties he had taken in accommodating the sacred text to the demands of his theory. In a work entitled *Archaeologiae Philosophicae,* he attempted to reconcile his account of earth history with the narrative in Genesis, but he never succeeded in throwing off the suspicion of heterodoxy aroused by his first speculations. Nevertheless, his *Sacred Theory* enjoyed a wide popularity. It went through several editions and established a vogue of scriptural geologizing which was to flourish for more than a century.

► THE MEANING OF FOSSILS

Burnet's theory of the earth was based largely on Scripture and Cartesian physics, with only casual reliance on observation. The Bible supplied the progression of events, Descartes the physical theory, observation and history the supporting detail.

Some of Burnet's contemporaries, however, were arriving at a concept of earth history by a more empirical route. Collecting fossils, or "figured stones," was becoming a favorite pastime among scholars and scientists, and a lively controversy developed concerning their nature and significance (5). In England the weight of scientific opinion favored regarding them as sports of nature, produced by some mysterious "plastic virtue" in the earth. Martin Lister and Edward Lhuyd, widely known as collectors of fossils, inclined to this view, though not without misgivings. The champion of the opposing view was Robert Hooke, the melancholy genius who conducted the experiments of the Royal Society in its early years and vied with Newton in the search for the true system of the world. Hooke was fascinated by the fossil specimens which came to his attention. He soon satisfied himself that they were the remains of real plants and animals, mostly marine species, and began eagerly to seek an explanation of their presence in the crust of the earth, often in regions far removed from the ocean which once had nourished them. In 1668, in a "Discourse on Earthquakes," he presented his solution of the problem to his colleagues in the Royal Society (6).

He began by dividing figured stones into two classes: mineral deposits, such as salts, crystals, and precious stones, and petrifactions, whether of bodies or of impressions of bodies. The petrifactions, he argued, must be derived from real plants and animals, since they resembled them in minute detail. Nature does nothing in vain, and what could be vainer than imitating living organisms in stone? No one would question their organic derivation except that it was hard to explain how the remains of once-living creatures had become embedded in mountain tops and in the bowels of the earth. This difficulty could be removed, said

Robert Hooke, 1635 · 1703

Hooke, by a very simple supposition: "That a great part of the Surface of the Earth hath been since the Creation transformed and made of another Nature; namely, many Parts which have been Sea are now

43

Land; and divers other Parts are now Sea which were once a firm land; Mountains have been turned into Plains and Plains into Mountains and the like" (7). History bore witness to changes of this kind, and daily observation suggested some of the causes by which they had been produced. The winds abraded the mountains; rivers and floods washed the land into the ocean; the sea attacked the land and formed and dissolved hills and valleys in its own depths; the force of gravity pro-' duced subsidences of the earth in some regions and corresponding elevations in others; earthquakes opened up great fissures and altered coastlines; and volcanic eruptions threw up mountains from the level plain.

Of these agents of terrestrial change, Hooke continued, the most powerful and effective had been earthquakes and eruptions. The Deluge was too brief to effect any considerable alteration of the surface of the earth. Earthquakes, on the contrary, had been felt in every part of the world at one time or another. They provided the most plausible explanation of the Deluge itself and of the original separation of the sea from the dry land in the first epoch of creation. To be sure, no earthquake known to sober history was sufficiently violent to raise up mountains like the Alps or the Andes, but volcanic eruptions had been known to throw up sizable mountains and to alter the relations of sea and land. That alterations of even greater magnitude had taken place in the first ages of the world seemed likely from Plato's account of the disappearance of Atlantis. Indeed, it was not improbable that the same earthquake which sank Atlantis heaved the British Isles from the sea, exposing great banks of oysters and other marine animals to the light of day. In those early times the crust of the earth was more fluid and plastic and the supply of subterranean fuel more abundant than today, hence the transformations wrought by Vulcan's agents were correspondingly greater.

If the theory of terrestrial instability were accepted, Hooke went on, change in the organic world followed as a necessary corollary:

> ...for since we find that there are some kinds of Animals and Vegetables peculiar to certain places, and not to be found elsewhere; if such a place have been swallowed up, 'tis not improbable but that those Animal Beings may have been de-

stroyed with them; and this may be true both of aerial and aquatic Animals: For those animated Bodies. . .which were naturally nourished or refresh'd by the Air would be destroy'd by the Water. . . .

[Likewise] there may have been divers new varieties generated of the same Species,. . .for since we find that the alteration of the Climate, Soil and Nourishment doth often produce a very great alteration in those Bodies that suffer it; 'tis not to be doubted but that alterations also of this Nature may cause a very great change in the shape, and other accidents of an animated Body. And this I imagine to be the reason of that great variety of Creatures that do properly belong to one Species; as for instance, in Dogs, Sheep, Goats, Deer, Hawks, Pigeons, &c. for since it is found that they generate upon each other, and that variety of Climate and Nourishment doth vary several accidents in their shape, if these or any other animated Body be thus transplanted, 'tis not unlikely but that the like variation may follow; and hence I suppose 'tis that I find divers kinds of Petrify'd Shells, of which kind we have none now naturally produced. . . (8) .

The disappearance of various species of plants and animals may very well have been accompanied, Hooke declared, by the extinction of an entire human civilization of a very high order, "a preceding learned Age wherein possibly as many things may have been known as are now, and perhaps many more." Noah's Flood, brought on by earthquakes, was probably the cause of these catastrophies in both the organic and the human world. What better means could the Omnipotent have adopted for the chastisement of mankind?

Such, in brief, were the views concerning the nature and significance of figured stones to which Hooke attempted to convert his colleagues in the Royal Society. The chief obstacle to the acceptance of his explanation was the unwillingness of his auditors to concede first, that fossils were the remains of real plants and animals, second, that the earth had really undergone the alterations he supposed, and third, that earthquakes and volcanic eruptions were capable of producing changes of the magnitude required by his theory. They were inclined to turn his argument around and maintain that figured stones *must* be sports of nature, since to suppose them real would require supposing that species could become extinct and that revolutions had taken place on the surface of the globe for which there was no evidence in reason; in Scripture, or in secular history. No adequate causes could be assigned for their production, least of all the agency of earthquakes and volcanoes. Hooke countered these objections in several ways. Again and again he came back to the fossil specimens, arguing that they *must* be the remains of plants and animals and that, if they were, profound changes *must* have taken place on the earth.

> There is no Coin can so well inform an Antiquary that there has been such or such a place subject to such a Prince, as these will certify a Natural Antiquary, that such and such places have been under the Water, that there have been such kind of Animals, that there have been such and such preceding Alterations and Changes of the superficial Parts of the Earth: And methinks Providence does seem to have design'd these permanent shapes, as monuments and Records to instruct succeeding Ages of what past in preceding. And these written in a more legible Character than the Hieroglyphicks of the ancient Egyptians, and on more lasting Monuments than those of their vast Pyramids and Obelisks (9).

The failure of recorded history to confirm the events implied by the fossil record was a serious objection, however, and Hooke did his best to overcome it. He made the most of Plato's account of the sinking of Atlantis and bolstered it with passages from the *Periplus* of Hanno the Carthaginian. He attempted, moreover, to show that Ovid's *Metamorphoses* contained a disguised account of earth history based on very early writings lost to posterity except for Ovid's rendition of them. He interpreted Ovid's narrative of the battle of the giants as a description of great earthquakes and eruptions. The story of Phaeton he interpreted as a mythological rendering of the origination of the present topography of the Mediterranean area "by a prodigious Catastrophy which Divine Providence...caused to be effected." The chronology of these events, Hooke conceded, was not apparent from Ovid's narrative, but perhaps it could be discovered by close study of the text. At any rate, it was very likely that the *Metamorphoses* contained a history of the earth from its first formation to the time when Ovid wrote.

Hooke also made use of the latest reports of earthquakes, volcanic eruptions, and fossil discoveries to substantiate his general hypothesis. In particular, he called attention to the effects of a great earthquake in the Antilles in 1690 and to the unearthing of giant elk horns in Ireland and huge "elephant bones" and "hippopotamus bones" in England, Pomerania, and Siberia. He even gave credence to the reported discovery of a ship with its crew aboard forty fathoms under the ground in Switzerland. In short, he used every available weapon to combat the prevailing belief in the stability of the terrestrial creation, drawing his arguments from observation, from history, from Scripture, and from the general analogy of nature. Concerning the last of these he wrote:

> For...we do find that all individuals are made of such a Constitution, as that beginning from an Atom, as it were, they are for a certain period of Time increasing and growing, and from thence begin to decay, and at last Die and Corrupt....As we see that there are many changings both within and without the body, and every state produces a new appearance, why then may there not be the same progression of the Species from its first Creation to its final termination? Or why should the supposition of this be any more a derogation to the Perfection of the Creator, than the other; besides, we find nothing in Holy Writ that seems to argue such a constancy of Nature; but on the contrary many Expressions that denote a continual decay, and a tendency to a final Dissolution; and this not only of Terrestrial Beings but of Celestial, even of the Sun, Moon and Stars and of the Heavens themselves, Nor have I hitherto met with any Doctrine among the Philosophers, that is repugnant to this Doctrine, but many that agree with it, and suppose the like States to happen to all the Celestial Bodies, that is, to the Stars and Planets that happen to the Individuals of any Species; and consequently if the Body of the Earth be accounted one of the number of the Planets, then that also is subject to such Changes and final Dissolution, and then at least it must be granted, that all the Species will be lost; and therefore, why not some at one time and some at another? (10).

In most of these arguments Hooke was a child of his time. He believed in the inspiration of Scripture, the tendency of nature to decline from original perfection, the sudden alteration of the surface of the globe by great upheavals, and the restricted time-scheme which went with this notion. He believed also in the doctrine of the creation and the argument from design, and in the general concordance of earth history and human history. But in his ideas concerning petrifactions and their significance for the study of earth history he was far ahead of most of his contemporaries. Accused of turning the world upside down for the sake of a few shells, he replied by citing the marvelous discoveries which had been made from much slighter evidence in astronomy and physiology. Surely, he declared,

> ...'tis a vain thing to make experiments and collect Observations, if when we have them, we may not make use of them; if we must not believe our Senses, if we may not judge of things by Trials and Sensible Proofs, if we may not be allowed to take notice of and to make necessary Consectaries and Corollaries, but must remain tied up to the Opinions we have received from others, and disbelieve everything, tho' never so rational, if our received Histories doth not confirm them; this will be truly *Jurare in verba Magistri,* and we should have no more to do but learn what they have thought fit to leave us: But this is contrary to the *Nullius in verba* of this Society, and I hope that sensible Evidence and Reason may at length prevail against Prejudice, and that *Libertas Philosophandi* may at last produce a true and real Philosophy (11).

In 1669 Hooke's cause received distinguished support from the publication in Italy of Nicolaus Steno's *Prodromus to a Dissertation Concerning a Solid Naturally Contained in a Solid.* Steno was a brilliant Danish physician and scientist residing at the court of the Grand Duke of Tuscany. He was well known to Hooke and his colleagues through his communications to the Royal Society on anatomical and physiological subjects. His new work, therefore, was promptly reviewed in the *Philosophical Transactions* and was translated into English by Henry Oldenburg, secretary to the Society (12).

Nicolaus Steno, 1638 · 1686

Like Hooke, Steno was convinced that fossils were organic remains. Not content with this general inference, he undertook to conceive how they had been produced, stating the problem thus: "Given a substance possessed of a certain figure, and produced according to the laws of nature, to find in the substance itself evidences disclosing the place and manner of its production."

In the case cf petrifactions he found the key to this problem in that the fossils were usually embedded in regular strata similar in appearance to the layers of sediment which muddy water deposits. From this analogy he concluded that the strata composing the earth's crust had been deposited in some kind of fluid medium and that the organic remains found in the strata had been incorporated during this process. In that case, he argued, the strata must have been formed successively, the lowest layer first and the uppermost last, each stratum spreading to cover the entire surface of the earth unless impeded by solid obstacles at the sides. The contents of each stratum must therefore indicate the conditions which prevailed when it was formed. Thus the strata composed of fine, homogeneous particles unmixed with foreign bodies must have been deposited at the time of the creation, when the waters covered the face of the earth and the upper strata had not yet been formed. Strata containing fragments of earlier formations and remains of marine creatures must necessarily have been formed under the sea at a later date. Still others, filled with grass, pine cones, tree trunks, and the like, must have been deposited by flooding rivers.

But if all the strata were originally deposited in horizontal layers, how did some of them come to be broken, twisted, and inclined at various angles to the horizon? To solve this question, Steno, like Hooke, supposed that earthquakes and volcanic eruptions had thrust up and broken the crust of the earth. He also conceived the "spontaneous slipping" of the upper strata owing to the disintegration of supporting formations by subterranean fire and water. Hills and mountains might be formed, he declared, either by upthrusting and faulting or by the tumbling down and erosion of higher eminences to form lower ones. Mountains of the latter type would necessarily be more disordered and mixed in composition than those from whose ruins they had been formed. Thus, some mountains were older than others, and the history of the earth could be read from the position and structure of its mountains and hills, plains and valleys.

To illustrate his thesis that "the present condition of any thing discloses the past condition of the same thing," Steno undertook to reconstruct the successive transformations which the surface of the earth had undergone in Tuscany. In the beginning, he declared, the tops of the highest mountains were covered with water. These elevated plateaus were subsequently broken into mountains and valleys by the downfall of the upper strata into huge underground caverns. The resulting uneven terrain was then submerged under water, and sandy strata containing organic remains were deposited. These were subsequently molded into sand hills and intervening valleys, swamps, and sunken plains by erosion, slipping of strata, and other geological processes. "Six distinct aspects of Tuscany we therefore recognize, two when it was fluid, two when level and dry, two when it was broken; and as I prove this fact concerning Tuscany by inference from many places examined by me, so do I affirm it with reference to the entire earth, from the descriptions of different places contributed by different writers" (13).

Up to this point, Steno had not mentioned the Bible. Now, however, he felt called upon to show that his version of earth history agreed with that recorded in Genesis, "in order that no one may be alarmed by the novelty of my view." The formation of the world in the bosom of the waters, the emergence of the dry land, and its subsequent flooding by a universal sea were all attested by Scripture, he said. The existence of vast plateaus after the waters receded was plainly indicated by natural phenomena, but neither sacred nor profane history spoke clearly on this point. The written record was equally unsatisfactory concerning the great terrestrial revolutions by which this level country was broken up into hills and valleys.

> But since the authors whose writings have been preserved report as marvels almost every year, earthquakes, fires bursting forth from the earth, overflowings of rivers and seas, it is easily apparent that in four thousand years many and various changes have taken place.... I should be unwilling to put credence in the mythical accounts of the ancients; but there are in them also many things to which I would not gainsay belief. For in those accounts I find many things of which the falsity rather than the truth seems doubtful to me. Such are the separation of the Mediterranean Sea from the western ocean; the passage from the Mediterranean into the Red Sea; and the submersion of the island Atlantis. The description of various places in the journeys of Bacchus, Triptolemus, Ulysses, Aeneas, and of others, may be true, although it does not correspond with present day facts (14).

Thus Steno, like Hooke, was hard pressed to explain the failure of history to record the changes which nature seemed to bear witness to.

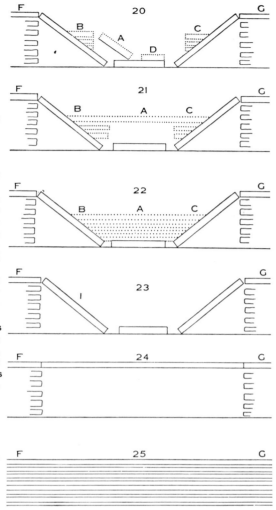

FIG. 3.1 — Six stages in the history of the earth, represented from the geology of Tuscany in Italy by Nicolaus Steno in 1669. "Fig. 25 shows the vertical section of Tuscany at the time when the rocky strata were still whole and parallel to the horizon. Fig. 24 shows the huge cavities eaten out by the force of fires or waters while the upper strata remained unbroken. Fig. 23 shows the mountains and valleys caused by the breaking of the upper strata. Fig. 22 shows new strata, made by the sea, in the valleys. Fig. 21 shows a portion of the lower strata in the new beds destroyed, while the upper strata remain unbroken. Fig. 20 shows the hills and valleys produced there by the breaking of the upper sandy strata." Reprinted by permission from **The Prodromus of Nicolaus Steno's Dissertation**, translated by John Garrett Winter. The University of Michigan, Ann Arbor, 1916.

Neither man seems to have guessed that earth history was much longer than human history. Hooke noted that the Egyptian and Chinese chronologers asserted the earth to be much older than Biblical scholars were willing to allow, but he was not inclined to desert the latter for the former. Steno was equally content with the traditional six thousand years. Indeed, he was mainly concerned to prove that the remains of plants and animals could survive for as many as four thousand years. He argued, for example, that the mollusks found in the stones of the Etruscan ruins at Volterra must go back at least to the Deluge, and that the "elephant bones" found in the fields of Arezzo must date back to the time of Hannibal's invasion of Italy.

In short, both Hooke and Steno glimpsed the possibility of deciphering the earth's history from evidence embedded in its crust. Steno boldly tried to conceive the processes by which the strata had been formed and deranged and to describe the successive transformations which geologic agents had wrought on the face of the globe. Hooke took a giant step farther when he inferred slow change in the organic creation from the instability of the terrestrial environment. But his assertion of the mutability of nature could not be called a theory of evolution, or even of progressive development. A century was to elapse before the notion of perpetual change in nature was linked to the idea of perpetual progress, before the time-scheme of earth history was separated from the time-scheme of human history.

▶ **THE OMNIPOTENT FLOOD**

The conjecturings of Burnet, Hooke, and Steno may seem inconclusive to us now, but in their own day they were sufficiently novel and disturbing to elicit a sharp protest in behalf of traditional views. Burnet's characterization of the earth as a "dirty little planet" ruined by the Deluge could not long go unreproved in a generation which produced and admired John Ray's *The Wisdom of God Manifested in the Works of the Creation*. In 1695, John Woodward, professor of medicine at Gresham College in London and an avid collector of fossils, came to the defense of the Creator with an *Essay Towards a Natural History of the Earth*. Woodward was willing to concede that the Deluge had revolutionized the face of the earth, but he denied indignantly that the alterations were haphazard and unfortunate. When man's nature became corrupted by sin, said Woodward, it became necessary to make corresponding alterations in his place of habitation. The Deluge was not simply a punishment for wickedness but a complete refitting of the globe to adapt it to the needs of fallen creatures. "How easy were it," exclaimed Woodward, "to shew that the *Rocks*, the *Mountains*, and the *Caverns*, against which these *Exceptions* are made, are of indispensable *Use* and *Necessity*, as well to the *Earth* as to *Man* and other *Animals*, and even to the rest of its Productions? that there

52

are no such *Blemishes,* no *Defects*: nothing that might have been *alter'd* for the *better*: nothing *superfluous*: nothing *useless,* in all the whole Composition: and so finally trace out the numerous *Footsteps* and *Marks* of the *Presence* and *Interposition* of a most wise and intelligent *Architect* throughout all this really wonderfull *Fabrick?"* (15).
Thus, Woodward saved the perfection of creation by arguing its perfect adaptation to imperfect human nature.

In general, however, he was less disturbed by Burnet's speculations than by those which inferred from the presence of marine fossils in the mountains that the earth had undergone unceasing alteration for countless ages. He was particularly shocked by the doctrine of the ancient atomists "that, in some Places, the *Sea* invaded the Land: in others, the Land got Ground of the Sea: that all *Nature* was in a *Hurry* and *Tumult*: and that as the *World* was first made, so should it be again dissolved and *destroyed* by *Chance.*" An enthusiastic collector of fossils, Woodward was well acquainted with the evidence which had prompted this view. He refused to admit, however, that wind, water, and volcanic fire could substantially alter the face of nature. These agents, he argued, served but to maintain the existing order of things. The soil washed into the sea by rivers was returned to the continents by rain. The relative positions of land and sea were approximately the same today as they had been in ancient times. In short, "the *Terraqueous Globe* is to *this Day* nearly in the same *Condition* that the *Universal Deluge* left it; being also like to *continue* so till the Time of its final *Ruin* and *Dissolution,* preserved to the *same End* for which 'twas first *formed"* (16).

John Woodward.
1665 . 1728

Despite these views, Woodward sided with Hooke and Steno rather than with Lister and Lhuyd [see page 89] in the controversy concerning the true nature of fossils. He recognized that they were organic remains, some of them different from any known living species, others similar to existing species but found in very different regions of the earth from their modern counterparts. He had, therefore, to explain their presence in beds of solid rock far removed from the sea. If the Deluge was to provide the explanation, a new conception of that catastrophe was obviously necessary. The crust of the earth must have been totally dissolved in the waters under the earth. Such a cataclysm could scarcely have been produced by the ordinary operations of nature; the laws of cohesion must have been temporarily suspended by divine fiat. Although the Deluge must have seemed like the return of Chaos to those who witnessed it, a more comprehending eye would have been able to discern God's hand at work in the whole affair, "directing all the several *Steps* and *Periods* of it to an *End,* and that a most *noble* and *excellent* one; no less than the *Happiness* of the whole Race of *Mankind* . . . which were to inhabit this ·Earth, thus *moduled anew,* thus suited to their present *Condition* and *Necessities*" (17). By this conception of the Deluge, Woodward managed to account for fossil deposits without conceding perpetual change on the earth's surface. Moreover, by making the Deluge the explanation of these puzzling phenomena he strengthened belief in the historicity of that event and in the full inspiration of Scripture. The Deluge explained the fossils, and the fossils proved that the Deluge had actually occurred. Nature and Scripture were in perfect accord.

► THE UNIFORMITY OF NATURE

Woodward had good reason to be concerned for the plausibility of the Deluge as an all-sufficient explanation of fossil phenomena. In Italy, the Abbé Moro, a churchman interested in geology, dismissed the Deluge as a supernatural event and attempted to explain the formation and subsequent alteration of mountains, hills, and valleys in terms of volcanic eruptions. In France, Bernard de Jussieu, demonstrator at the Botanical Garden in Paris, in attempting in 1718 to account for the impressions of tropical plants which he had observed in the strata at St. Chamond, was content to suppose a great flood from the south-east, but his colleagues in the Royal Academy were more sceptical. The Secretary of the Academy, Bernard de Fontenelle, and the great entomologist Réaumur were convinced that the Deluge could never have produced these impressions, still less the great banks of oyster shells in Turenne more than one hundred miles from the sea (18). Clearly, the scientific world was ready for a theory of the earth which would dispense with the Deluge and seek the key to fossil phenomena and earth history in nature itself, as Hooke and Steno had tried to do.

54

Buffon's theory of the earth, contained in the first volume of his *Natural History* (1749), had precisely that aim. The time had come, said Buffon, to put aside Scripture and to view the surface of the earth as a system of matter in motion. Burnet, Whiston, and Woodward had erred in making the Deluge the prin-   Buffon, cipal explanation of terrestrial phenomena. This   1707 · 1788 event, being miraculous, could not be explained by natural causes. Like all great floods, the Deluge took its toll of life and property, but it could not and did not alter the great structures of nature. The present appearances on the surface of the earth must be explained, not as the result of one great catastrophe, but as effects of natural processes operating daily within plain observation.

> ...to give consistency to our ideas, we must take the earth as it is, examine its different parts with minuteness, and, by induction, judge of the future from what at present exists. We ought not to be affected by causes which seldom act, and whose action is always sudden and violent. These have no place in the ordinary course of nature. But operations uniformly repeated, motions which succeed one another without interruption, are the causes which alone ought to be the foundation of our reasoning (19).

So saying, Buffon proceeded to put his precept into practice. In successive chapters on the formation of strata, on their fossil contents, on rivers, seas and lakes, tides, ocean currents, winds, volcanoes and earthquakes, on new islands, on erosion, and on the changes of land into sea and sea into land, he presented a multitude of facts drawn from his reading and observation, along with a variety of opinions and interpretations, both his own and those of the authors he quoted. The result was a volume which, for all its theoretical and factual errors, provided a systematic, informative, and readable introduction to the study of the earth's surface. The theory of the earth was placed on a new footing, and a wide field for future investigation was opened to view.

Yet when Buffon turned from a general view of the new science of the earth to the particular problems posed by terrestrial phenomena, his contributions were less notable, and he did not always practice what he preached. Like several naturalists before him (notably Benoit Demaillet [20], he conjectured that the main features of the earth's surface had been formed in the bosom of the sea by deposition of sediments and by the action of tides and ocean currents. To the evidences which others had cited in favor of this idea he added his own observations concerning the uniform "correspondence of angles" of hills and ridges on the opposite sides of valleys. "Every traveller," he declared, "may remark this correspondence in opposite hills. When a hill makes a projection to the right, the opposite one uniformly recedes to the left. Besides, in opposite hills separated by valleys, there is rarely any difference in their height. The more I observe the contours and elevations of hills, I am the more convinced of the correspondence of their angles, and of their resemblance to the channels and banks of rivers" (21). Buffon did not conclude, however, that these valleys had been cut by rivers. If he had, his time perspective would have been greatly enlarged. Instead, he assumed that the features of the continents had been sculptured by ocean tides and currents while the entire globe was covered with water and the strata were still being formed. The great mountain ranges of both hemispheres he attributed to the general motion of the tides, which he supposed had heaped the undersea sediments to great heights. To more local motions, "occasioned by winds, currents, and other irregular agitations of the sea," he attributed the formation of smaller mountains and hills. He conceded that all these formations had been wasted by wind, water, and volcanic fire since the time of their elevation above sea level, but he considered these alterations relatively unimportant. "The ancient form is still recognizable; and I am persuaded, that every man may be convinced, by his own eyes,. . . that the surface which we now behold, received its configuration from the currents and movements of the sea" (22).

But how had the mountains formed in the depths of the ocean become elevated above sea level? Buffon admitted the difficulty of the question. He never doubted, however, that a satisfactory answer would eventually be found. Volcanoes and earthquakes he considered too superficial and restricted in their scope to elevate mountains from the sea. As an alternative he suggested that the sea was moving gradually from east to west, slowly swallowing up the lands to the west as it abandoned those on the east. Or possibly some great cavern had collapsed in the interior of the earth, altering the contours of its surface and the distribution of sea and land. Such a revolution, Buffon hastened to add, would probably not occur suddenly but by gradual stages over a long period of time.

Thus Buffon returned to his original principle that the explanation of terrestrial phenomena must be sought in the everyday workings of nature, not in cataclysmic events. This was the basic principle of *uniformitarianism*, the foundation stone of modern geology, as distinguished from *catastrophism*, the belief that sudden upheavals produced by causes no longer operating were required to account for the evidences of profound change on the earth's surface. On the whole, Buffon adhered to the principle of uniformity he had announced. The origin of the earth itself he derived from a cataclysm -- the collision of a comet with the sun — but once he had his planet he attempted to see how it could turn into a habitable world by the steady operation of the laws of nature. The flowing mass of molten matter gradually cooled, the vapors around it condensed to form the primeval ocean, sediments were deposited on the vitrified floor of the ocean and formed into hills and valleys by the action of tides and currents, and the whole was eventually raised above the water, there to be attacked by wind, water, and fire, thus providing the detritus for still new formations under the sea.

> From what has been advanced [said Buffon] we may conclude, that the flux and reflux of the ocean have produced all the mountains, valleys, and other inequalities on the surface of the earth; that currents of the sea have scooped out the valleys, elevated the hills, and bestowed on them their corresponding directions; that the same waters of the ocean, by transporting and depositing the earth, &c. have given rise to the parallel strata; that the waters from the heavens gradually destroy the effects of the sea, by continually diminishing the height of the mountains, filling up the valleys, and choaking the mouths of rivers; and, by reducing every thing to its former level, they will in time, restore the earth to the sea, which, by its natural operations, will again create new continents, interspersed with mountains and valleys, every way similar to those which we now inhabit (23).

In this passage Buffon seemed to anticipate Hutton's conception of the earth's surface as a system of matter in motion in infinite time, "with no vestige of a beginning, no prospect of an end." In truth, Buffon made a great point of the relativity of human conceptions of time and space, observing that "human life is only a point of time, a single fact in the history of the operations of God." He thought in terms of a time-scheme much larger than the traditional six thousand years allowed by Archbishop Ussher. He even adopted Halley's suggestion that the rate of increase of salinity in the sea would provide a chronometer of earth history. But he stopped short of a true conception of the vast antiquity of the earth, partly because, unlike Hutton, he had a theory of the *origin* of the earth and the solar system, and partly because he believed that the formation of the continents had taken place under the sea while the strata were still in a plastic condition. Moreover, he shared Hooke's notions about the superior vigor of nature in her pristine state and speculated that the changes which took place on earth immediately after the separation of the earth from the sun were much greater and much quicker than those now observable. "For, as terrestrial substances could only acquire solidity by the continued action of gravity, it is easy to demonstrate, that the surface of the earth was at first much softer than it is now; and, consequently, that the same causes, which at present produce but slight and almost imperceptible alterations during the course of many centuries, were then capable of producing very great revolutions in a few years" (24). Hence, although Buffon did much to spread the idea that nature was changeable, although he began to separate human history from earth history and expanded the traditional time-scheme considerably, he fell short of a complete uniformitarianism and the vast time-perspective it implied. His views were sufficiently unsettling to provoke censure by the clergy nevertheless. The first three volumes of the *Natural History* had slipped through the royal censorship in 1749, but Buffon was forced to preface the fourth volume, published in 1753, with a formal retraction of the heretical opinions expressed in the first, particularly those contained in his theory of the earth. In answer to the charges leveled against him by the theological faculty of Paris, Buffon made a solemn declaration: "That I have no intention of contradicting the text of Scripture; that I believe firmly everything related there concerning the creation, whether as to the order of time or as to the actual circumstances, and that I abandon whatever concerns the formation of the earth in my book, and in general everything which could be contrary to the narration of Moses, having presented my hypothesis concerning the formation of planets only as a pure supposition of philosophy" (25). The theologians were apparently satisfied with this, but they were to discover twenty-five years later, when Buffon published his *Epochs of Nature,* that the heretic had never really abandoned his theory of the earth.

► THREE KINDS OF MOUNTAINS

During the quarter century between Buffon's publication of a program for the study of the earth and his attempt in 1778 to delineate the epochs of nature, the nascent science of geology made great strides. In Germany, where mining and metallurgy were of great economic importance, the successive strata composing the crust of the earth were painstakingly described, and efforts were made to reconstruct imaginatively the series of events which had produced them. Johann Gottlob Lehmann's *Investigation into the History of Stratified Mountains* (26), published in 1756, was one of the earliest of these pioneering ventures into the field of stratigraphical geology. A teacher of mining and mineralogy in Berlin, Lehmann strongly emphasized the utility of his researches, but he was also interested in general theory. After reviewing the theories of Whiston, Woodward, Moro, and others, he proposed a classification of mountains based on their position, structure, and mineral content and tried to account

Johann Gottlob Lehmann, d. 1767

for the differences among these classes by conjectures as to their order and mode of formation. He recognized three types of mountains. The first comprised those lofty ranges, such as the Alps and Carpathians, whose strata, reaching deep into the earth and inclining at sharp angles to the horizon, were of relatively homogeneous composition, devoid of organic remains but abounding in precious and useful metals. These he called *Gang-Gebirge,* that is, ore-bearing mountains. Superimposed on the flanks of these great ranges and extending laterally from them were the *Floetz-Gebirge,* or stratified mountains, whose horizontal or only slightly-tilted strata bore remains and impressions of sea creatures. From these mountains came coal, marble,

FIG. 3.2 — A very early stratigraphical section of the earth's crust, from
J. G. Lehmann's **History of Stratified Mountains** (1756). The terrain lies in the
vicinity of Nordhausen in central Germany. The strata are numbered and
described in the accompanying text by number. Thus, No. 5., indicated at lower
right, is "common limestone, which effervesces when treated with an acid solvent,
and which miners call **Zechstein**."

and limestone for man's use. Finally, there were miscellaneous other
mountains produced by various accidents of nature, such as earthquakes and volcanic eruptions.

The characteristics and relative locations of these three types of
mountains could be explained, Lehmann suggested, by supposing
them to have been formed at different epochs in the earth's history.
The ore-bearing mountains were obviously the oldest, since they
underlay the other types and at the same time overtopped them.
Moreover, since they contained no organic remains, they must have
been formed before life appeared on the earth, or, in other words, in
the very beginning. Scriptural references to the art of mining showed
that the primitive earth had had its ore-bearing mountains, and the

Biblical account of creation suggested the general manner in which these mountains had taken form in the depths of the primeval ocean. The face of the primitive earth was probably altered considerably at the time of the Fall, Lehmann continued, but thereafter it suffered few changes until the Deluge. In that catastrophe, however, the aspect of nature was radically changed, and new mountains were formed on the sides of the old.

> When the water fell, it mixed together the earth and its plants, animals, trees, and the like and deposited them at the feet of the great mountains as it gradually subsided: thus arose new mountains composed largely of horizontal strata of one kind or another. The surface of the earth thereby acquired a new form in these regions. Changes even took place in the interior of the earth. For, wherever the waters acted upon soluble earths and minerals, such as limestone mountains, they were able with little difficulty to dissolve these and to carry away the dissolved parts, thus creating caverns, sink holes, canyons, etc. The high mountains were denuded of their fertile soil, and hence are seen upon them really high peaks and many bare rocks and crags. In the course of time, the substances buried under the newly created mountains underwent alteration. A part of them rotted. Others were subject to other changes. Thus we find that some substances were petrified: for example, trees, bones, shells, snails, etc. Others, although they rotted away, impressed their forms in the soft clay matrix in which they had been deposited before the matrix hardened; such are fishes, crabs, plants, flowers. Still others were penetrated by certain kinds of earths in the course of time: for example, the coal-wood fossils such as are found frequently in England, France, Germany, Bohemia, Poland, and Silesia. Still others were infiltrated by minerals, as in the case of the horns of Ammon, the belemnites, and other objects saturated with silica. Still others were converted into ores, as, for example, the mussels turned into iron ore at Freienwald, the wood changed into iron at Orbissau in Bohemia, etc. Still others, on the contrary, were completely destroyed (27).

Such was the origin of the *Floetz-Gebirge,* or stratified, fossiliferous mountains. The mountains of the third kind Lehmann assigned to the post-diluvian period, attributing their formation to landslides, violent storms, sudden inundations of the sea, volcanic eruptions, earthquakes, and other accidents of nature. These were the smallest and most heterogeneous of all, having been formed, in many cases, from the debris of older mountains. Thus, the characteristics of Lehmann's three types of mountains were explained by his theory of their dates and modes of origin.

Lehmann's scheme of earth history was little different in many respects from that offered by Whiston many years earlier. There was the same confident reliance on Scripture, the same willingness to interpret the sacred text broadly when reason and observation seemed to require it. There was the same assumption of the stability of the great structures of nature, and the same tendency to synchronize the revolutions on the surface of the globe with the great epochs of human history delineated in the Bible. But there were differences, too. Lehmann's classification of mountains was founded on careful observation, hence it did not stand or fall with his theory of origins. Moreover, although his first two types of mountains were conceived as originating under circumstances not likely to be repeated, the mountains of the third type were attributed to natural events which, though somewhat unusual in character, were likely to happen again, producing similar effects. There was, in short, some appreciation of the instability of the earth's surface in Lehmann's book, but not enough to overcome the long-standing presumption that structure was fundamental and change only superficial in the economy of nature.

► **BAPTISM BY FIRE**

Old habits of thought persisted stubbornly, but new evidence raised evernew doubts concerning them. In 1751, five years before Lehmann described the mountains of Germany, a startling discovery was made in the French province of Auvergne by Jean Étienne Guettard, one of the founders of mineralogical cartography. In 1746 Guettard had communicated to the Royal Academy a memoir on the distribution of rocks and minerals, with accompanying maps. A few years later he made a journey through central France to see at first hand a region which he had described in his memoir but had never visited. Passing through Moulins, he was astonished to see milestones which seemed to be composed of volcanic lava. Volcanoes in central France? Surely that was impossible, yet here were stones which seemed un-

FIG. 3.3 — Extinct volcanoes of central France, first recognized as such by Jean Étienne Guettard in 1751. There were no historical records of volcanic activity in this region. The eminence marked **h** in the far background (right center) is the famous Puy-de-Dôme, volcanic itself, from which Guettard surveyed the chain of extinct volcanic cones. Up its slopes Blaise Pascal had a barometer transported a century earlier, to confirm the theory that the column of mercury in the Torricellian barometer was supported by the pressure of the atmosphere. The illustration is from G. P. Scrope, **The Geology and Extinct Volcanos of Central France**, (2nd ed.; London: 1858), Plate III.

mistakably of volcanic origin. The very buildings of the region were constructed of them. But where were the volcanoes? With mounting excitement Guettard followed the trail of mineralogical evidence until it led him to a mountain rising above the little

Jean Étienne Guettard,
1715 . 1786

village of Volvic in Auvergne. Everything connected with the mountain announced that it had once been a volcano: its conical shape, the gray-white and pale rose granite at its base, the disorderly heaps of pumice stones, the scorified rocks and spongy slag on the slopes, and, most dramatic of all, the funnel-like vent near the summit of the mountain. Nor was this all. There were other peaks in the vicinity, notably the Puy-de-Dôme, near Clermont, which strikingly resembled the mountain of Volvic. A visit to the Puy-de-Dôme removed the last vestige of doubt from Guettard's mind. The view from its summit was as terrifying as it was beautiful to one who knew the secret of the silent, conical peaks stretching away into the distance.

Guettard lost no time in reporting his discovery to the Academy of Sciences. It was certain, he told his colleagues, that many of the mountains of Auvergne and Dauphiné had once been active volcanoes. It was even possible that they were not completely extinct, for hot springs issued from many, and tremblings of the earth had been felt in their vicinity in recent times. Perhaps the French in Auvergne were no safer from Vulcan's wrath than the Italians in Catania! It was an odd circumstance, Guettard added, that the ancient eruptions so plainly attested by nature had escaped the notice of historians. Earthquakes had been recorded in great numbers by the early chroniclers, but of volcanic eruptions in the Auvergne region there was not the slightest hint. To what a remote age, then, must these terrible events be assigned! (28).

The existence of extinct volcanoes once surmised, traces of them were found everywhere. In France, Nicolas Desmarest took up the trail which Guettard had blazed. Desmarest was especially interested in the columnar basalt formations found in various parts of Europe, notably in the "Giant's Causeway" in county Antrim, Ireland. Touring Auvergne in 1763, to see for himself the volcanic phenomena which Guettard had described, he came upon some basalt columns much like those reported in Ireland, prismatic in form and of the same grain, color, and hardness. From their appearance and their situation on a bed of scorified earth he concluded that these columns, like the supporting formations, had originated from volcanic lavas. By studying and mapping the region from Volvic to Mont D'Or, Desmarest was able to show the relation of the columnar basalts to the lava flows of extinct volcanoes and to display a gradation of basaltic formations, including regular polygonal columns, irregular columns, truncated pyramids, trapezoids, and faceted stones, all of the same grain and color, hence all basaltic lavas. Their differences in appearance, situation, and form he explained by reference to the varying conditions which had prevailed at the time of their solidification and to the different "degradations" which they had suffered since then.

Nicolas Desmarest, 1725 · 1815

But if the basalts of Auvergne were of volcanic origin, then why not those of other regions? There were, said Desmarest, at least ten districts in Europe known to contain prismatic basalts: one in Ireland, two in Germany, four in France, two in Italy, and one in Sicily. Non-prismatic basalts had been found even more widely, in Tahiti and New Zealand, for example. If the rest of the globe contained as many evidences of volcanic activity as those parts which had already been examined attentively, it could not be doubted that subterranean fire had played a much greater part in its history than had commonly been supposed. The basalts of Europe, Desmarest added, had unquestionably been formed at different times and in differing circumstances. Some of them, for example, contained tiny fragments of calcareous rocks, presumably the debris of strata shattered by the eruption of underground lavas. In other cases calcareous strata lay superimposed on basalts which contained none of these gray and white fragments; these must have solidified before the calcareous strata were deposited. From similar traces and inferences, Desmarest predicted, the history of nature's operations on the surface of the globe would eventually be deciphered. In that history, volcanic eruptions would be found to have played a major role, not only within the memory of man but also in remote ages far antedating recorded history (29).

In support of these contentions, Desmarest undertook to establish the relative antiquity of the various formations of basaltic lava in the regions he had examined. The most recent, he reasoned, were those which could be traced without any difficulty to an existing volcanic cone, either active or extinct. These formations, though weathered by wind and water, still preserved the continuity of the original flows. The lava sheets of the second epoch, on the contrary, had been cut to pieces by erosion, only the most resistant patches remaining in detached situations over the countryside. The volcanic cones which poured them forth had long since disappeared. Still more ancient were the lavas which underlay sedimentary strata or alternated with them. Since the sedimentary strata were undoubtedly marine deposits, the basaltic formations beneath them must once have been submerged under the sea. In some cases submarine eruptions must have alternated with periods of sedimentation. Thus, from the position and appearances of the various basalts Desmarest traced three epochs of nature. That they had been of vast duration he had no doubt, but he considered it unwise or unprofitable to attempt to estimate the total time elapsed. His aim was to fix the relative, not the absolute, chronology of events, but to do so on the assumption that "Nature has been constrained to the same mode of operation in the most remote ages as in the most modern times" (30).

The search for traces of extinct volcanoes was not confined to France. In Italy, Sir William Hamilton, British ambassador at Naples, found his attention drawn from Vesuvius itself to the evidences of earlier volcanic activity in Sicily and southern Italy. From the top of Mount Etna he descried forty-four conical mountains on the Catanian plain, each with its own crater. Examining the soil around Naples, he concluded that all the high grounds around that city, as well as the islands of Procida and Ischia, were of volcanic origin. Farther north, along the road between Naples and Florence, he found traces of other eruptions. Comparing these phenomena with those reported in other parts of the world, he concluded that volcanoes had played a large role in the history of the earth. "Surely," he wrote to the President of the Royal Society, "there are at present many existing volcanoes in the known world; and the memory of many others have [sic] been handed down to us by history. May there not therefore have been many others, of such ancient date as to be out of the reach of history?" (31).

In Germany, Vulcanism, the doctrine that volcanic action had been the chief agent in giving the earth its present form, found a stout champion in Rudolph Raspe, an amateur geologist best known to history as the author of *Baron Munchausen's Travels*. In a *Supplement* to Sir William Hamilton's *Observations on Italian Volcanoes*, published in 1776, Raspe advanced the thesis that the Dornberg and the Habichwald mountains, enclosing the valley of Cassel, were of volcanic origin. These peaks, he declared, were of a different nature from the calcareous hills on which they rested. The latter were composed of parallel strata bearing organic remains, the former of miscellaneous clays, ashes, marls, sands, black vitreous rocks, slags, and the like — the very kinds of materials contained in the volcanic mountains described by Italian observers. Could there be any doubt, then, that the German mountains were of similar origin?

Their remarkable elevation, their large extent of at least 20 English miles square, and their present exterior appearance, seem clearly to indicate, that subterraneous fermentations, heat, and fire, worked many centuries to raise and accumulate them by many eruptions on a marine ground, and perhaps in the midst of an ancient sea; but that water, rain, frost, and the inclemencies of the atmosphere since times immemorial, have been at work to destroy and to level them again. Isolated, and for the most part conical ridges and points of basalts, of a like origin, and of the same remote antiquity, are dispersed all over Hesse; and there seems to be in Lower-Hesse in the neighbourhood of Cassel a chain of volcanic hills, running through Upper-Hesse and the Wetterau to the Mayn and to the Rhine, nay perhaps through Thuringia and Franconia to the Saxonian and Bohemian mountains, which partly are known to be of a volcanic structure (32).

How long ago these German volcanoes had been active Raspe did not venture to say. History was silent on the subject, but it was silent on everything connected with Germany until fairly recent times. The record of nature was all that there was to go by. The sandstone strata, being lowest, were presumably primordial. The calcareous strata above them were probably deposited at the bottom of an ancient sea and subsequently fractured and inclined by earthquakes and eruptions, perhaps the very eruptions which formed the Dornberg and the Habichwald. Of still more recent origin were the beds of volcanic detritus which streams and rivulets had torn loose from the mountainsides and deposited in the valley of Cassel. How long it had taken the running waters to cut the Druseltal through the high rocks of the Habichwald and to roll the detached fragments to the gates of Cassel and grind and blunt them into smooth pebbles could only be guessed. The relative antiquity of these various operations of nature was clear, but their absolute chronology was uncertain. With respect to the latter question Raspe left his readers to form their own conclusions.

▶ RELATIVE VS. ABSOLUTE CHRONOLOGY

Now, gradually, a new time sense was being formed. But the general theory of the earth responded slowly. Lehmann had found more agreement than disagreement between Genesis and geology.

Hamilton and Raspe were content to hint at the remote antiquity of the globe without pushing the matter farther. Desmarest had traced the history of volcanic activity into a remote past, but had had no use for idle speculation concerning the history of the globe. Nevertheless, events were pressing toward a radical transformation of human conceptions of earth history, and prescient minds reached out to anticipate the change. One of the boldest of these forerunners of the modern view of earth history was Giovanni Arduino, professor of mineralogy

Giovanni Arduino, 1714 • 1795

at Padua. Like Lehmann, Arduino divided mountains into three classes, which he named primary (or primitive), secondary, and tertiary. The first two classes corresponded roughly to Lehmann's ore-bearing and stratified mountains, the third comprised "low mountains and hills which seem to be composed of gravels, sands, and limose, argillaceous and marly earths, materials almost always mixed copiously with marine shells." These three types of mountains Arduino considered to have been formed successively in the order suggested by their names, but he did not, like Lehmann, assign them to definite epochs in the history of the earth. In any given situation, said Arduino, the primary mountains could be shown to be older than the secondary, and the secondary could be proved to antedate the tertiary. But this was not to say that all primary mountains had been formed at one time — at the creation, for example. "My divisions are extremely general and concern only the principal distinct epochs of each series of those events on which apparently depend the production, the native essential properties, the modifications and changes of materials and aggregates which taken together constitute each one of the aforesaid orders in particular" (33).

Here was a marked change in point of view. "Primitive" no longer meant primigenial, dating from the creation of the earth; "secondary" no longer referred to the supposed effects of the Deluge; "tertiary" mountains and hills were not necessarily post-diluvian. An absolute time-scheme had been replaced by a relative one. Some strata could be shown to be older than others, but none could be considered primordial. The task of the geologist, therefore, was to determine the relative antiquity of the various formations and to try to grasp the processes by which they had been formed and modified. Thus, according to Arduino, the primitive mountains could be divided into two subdivisions. The first and oldest, composed of vitrescent rocks — schist, granite, porphyry, basalt, and trap rock — bore unmistakable signs of a melting agent in their formation and subsequent alteration. The second, superimposed on the roots and flanks of the first, owed their origin to the action of both fire and water. The calcareous mountains in the Alps, on the other hand, were plainly the work of the sea, their strata having been formed in successive periods of time, each layer bearing its distinctive fossil remains. Moreover, all of these formations, whether igneous or aqueous in origin, had been altered and deranged by the forces of nature in the course of time. Streams and rivers had cut deep valleys in the mountains and sculptured the hills. Volcanic eruptions and subterranean fires had deposited lava, tufa, pumice, and ashes, had lifted and distorted enormous beds of rock, and had left extinct craters as a silent witness of their operations.

> These [declared Arduino] are the principal means employed by nature in our age to diversify here and there and give new forms to the face of the earth, here raising new mountains, there altering the old horizontal strata of marine deposits in a thousand ways, here uplifting islands and reefs from the deep abyss and submerging those already existing, there engulfing plains and mountains and forming new lakes and deep gulfs. Of similar events we have all too recent horrible memories: and of those happenings in the remotest ages we see the effects abundantly in every part of the globe (34).

Arduino had, in effect, arrived at a concept of the uniformity of nature's operations on the earth's surface. Gone was all pretense of reading the history of the earth from the pages of Scripture, gone the attempt to explain terrestrial phenomena as products of one or two great catastrophes, gone the assumption that earth history was synchronous with human history. The working hypothesis was "the celebrated supposition of Leibniz and other learned men to the effect that our earth does not possess its original aspect, and that that which it now has is a result of innumerable complicated effects of fire and water." The plain implication was that the forces of nature had required millions of years to produce these effects, but Arduino preferred to remain silent on this point, perhaps for prudential reasons. It remained for Hutton to portray the agencies at work on the surface of the globe as a self-balancing system of natural processes operating perpetually to destroy and perpetually to recreate the face of the earth throughout endless ages.

► WATER IS GREATER THAN FIRE

Arduino's views, outlined in 1760 and amplified in 1774, set the stage for a forthright exposition of geologic uniformitarianism, but the scientific world was not ready for so bold an act. In the decade which was to elapse before Hutton threw down the gauntlet to the catastrophists it was the issue between Vulcanists and Neptunists, not that between uniformitarians and catastrophists, which agitated the geological world. In Germany, Abraham Gottlob Werner threw the weight of his growing prestige on the side of the Neptunists, who regarded water as the master agent in the history of the earth. In 1775 when he assumed his professorial duties at the Freiberg School of Mines, he found the Vulcanist star in the ascendant. Basalt and most of the so-called primitive rocks were regarded as volcanic in origin. Werner could not agree. From his own examination of the basalts of Saxony he became convinced that neither basalt nor the primitive rocks were of igneous origin but that, on the contrary, nearly all of the formations on the surface of the globe had resulted from crystallization or deposition in a watery medium. In his *Short Classification and Description of Various Types of Mountains,* published in 1787, Werner divided rocks into four types according to their nature and origin: namely, primitive, sedimentary, volcanic, and alluvial. To only one of these types, the volcanic, did he assign an igneous origin (35).

Abraham Gottlob Werner, 1749 • 1817

Werner did not propose a general theory of the earth in this work nor in any subsequent publication, but his ideas were made known to the public by his many disciples gathered from all parts of Europe. According to them, Werner supposed that a vast primeval ocean had enveloped the globe at the creation, diminishing in volume as the centuries passed and rocky strata were formed under its waters. In the early stages of the ocean's subsidence the *primitive* rocks were formed by crystallization or chemical precipitation. Then, about the time that the first types of life began to appear, the *transition* rocks were formed above the primitive, partly by chemical processes and partly by mechanical deposit of sediments. Next came the *floetz*, or stratified, formations, filled with remains and impressions of plants and animals. Finally, after the continents had been exposed by the continued retreat of the sea, *alluvial* strata were deposited by streams and rivers, and volcanoes spewed forth lava and volcanic ash.

To account for certain anomalies and transpositions in the succession of strata, Werner found it necessary to suppose temporary resurgences of the ocean separated by long periods of time to permit weathering of the strata by the elements. The superposition of floetz and trap rocks on the wasted remains of earlier formations convinced him, said one of his disciples,

> ... that the globe of the Earth is of remote antiquity; that its surface was inhabited by animals, and covered with vast forests, when it underwent a great revolution, perhaps the last of several which it has experienced; that this revolution occasioned the disintegration of many of the rocky masses already existing, — the total destruction of the forests, — and was followed or accompanied by a mighty inundation, which rose to a height equal perhaps to that of the highest mountains; that this immense and necessarily raging sea produced accumulations of gravel and sand, over which, when it had somewhat abated of its agitation, were deposited the earthy, clayey, and bituminous particles with which it was charged: that as the water became more and more tranquil and pure, the precipitates had become less earthy, and the union between their particles more intimate; wacke, basalt, greenstone, and porphyry slate, being successively produced, as it approached to that state of calm and purity favorable to crystallization (36).

Thus, Werner depended upon regular processes of nature for the building of rock formations and their subsequent weathering, but he had no convincing explanation of the gradual retreat of the sea, and he was forced to introduce temporary resurgences to account for phenomena inconsistent with his general hypothesis. The resulting panorama of earth history differed in many ways from the old picture of creation, deluge, and final conflagration, but it was not hopelessly at odds with it. Although Wernerians spoke of the "remote antiquity" of the globe, their time-scheme remained vague and uncertain. Great revolutions were declared to have taken place on earth, but there was little suggestion that the forces which had produced them were still in operation. The pattern of earth history was drawn from nature rather than from Scripture, but it could be and was reconciled with the narrative in Genesis by a modified interpretation of the six days of creation. The occurrence of mighty terrestrial revolutions, the gradual appearance of living forms, even the destruction of some species could be admitted without disturbing in the least the cheerful conviction that the earth had now attained its divinely appointed form as a stage for the activities of human beings.

► EPOCHS OF NATURE

Werner's Neptunism was tinged with catastrophism, but catastrophism was not confined to the Neptunian camp, as the examples of Pallas and Buffon will show. Peter Simon Pallas was a German naturalist of international reputation. In 1768 he was invited by

Peter Simon Pallas, 1741 · 1811

Catherine the Great of Russia to fill the chair of natural history at the Imperial Academy in St. Petersburg and to organize scientific exploration in her domains. His reports of these expeditions gave the scientific world its first general view of the topography, geology, zoology, botany, and ethnology of the vast Russian Empire. His *Observations on the Formation of Mountains,* published in 1777 and again in 1782, provided a welcome confirmation of the findings of Arduino, Saussure, Lehmann, and other observers. In Russia, as in western Europe, the substratum proved to be granite, followed successively by schistose formations, sedimentary rocks bearing marine fossils, and alluvial strata filled with "elephant bones," tree trunks, and similar debris. In explaining the genesis of these formations, Pallas made liberal use of both fire and water. He had no theory of the origin of the granite core of the earth or of the primeval ocean which he thought must have covered all but its most elevated peaks and plateaus. He was sure, however, that the formation of the schistose mountains and the fracture and elevation of the calcareous strata could be explained by assuming extensive volcanic activity in every quarter of the globe at some early period in its history. Traces of these early volcanoes were hard to find, he conceded, but new facts were coming to light every day, and the volcanic hypothesis explained the apparent phenomena better than any other. "These operations of volcanoes," he declared, "have continued in different places, especially in the vicinity and at the bottom of the seas, up to our own day. It is by their agency that new islands have been seen to rise from the depths of the ocean; it is probably they which raised all those enormous calcareous Alps, formerly coral rocks and beds of shells, such as are still found today in the seas which foster these productions" (37).

Deluges played a part in Pallas' version of earth history, too. Noting that the slow and regular processes of erosion, deposition, and solidification would require millions of years to build the rock-ribbed continents at the expense of the sea, Pallas sought a more rapid and efficient cause. Suppose, he conjectured, that a great submarine eruption off the coast of China should have elevated the Japanese and Philippine Islands above the surface of the ocean and sent a series of tidal waves rolling northward and westward into Asia and Europe. Such a flood might well have swept elephants, rhinoceri, tropical plants, and much other debris into northern Europe and Asia. Attested by Scripture and by the traditions of every nation, the Deluge provided, said Pallas, a far better explanation of the facts of natural history than Buffon's supposition of the formation of the continents by ocean tides and currents.

Pallas' reference to Buffon was apropos, for Buffon was just then bringing forth his *Epochs of Nature*, a grand synthesis of the history of the earth from its origin as a lump of molten matter to its present condition as a theater of human life. Buffon explained his project in glowing language calculated to expand the reader's conception of the amplitude of nature's operations on the globe.

Buffon, 1707 • 1788

> Nature being contemporary with matter, space, and time, her history is that of every substance, every place, every age; and although it appears at first glance that her great works

7 3

never·alter or change,....one sees, on observing more closely, that her course is not absolutely uniform; one realizes that she permits considerable variations, that she undergoes successive alterations, that she lends herself to new combinations, to mutations of matter and form; finally, that the more fixed· she appears in her entirety, the more variable she is in each of her parts; and if we comprehend her in her full extent, we cannot doubt that she is very different today from what she was in the beginning and from what she became in the course of time: these are the different changes which we call her epochs (38).

Of these epochs of nature Buffon named six: the *first* when the globe was fused by fire, the *second* when the molten mass was consolidated and vitrified rocks were formed, the *third* when water covered the whole earth and calcareous beds containing the remains of marine life were deposited, the *fourth* when the waters withdrew from the present land masses, the *fifth* when the elephant, the hippopotamus, and other tropical animals inhabited the northern regions of the globe, the *sixth* when the two hemispheres were separated from each other and man began to alter the face of nature. To calculate the time required for these developments Buffon attempted to estimate the rate of cooling of a molten mass such as the earth had once been. From experiments on the refrigeration of various metals he concluded that perhaps seventy thousand years had elapsed since the earth began to cool. He even attempted to estimate the duration of the several epochs, allowing, for example, about fifteen thousand years for the fifth epoch.

It was no mere coincidence that the number of epochs in Buffon's scheme of earth history matched the number of days allotted for the creation of the world in the first chapter of Genesis. Buffon had not forgotten his rebuff by the clergy a quarter of a century earlier. He ventured to suggest that the six days of creation might reasonably be regarded as six periods of indefinite length and that a long period of time probably intervened between the first creation of matter and its fashioning into an orderly world. These interpretations were volunteered, he declared, in a sincere effort to reconcile science and theology. If they should prove unacceptable to liberal-minded persons, he begged them to "...judge me by my intention and to consider that since my system concerning the epochs of nature is purely hypothetical, it cannot injure revealed truths, which are immutable axioms independent of all hypothesis, to which I have submitted and do submit all my thoughts" (39). Buffon's precautions were unavailing. The professors of the Sorbonne would be satisfied with nothing short of outright retraction. Only the favor of the king and the temper of the times saved Buffon from renewed humiliation.

Shocking as it was to accepted opinion, Buffon's geologic time-scheme fell far short of the perspective required for the development of a theory of organic evolution. Despite his great imagination and his broad command of the field of natural history, Buffon was still not ready to commit himself unreservedly to a uniformitarian view of nature. In this respect he was less far advanced in 1778 than he had been in 1749. At that time he had doubted whether any truly primitive formations were to be seen on the earth. Now, however, he was convinced that the granitic mountains constituted part of the original globe formed by the rotation and cooling of the molten matter derived from the sun. The sedimentary strata, he believed, were deposited on this uneven vitrified core of granite after it had cooled enough to condense surrounding vapor which thus formed a great ocean covering most of the globe. During their formation these strata were molded by ocean tides and currents. Once hardened, they were altered further by earthquakes, volcanic eruptions, the collapse of underground caverns, and, after they had been exposed above sea level, by wind and water erosion. Thus, in the *Epochs,* Buffon invoked the slow but steady action of everyday processes of nature chiefly to account for the formation of the sedimentary strata. The inclination and fracture of these strata he attributed either to the underlying contours of the original core of the globe or to accidental causes, mostly of a catastrophic nature. Volcanoes were given a more prominent role than formerly but were still regarded as superficial explosions produced by the seepage of ocean water into contact with subterranean fires. In the last analysis, Buffon's brave words concerning the mutability of nature and the amplitude of time at her disposal were qualified by his belief that the fundamental structures of the earth's surface were formed under circumstances not likely to be repeated and that nature's forces, though operating continually to modify these structures, had not had time enough to erase the original features of the globe.

Moreover, since he regarded the gradual cooling of the globe as the controlling factor in earth history, he looked forward to the slow decline and ultimate extinction of all living forms. Despite the overtones of progressive improvement in his account of the formation of a habitable world, Buffon remained deeply influenced by the old notion that change consists in decline from the pristine vigor of nature.

► NO VESTIGE OF A BEGINNING,
NO PROSPECT OF AN END

The year Buffon died (1788) a new theory of the earth appeared. It was so uncompromising in its adherence to uniformitarian principles, so comprehensive in its synthesis of geologic processes, and so breathtaking in its time-perspective as to challenge immediate attention. The author was James Hutton, a member of the brilliant literary and scientific circle which adorned Edinburgh society in the late eighteenth century. The son of an Edinburgh merchant, Hutton pursued a varied course of studies at home and abroad before retiring to the country to enjoy the delights of agriculture, science, and philosophy. Geologic excursions became his favorite pastime and the unlocking of the secret of the earth's history his consuming ambition. In 1785, having moved to Edinburgh to enjoy its stimulating intellectual atmosphere, he presented to his colleagues in the Royal Society of Edinburgh the result of thirty years' research and reflection: "Theory of the Earth; or an Investigation of the Laws Observable in the Composition, Dissolution, and Restoration of Land upon the Globe" (40). When published in 1788 it did not long go unnoticed. Jean Deluc, a well-known Swiss naturalist attached to the royal court in London, attacked Hutton's uniformitarian doctrine in a series of letters published in the *Monthly Review*. Richard Kirwan, "the Nestor of English chemistry," joined the assault on Hutton in 1793, charging him with an infidel purpose to subvert the credit of Scripture. Hutton replied by expanding his original paper into a two-volume work, *Theory of the Earth*, published in 1795 (41). The controversy thus begun continued long after Hutton's death. As a matter of fact, the outburst which greeted Sir Charles Lyell's *Prin-*

James Hutton,
1726 · 1797

ciples of Geology in 1830 was but another engagement in the same contest, brought about by Lyell's reaffirmation of Huttonian principles [see page 249].

Hutton began by taking seriously a principle which Buffon had announced but had not always observed: the principle that geologic phenomena should be explained in terms of everyday processes of nature.

> In examining things present [Hutton declared] we have data from which to reason with regard to what has been; and, from what has actually been, we have data for concluding with regard to that which is to happen hereafter. Therefore, upon the supposition that the operations of nature are equable and steady, we find in natural appearances, means for concluding a certain portion of time to have necessarily elapsed, in the production of those events of which we see the effects (42).

It was no valid objection to this method of explaining natural phenomena that it required the assumption of millions of years of earth history. "Time, which measures everything in our idea, and is often deficient to our schemes, is to nature endless and as nothing." Nor need anyone be disconcerted because the uniformitarian hypothesis implied the mutability of the everlasting hills. Change and process were in no way inferior to permanence and stability. "Every material being exists in motion, every immaterial being in action and passion; rest exists not anywhere; nor is it found in any other way, except among the parts of space" (43).

In general, Hutton continued, three kinds of agencies determined jointly the aspect of nature on the globe. First, regular strata were formed by gradual deposition in the ocean; second, these strata were consolidated, elevated, and altered by subterranean heat and pressure; thirdly, the face of the earth was worn away by wind, water, and organic decay. These three types of processes, viewed in relation to each other, formed a self-regulating and self-preserving system of matter in motion. From the present state of the system its previous states could be inferred and its future states conjectured.

This concept of the surface of the globe as a law-bound system of matter in motion Hutton regarded as his great achievement. From it

7 7

he drew a revolutionary conclusion: that the operation of the terrestrial system had produced a long succession of worlds on the face of the globe and must continue to do so in ages to come. Hence, no mountains were truly primitive; every rock formation bore evidence of having been formed under the sea by solidification and fusion of the detritus of previous worlds, and everything indicated that a new earth must now be taking form under the ocean.

But what of the globe itself? What of its origin? Buffon's answer to this question did not please Hutton. The perfect equilibrium and regularity of the terrestrial system proved, he declared, that it was divinely created as a means for perpetually maintaining the surface of the earth as a fit habitation for living beings. Although the system of the globe might not be so simple and regular as scientific convenience required the geologist to suppose, nevertheless there were systems in nature toward the discovery of which the naturalist could move by constructing hypothetical models and comparing the consequences deduced from them with the observable facts of nature. But this method could not be used to account for the systems themselves; there were no data for such an inquiry. Instead of wasting time and thought in fruitless speculation, therefore, would it not be better to acknowledge the evidences of wisdom and benevolence in the construction of these systems?

> For having, in the natural history of this earth, seen a succession of worlds, we may from this conclude that there is a system in nature; in like manner as, from seeing revolutions of the planets, it is concluded, that there is a system by which they are intended to continue those revolutions. But if the succession of worlds is established in the system of nature, it is in vain to look for anything higher in the origin of the earth. The result, therefore, of this physical inquiry is, that we find no vestige of a beginning, — no prospect of an end (44).

Once the naturalist had discovered a *system* in nature, his next task, according to Hutton, was to discern the purposes the system was meant to serve and its fitness for those ends. The geologist, for example, should not stop at explaining terrestrial phenomena as the outcome of a system of natural processes; he must go on to show how these processes worked together to maintain a stage for human activity.

Not only are no powers to be employed that are not natural to the globe, no actions to be admitted of except those of which we know the principle, and no extraordinary events to be alleged in order to explain a common experience, [but also] the powers of nature are not to be employed in order to destroy the very object of those powers; we are not to make nature act in violation of that order which we actually observe, and in subversion of that end which is perceived to be in the system of created things (45).

The idea of an end for which the system of the earth was created was by no means a pious afterthought for Hutton. On the contrary, his a priori conviction that terrestrial operations had been designed to maintain a system of life predisposed him to look for processes counteracting those by which the continents were daily washed into the sea. He had no use for Burnet's idea that the deterioration of the mountains was but a manifestation of the tendency of all things to decline from their pristine perfection. To Hutton the "poetic fiction of a golden age" seemed unworthy of the wise Creator revealed in Newton's demonstration of the motions of the solar system. Surely God would not permit a system He had created to run down and become unsuited to the purposes for which it was made. There must be some provision in the system of the globe for *repairing* the ravages of wind and water. This provision Hutton found in the formation of new continents under the sea and in their subsequent elevation above the waters. But why must the existing continents be wasted? Why this perpetual destruction and . creation of land masses? Because, answered Hutton, the dissolution of rock formations was essential to prepare soil for the growth of plants. Plants provided food for animals, and both plants and animals served the needs of man. Thus all the processes of the globe were related to each other and to the end for which they had been contrived. The harmony of the whole ministered to man's intellectual and spiritual as well as to his physical needs by leading his inquiring mind to a knowledge of his Creator.

What a comfort to man, for whom that system was contrived, as the only living being on earth who can perceive it; what a comfort, I say, to think that the Author of our existence has given such evident marks of his good-will towards man, in this progressive state of his understanding! What greater security can be desired for the continuance of our intellectual existence, — an existence which rises infinitely above that of the mere animal, conducted by reason for the purposes of life alone (46).

Although Hutton was content to have made room for final causes in the science and system of the globe, most of his contemporaries were not so easily satisfied. The desperate ingenuity with which respected geologists such as Kirwan, Deluc, and Professor Jameson of Edinburgh sought to avoid the force of Hutton's reasoning betrayed their determination to save the traditional time-perspective and the stability of nature at all costs. In his reply to Kirwan, Hutton went to the root of most of the nontechnical objections to his theory. "He accuses me," said Hutton, "of giving this world a false or imperfect constitution...for no other reason, that I can see, but because this may imply the formation of a future earth, which he is not disposed to allow; and he is now to deny the stratified construction of this present earth to have been made by the deposit of materials at the bottom of the sea, because that would prove the existence of a former earth, which is repugnant to his notion of the origin of things, and is contrary, as he says, to reason, and the tenor of the Mosaic history" (47). So far as reason was concerned, Hutton answered, the notion of a succession of worlds did not imply the eternity of the world. Millions of years did not add up to eternity. However long the system of the globe might have been in operation, it must have been created, since it bore the marks of intelligent design.

The friends of revelation would not be put off with a nice distinction between time and eternity. Kirwan returned to the attack with a volume of *Geological Essays* in 1799. He was induced to compose these, he explained, by observing the pernicious influence of false but ingenious theories on the reading public and by his own discovery of the remarkable coincidence between the Mosaic narrative of creation and the history recorded in the rocky strata of the earth's crust. Dismissing Buffon as a wild visionary, Kirwan sketched the history of creation along Mosaic and Wernerian lines, contributing the novel suggestion that the light mentioned in the second verse of the first chapter of Genesis came from erupting volcanoes. After describing the primitive state of the globe, he explained how a miraculous flood from the south overwhelmed the continents, scattering the debris of plants and animals far and wide. Next he undertook to demonstrate the recent origin of volcanic phenomena, observing that nothing had been so injurious to the credit of Scripture as the assertion of the great antiquity of the globe. In the final essay, devoted entirely to Hutton's theory of the earth, he labored with great ingenuity to prove that mountains were *not* washed into the sea, that metamorphosed rocks were formed by *chemical* precipitation rather than by fusion, that weathering was *not* essential to provide soil for vegetation — in short, that the idea of a succession of worlds was unscientific and preposterous (48).

Hutton died in 1797, but his theory found an ardent champion in John Playfair, professor of mathematics and natural philosophy at the University of Edinburgh. In 1802, in a work entitled *Illustrations of the Huttonian Theory of the Earth,* Playfair presented the new geology with great ability and defended it against scientific and religious objections. The Bible, he argued, was intended as a guide to man's destiny and duty, not as a key to the understanding of physical nature; consequently, its pronouncements in scientific matters were not to be taken literally. The attempt to squeeze the history of the earth into six thousand years had produced one absurdity after another. Why this attachment to the idea of six thousand years? Did not Hutton's conception of the surface of the globe as a system of processes designed to provide a habitation for living creatures throughout innumerable generations enlarge and ennoble man's concept of divine wisdom and power? Did not the whole analogy of nature — the endless succession of generations in animal species and the successive states of the solar system — support the notion of a time series extending indefinitely into the past and into the future? Knowledge of the origin of things was not essential to man; it was enough for him to recognize the handiwork of his Creator in the design of the creation. "It is but reasonable, therefore, that we should extend to the geologist the same liberty of speculation, which the astronomer and the mathematician are already in possession of; and this may be done, by supposing that the chronology of Moses relates only to the human race" (49).

The strategy of granting the recent origin of man in order to gain acceptance for the indefinite antiquity of the earth was a shortsighted one. So long as the earth and man were assumed to have been created

John Playfair, 1748 · 1819

81

at approximately the same time, it was easy and natural to regard the earth and its flora and fauna as designed for the use and instruction of man. The steadily accumulating evidence pointing to a progress from lower to ever higher forms of life in the geologic series could not shake the traditional view of the relation of the inorganic world to the organic and rational worlds so long as the progression was thought to have taken place in a few thousand years. But if Hutton was right and the earth millions of years old, whole economies of vegetable and animal life must have flourished and expired long before man made his appearance. How, then, one might logically ask, could the maintenance of human life be regarded as the final cause of these vast revolutions on the globe? Were not the means out of proportion to the end?

If Playfair was aware of this problem, he did not indicate it. As for Hutton, a passage in his *Theory of the Earth* suggests that he would rather grant the antiquity of man than face the consequences of the opposite view. In this passage he spoke of the system of the globe as being "calculated for millions, not of years only, nor of the ages of man, but of the races of men, and the successions of empires" (50). The study of the earth's crust, Hutton seemed to suggest, would eventually disclose a history of human races and empires comparable in duration and vicissitudes with the globe's turbulent history. Playfair abandoned the antiquity of man in order to save geology by saving the credit of Scripture; Hutton assumed man's antiquity in order to save belief in the wise contrivance of nature.

Although Playfair devoted special attention to the Scriptural issue, he did not imagine that it was the only, or even the main, obstacle to the acceptance of Hutton's theory. He took full account of the preconceptions which opposed it:

> The greatness of the objects which it sets before us, alarms the imagination; the powers which it supposes to be lodged in the subterraneous regions; a heat which has subdued the most refractory rocks, and has melted beds of marble and quartz; an expansive force, which has folded up, or broken the strata, and raised whole continents from the bottom of the sea; these are things with which, however certainly they may be proved, the mind cannot soon be familiarized. The change and movement also, which this theory ascribes to all

that the senses declare to be most unalterable, raise up against it the same prejudices which formerly opposed the belief in the true system of the world...Even the length of time which this theory regards as necessary to the revolutions of the globe is looked on as belonging to the marvellous; and man, who finds himself constrained by the want of time, or of space, in almost all his undertakings, forgets, that in these, if in any thing, the riches of nature reject all limitation (51).

Playfair saw no reason for despair in all of this. Astronomy had come into its own despite similar objections. There had been a long succession of false theories before Newton discovered the true system of the world. If Hutton had now discerned the true system of the earth, future geological investigation would serve but to corroborate, correct, and extend his principles, as the progress of astronomy had confirmed Newton's.

The comparison between Hutton and Newton was more just than Playfair knew. Both men had discovered and demonstrated the existence of a natural system. Both had done so in the firm conviction they were tracing out mechanisms contrived by God to provide a suitable stage for human activities. Both had substituted a general providence operating by law for particular providences proceeding from the momentary purposes of the Deity. But over both men hung the possibility that the system described might prove to be an incidental product of operations in a larger system of matter in motion. It is significant in this respect that Playfair admired Laplace for his synthesis of celestial mechanics but not for his nebular hypothesis. In reviewing the *Celestial Mechanics,* Playfair expressed regret that Laplace had failed to cite the stability of the solar system and the rotation of all its planets and their satellites in the same direction and plane as proof of its divine origin. For Playfair, as for Hutton, the business of the scientist was to discover natural systems, not to account for their origin.

Hutton's theory of the earth had an important bearing on the controversy between the Neptunists and the Vulcanists. Heretofore the Vulcanists had conceived the operations of subterranean fire largely in terms of volcanic eruptions, attributing the origin of nearly all igneous rocks to that source. But Hutton regarded volcanoes simply as escape vents for a vast sea of molten rock generated deep under the earth by the enormous heat and pressure there. It was these forces, he supposed, which fused the materials deposited on the floor of the ocean into beds of solid rock, intruded veins of molten rock into their cracks and crevices, and elevated them high above sea level. Thus, the regular operations of subterranean heat and pressure, not the occasional outbursts of volcanoes, were central in Hutton's conception. "Plutonism," suggested Playfair, was a better name than "Vulcanism" for this new theory of igneous agency.

Although Hutton's theory was attacked chiefly because of its extreme emphasis on subterranean heat as a geologic agent, there was a strong undertone of hostility to his uncompromising uniformitarianism and to the notion of a long succession of terrestrial worlds, each formed from the debris of its predecessor. Many confirmed Vulcanists parted company with Hutton on these points. Desmarest, for all his staunch advocacy of the igneous origin of basalt and the importance of volcanic activity in earth history, could not accept Hutton's inclusion of granite among the igneous rocks, his theory of the consolidation and elevation of strata by subterranean heat and pressure, or his idea of a perpetual succession of worlds. Though willing to concede the "remote antiquity" of nature's operations on the surface of the globe, Desmarest still conceived them as a series of modifications of a "primitive earth" which had been formed in the depths of a primordial ocean (52). Likewise Dolomieu (the French geologist whose pioneer work in volcanic studies was commemorated in the naming of the mineral *dolomite*), remained more Wernerian than Huttonian in his general view of earth history (53). The Wernerians, for their part, were firmly committed to the doctrine of a linear progression of revolutions aimed at fashioning the primitive earth into a residence for man. They were willing to allow "countless ages" for this work of nature, but they could see no necessity for accepting the staggering time-perspective implied in Hutton's theory. The Wernerian explanation of the succession of strata required no such infinity of time. "It presumes," wrote one of Werner's Scottish disciples, "not to carry its researches past the commencement of the present world, or to extend them beyond its termination; it is satisfied with endeavouring to trace the causes of the appearances which at present exist; and the characters of its explanations are these of fair and legitimate deduction" (54).

The main difficulty in Wernerian catastrophism was the problem of accounting for the supposed subsidence of the primeval ocean and for its mysterious resurgences from time to time. Many Wernerians threw up their hands, trusting to the future to provide a solution of the difficulty. In other quarters the search for a general cause of the revolutions which had racked the globe led to hypotheses drawn from astronomy. According to Desmarest, writing in 1794, the elevation of the continents above sea level had probably resulted from the nodding motion of the earth's axis and from a change in the inclination of the ecliptic to the plane of the equator, occasioned by a change in the earth's center of gravity.

> These events, recognized by most naturalists, have been sufficient to produce the most marked alterations on the surface of our globe: they have not only caused the sea to leave places which it once occupied and to submerge others, but also they must have altered the total position of the globe relative to the sun and consequently caused a total change in climates and influenced the individuals living there. That appears to furnish us a natural explanation of a great many phenomena which present themselves in the strata of the earth (55).

Among the advocates of this hypothesis Desmarest mentioned specifically Guillaume Rouelle, lecturer in chemistry at the Jardin des Plantes. Dolomieu was also inclined toward an astronomical explanation of the great events of earth history. In a letter written in 1788 he confessed his liking for Whiston's system, "with many modifications, however" (56). Eight years later he reported to Pierre Picot, professor of theology in Geneva, the result of some conversations with Laplace on this subject. "He told me," wrote Dolomieu, "that among causes exterior to our globe, there were none which could sensibly change either the center of gravity nor the level of the seas and that all the forces acting constantly on the earth concur to maintain it in its present state, but he did not know whether there might not be internal causes which, *accidentally,* could act and influence events on its surface; such things as those which might produce immense volcanic explosions. He thinks also that the *possible* shock of a comet may serve as a foundation for some geological systems" (57).

While Dolomieu toyed with the idea of an astronomical explanation of the history of the globe, new light was being thrown on the subject by meticulous examination of the organic remains contained in the strata composing the earth's crust. In the British Isles it was William Smith, a humble surveyor and self-made engineer, who developed the idea and technique of identifying and tracing geological strata by their fossil contents. Although his discovery was of major importance, Smith did not examine its bearing on the general theory of the earth. Instead, he confined himself to perfecting his stratigraphical maps of the British Isles, leaving it to his friend and associate, the Reverend William Townsend, to argue the agreement between Genesis and stratigraphical geology (58). Across the Channel the great Cuvier and his associate Alexandre Brongniart were undertaking a paleontological examination of the stratigraphy of Paris and its environs. The Paris Basin, they explained in their report of researches in 1811, was peculiarly well suited to investigations of this kind.

William Smith, 1769 · 1839

> The region in which this capital is situated is perhaps one of the most remarkable yet observed in respect to the series of different terrains composing it and the extraordinary remains of former beings which it contains. Millions of marine shells alternating regularly with fresh-water shells compose its principal mass; bones of terrestrial animals unknown even as to their genera fill certain parts; other bones of species of considerable size, whose modern counterparts are found only in far distant countries, are scattered about in the uppermost strata; a very marked indication of a great irruption coming from the southeast is imprinted in the forms of the promontories and the directions of the principal hills; in a word, there is no canton better suited to instruct us concerning the last revolutions which terminated the formation of our continents (59).

From their examination of these cantons Cuvier and Brongniart were able to distinguish eleven distinct geologic *formations*, using that word in Werner's sense of "an ensemble of strata of the same or different nature but formed at the same epoch." After listing these formations, they explained the technique of differentiating between them, then hazarded some conclusions as to the manner in which the calcareous strata had originated. They must have been formed sometime after the chalk and clay formations, since they rested upon them and differed widely from them in composition and content. Probably they were deposited slowly in a tranquil sea, since the strata were regular and the fossil shells perfectly preserved. Moreover, a considerable time must have elapsed in the process, for the fossil content varied from layer to layer, the number of species diminishing progressively until none were to be found in the upper strata. "The waters which deposited these strata

Georges Cuvier, 1769 • 1832

Alexandre Brongniart, 1770 • 1847

either no longer contained shellfish or had lost the property of conserving them" — a very strange circumstance, the authors added, since the modern species of shellfish seem identical with those known to classical antiquity. Finally, both the calcareous and the primitive strata bore marks of having suffered intensive alteration between the time of their formation and the period in which they were covered with new deposits. Many of the calcareous beds contained deep pits filled with foreign materials. These pits must have been excavated by some geologic agent and filled with iron-bearing clay, sand, and stones before the next layer was deposited over them. These operations must have required a considerable period of time. It was unfortunate, said the authors, that there were no data by which to form even a rough estimate of the time required.

Had Cuvier and Brongniart been willing to adopt Hutton's uniformitarianism, the problem might have appeared in a different light, but they were not ready for so bold a theory. It had been suggested, they noted, that the deep valleys in the secondary formations had been eroded by running water. But what a vast volume of water would have been necessary to cut through these strata of solid rock! Cuvier and Brongniart drew back from the abyss of time which seemed to open before them. "In geology," they concluded, "we must limit ourselves to the observation of facts, since the hypothesis which appears the simplest and most natural is subject to insoluble difficulties for the present" (60).

Such is the economy of nature, that no instance can be produced of her having permitted any one race of her animals to become extinct....

THOMAS JEFFERSON, 1785

...we are forced to submit to concurring facts as the voice of God — the bones exist, the animals do not!

REMBRANDT PEALE, 1803

4.

Lost Species

As the organic remains in the earth's crust were collected and compared with each other and with living plants and animals, an astonishing fact came to light — some of the fossil species were quite unlike any existing ones. John Ray was well aware of this fact and its implication that some links in the chain of creation had been lost. In his account of his travels through Europe, he noted the presence in every country he visited of petrified shells of various kinds, many of them different from any modern species. "Such are, for example, the *Serpentine* stones or *Cornua Ammonis* supposed originally to have been *Nautili*, of which I myself have seen five or six distinct species, and doubtless there are yet many more." "If it be said," he added, "that these species be lost out of the world: that is a supposition which philosophers hitherto have been unwilling to admit, esteeming the destruction of any one species to be a dismembring the universe and rendring it imperfect, whereas they think the divine providence is especially concerned to preserve and secure all the works of the creation." Ray spoke for himself as well as others here. Hard pressed by the assertion of his friend Dr. Martin Lister "that when he particularly examined some of our *English* Shores for Shells, and also the fresh Waters and the Fields, that he did never meet with any one of those *Species* of Shells found at *Adderton* in *Yorkshire, Wansford-Bridge* in *Northamptonshire,* and about *Gunthorp* and *Beavoir-Castle,* &c. any where else, but in their respective Quarries," Ray took the easy way out. "Why," he answered, "it is possible that many Sorts of Shell-Fish may be lodged so deep in the Seas, or on Rocks so remote from the Shores, that they may never come to Sight." (1).

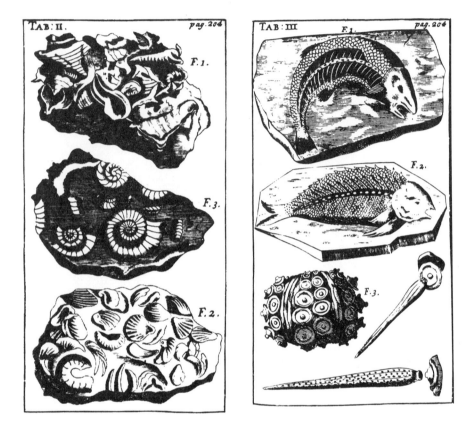

FIG. 4.1 — Organic remains embedded in stone, figured in the 3rd ed. of John Ray's **Three Physico-Theological Discourses.** Fig. 3 on the left shows the **Cornua Ammonis,** or "horns of Ammon," of which the remains of five or six distinct species, but no living species, were known to Ray. Some of the fossil specimens, Ray noted, were "about a foot in diameter, far exceeding the bulk of any shell-fish now breeding or living in our seas." Fig. 3 on the right represents: "A Sea-Urchin petrify'd with its Prickles broken off, which are a Sort of **Lapis Judaicus,** or **Jew-Stones;** their Insertions on the Studs or Protuberances of the Shell are here shown." Ray was extremely reluctant to admit that species might become extinct.

The question of nature's integrity presented itself most dramatically in repeated discoveries of huge bones, teeth, and tusks in both the New and the Old World. In 1706 Governor Joseph Dudley of Massachusetts was astonished to receive from two visiting Dutch colonists a huge tooth and several enormous bones which they said had been found under the bank of the Hudson River not far from Albany. They added that the ground where these objects were found was discolored for a distance of seventy-five feet and that all who saw the place judged that the discoloration measured the length of the creature whose bones were interred there. Dudley hastened to inform the learned Cotton Mather of this prodigy. The tooth, he wrote, was exactly like a human eye-tooth except for its size. It was nearly six inches high and thirteen inches in circumference and weighed two pounds three ounces. It resembled two other huge teeth recently discovered, one of them presented to Lord Cornbury, the other exhibited at Hartford. There could be no question, said Dudley, that the tooth was a human tooth. Perhaps its owner had belonged to one of the races of giants mentioned in Scripture and in Jewish tradition.

> The distance from the sea takes away all pretension of its being a whale or animal of the sea, as well as the figure of the tooth. Nor can it be the remains of an Elephant: the shape of the tooth, and admeasurement of the body in the ground, will not allow that.
> There is nothing left but to repair to those antique doctors for his origin, and to allow Dr. Burnet and Dr. Whiston to bury him at the Deluge; and if he were what he shows, he will be seen again at or after the conflagration, further to be examined (2).

In 1712 Cotton Mather, adopting Dudley's idea that the bones and teeth were those of an antediluvian giant, sent an account of them to Professor John Woodward, whose theory of the earth was well known in America. "It were to be wished," lamented Woodward in communicating the account to the Royal Society, "the Writer had given an exact Figure of these teeth and Bones" (3).

The Royal Society had already received an account of some "elephant bones" found near Gotha in Germany. In 1728 the newly elected President of the Society, Sir Hans Sloane, attempted to bring together the scattered bits of evidence relating to elephant bones and teeth found in the earth. From his own collection he described several specimens, including a huge tusk brought back from Siberia by the traveller John Bell. He then quoted from other travellers in Siberia concerning a flourishing trade in the tusks of a creature known as the *mammoth* in Russian folklore and represented as living underground. Next came a letter from Basilius Tatischow, Director-General of Mines in Siberia, describing the discovery of three great tusks, one of which had been presented to the Czar. These testimonies showed, said Sloane, that something more was involved than the remains of a few elephants brought into the north by Greek and Roman armies. The most probable conjecture was that the tusks and bones in question were those of elephants overwhelmed and swept away in the universal Deluge. "It is to be hoped that this Matter will one Time or other be set into a still clearer Light, particularly after the order his late Czarish Majesty [Peter the Great] was pleased to give to the Governor-General of Siberia, to spare no care nor Cost to find a whole Skeleton of this Animal, and to send it to Tatischow" (4).

In a second article, Sloane culled from ancient and modern authors a great number of references to discoveries of large teeth and bones supposed to be those of human giants. These remains, he surmised, were probably those of elephants and whales. The errors committed with respect to them should bring home to naturalists the importance of studies in comparative anatomy — "I mean to examine, with more Accuracy than hath been hitherto done, what proportions the Skeletons and Parts of Skeletons of Men and Animals bear to each other, with Regard either to Size, or Figure, or Structure, or any other Quality. This would doubtless lead us to many Discoveries, and is otherwise one of those Things, which seem to be wanting to make Anatomy a Science still more perfect and compleat" (5).

Among those dispatched by Peter the Great to explore the natural history of Siberia was a physician named Daniel Gottlieb Messerschmidt. Hearing of the discovery of a huge skeleton in a bank of the river Indigirska, Messerschmidt located one Michael Wolochowicz who had carried away the skull, a tusk, a molar tooth, and a thigh bone. Messerschmidt made drawings of these and took them with him to his home in Danzig, where he showed them to his colleague Dr.

John P. Breyne. Breyne had already read a paper before the local learned society on the subject of Siberian bones and teeth, declaring them to be the remains of elephants drowned in the Flood and swept by its waves into the northern regions of the globe. He encountered Sir Hans Sloane's discussion of the problem in the *Transactions* of the Royal Society shortly thereafter, and immediately sent Sloane a copy of his own dissertation accompanied by Messerschmidt's drawings and descriptions. These were printed in the Society's *Transactions* in 1737 (6).

FIG. 4.2 — The cranium of a woolly mammoth found in a river bank in Siberia by a Russian soldier and figured by Daniel Gottlieb Messerschmidt, a German physician and naturalist in the employ of Peter the Great. Messerschmidt's drawing was used by Cuvier in 1799 to establish the woolly mammoth as an extinct species of the elephant genus. A full skeleton of a woolly mammoth secured in Siberia in 1806 may be seen in Figure 4.11 (p. 118).

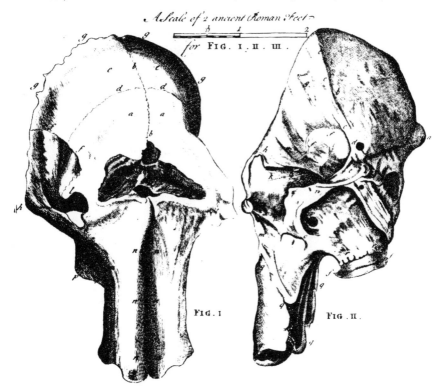

The scene now shifts to North America, where in 1739 the Baron de Longueuil, commandant of the French fort at Niagara, going down the Ohio to join forces with Lemoyne de Bienville against the Chickasaws, came upon some "elephant bones" at the edge of a marsh not far above the falls of the river. Longueuil collected some of these remains, including a tusk, a thigh bone, and several molars. When he returned to France the following year, he deposited these curiosities in the Royal Cabinet, where they were examined by Buffon and Daubenton and compared with similar objects brought back from Russia by Joseph Delisle, astronomer to Catherine the Great. Not until 1761, however, did Buffon introduce the readers of his *Natural History* to "the *prodigious mammoth*, a quadruped animal whose enormous bones we have often beheld with astonishment and which we have judged to be at least six times larger than the strongest elephant." Of this tremendous animal, said Buffon, not a trace was left except the bones, tusks and teeth found in great numbers in both the New and the Old World. If this, the largest of all terrestrial animals, had disappeared from the face of the earth, how many smaller species must also have perished in the long course of nature's vicissitudes! (7).

Buffon's anatomist, Jean Louis Marie Daubenton, was of a more cautious temperament. In 1762 he undertook to prove before the Royal Academy of Sciences that the Siberian mammoth and the creature of the Ohio country belonged to the same species as the common elephant of Asia and Africa (naturalists had not yet distinguished the Asian from the African species). To this end he presented a drawing of three thigh bones: the one found by Delisle in a Russian monastery, the one brought from America by Longueuil, and a third from an elephant formerly in the Royal Menagerie. He conceded that the bones differed in several respects, especially in size and thickness, but argued that these differences were no greater than might be expected to result from differences in age and sex within the same species. He then compared the tusks and found them similar in composition and substance though differing in size. There was but one jarring element in the comparative picture: the teeth from America were not composed like the others of successive layers of bone and enamel; instead the enamel surrounded an interior bony substance, as in the case of the pig family. This anomaly could be accounted for, said Daubenton,

FIG. 4.3 — The thigh bone of an American mastodon (Fig. 1, middle) compared with that of a Siberian mammoth (Fig. 2, bottom) and that of a modern elephant (Fig. 3, top), drawn by Jean Louis Daubenton for the **Memoirs** of the Royal Academy of Sciences in Paris, 1764. The mammoth bone was brought to Paris from Russia by a French astronomer; the mastodon bone from the Ohio Valley by a French officer stationed there. Daubenton argued incorrectly that the three bones belonged to animals of the same species, that is, that they were all "elephant bones."

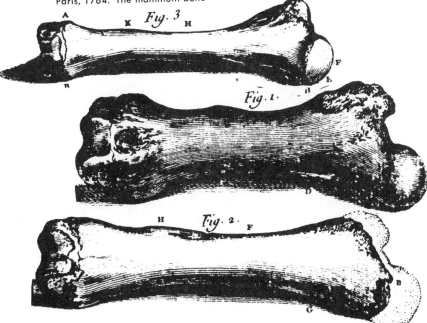

by supposing these teeth to have belonged to a giant hippopotamus whose remains became intermixed with those of the American elephant. Surely the Indian savages who brought the teeth and bones to Longueuil had taken little care to prevent the mixing of separate finds. These arguments apparently persuaded Buffon, for in 1764 he adopted Daubenton's view that the tusks and bones found in the northern regions of both hemispheres were those of elephants. The troublesome "hippopotamus teeth" were passed by in silence (8).

Meanwhile, scientific curiosity concerning the "elephant bones" of the Ohio had been aroused in America. In 1762 John Bartram, the Philadelphia Quaker who supplied Linnaeus and other European naturalists with descriptions and specimens of American plants and animals, heard that a large tooth and a fragment of a tusk had been brought to Fort Pitt by some Indians. He requested his friend James Wright to make inquiries among the Indians concerning the place where these objects had been found. Through an interpreter, Wright secured an account of the site which came to be known as Big Bone Lick. According to Wright's informants, the lick contained five entire skeletons, the heads pointing toward a common center. The bones were of enormous size and were accompanied by tusks ten or twelve feet long. No such creatures as these had ever been seen alive by the Indians, but legend said that they had once been hunted through the forests by men of gigantic stature and that when the last of these men had died, God had destroyed their mighty prey in order to protect the present race of Indians (9).

Among those best acquainted with the Ohio country at this time was George Croghan, deputy to Sir William Johnson in Indian affairs. Croghan, too, had heard of the salt lick strewn with giant bones; perhaps he had seen the specimens the report of which had aroused Bartram's curiosity. At any rate, he resolved to visit the lick at the first opportunity. In the spring of 1765 he was sent by General Gage to negotiate with the Indians of the Illinois country. He proceeded down the Ohio, and on the morning of May 31 left the river to find the lick. Passing through an open woodland, he and his party came upon a broad, hard-beaten buffalo trail which led them straight to their destination. After collecting an assortment of bones and tusks, they returned to their boats and went on down the river. A week later they were set upon by hostile Indians and the whole party was killed or captured. Croghan was wounded by a tomahawk, but he recovered and, obtaining his freedom, found his weary way back to Fort Pitt. His collection of bones was gone, but his interest in securing some of the giant fossils was undiminished. On a second expedition into the Illinois country in the following year he made a new and larger collection and managed to send it safely down the Mississippi and thence to New York. There the treasure was divided into two shipments consigned to London — one destined for his friend Benjamin Franklin, the other for his superior, Lord Shelburne (10).

The arrival of these shipments in London in February, 1767, created a great stir among the literati. In November and December of that year Peter Collinson, the Quaker naturalist who subsidized John Bartram's researches into American natural history, presented an account of Croghan's "tusks and grinders" to the Royal Society. The tusks, he informed the Society, were like those of an elephant. The grinders, on the other hand, had a peculiar knobby surface which seemed adapted for browsing on trees and shrubs. The creature possessing these grinders, "wherever it exists," must be herbivorous. It might be a new species of elephant, or possibly a tusked animal quite different from the elephant in size and shape. If an elephant, it presented a difficult problem for naturalists. The Deluge served to explain the presence of elephant teeth and bones in Siberia: "But what system or hypothesis, can, with any degree of probability, account for these remains of elephants being found in America, where those creatures are not known ever to have existed, is submitted to this learned Society" (11).

The problem posed by Collinson was taken up by the celebrated anatomist William Hunter. Hunter ransacked London for fossil bones and teeth and turned up a considerable number in the British

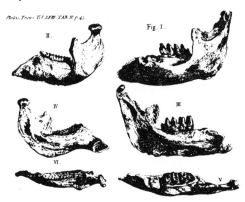

FIG. 4.4 — The lower jaw of an American mastodon (Figs. I, III, V, on the right) compared with that of an African elephant (Figs. II, IV, VI, on the left) by the British anatomist William Hunter in the **Philosophical Transactions** of the Royal Society of London in 1768. The mastodon jaw came from the collection of bones and teeth gathered by the American trader George Croghan at Big Bone Lick in Kentucky and shipped by him to Franklin and Lord Shelburne in London.

Museum, the Tower of London, and in private collections. He inspected the Croghan shipments and compared his findings with those reached by Daubenton and Buffon in their examination of the Longueuil collection. On February 25, 1768, he presented his conclusions to the Royal Society. The supposed American elephant he declared to be "a *pseud-elephant, or animal incognitum.*" In support of this position, he called attention to the many differences observable in Daubenton's drawing of the three thigh-bones — differences in the shape and proportion of the "head" and in the length and direction of the "neck," as well as in the overall size of the bones. He then presented his own comparison of the lower jaw in Lord Shelburne's collection with the lower jaw of an ordinary elephant. These two jaws could not possibly belong to animals of the same species, he declared. The American creature was plainly carnivorous. In size of tooth and bone it could be approached only by the Siberian incognitum. The American and Siberian forms would probably prove to be of the same species. If so, concluded Hunter, "it should follow that the *incognitum* in former times has been a very general inhabitant of the globe. And if this animal was indeed carnivorous, which I believe cannot be doubted, though we may as philosophers regret it, as men we cannot but thank Heaven that its whole generation is probably extinct" (12).

It need hardly be said that Franklin took a lively interest in the discussion precipitated by Croghan's tusks and grinders. On first inspecting the teeth, he thought that they must belong to a carnivorous animal. Later he adopted Collinson's view that the creature was too heavy and slow to prey on other animals and that the protuberances on the grinders would serve just as well for champing branches and shrubs as for chewing flesh. He could see no reason for making a separate species of the "American elephant," but he was puzzled by the fact that its remains were found in latitudes much too cold for elephants. Perhaps, he suggested to one of his correspondents, the climate of the various parts of the globe had been altered by a shift in the position of the earth's axis (13).

Collinson lost no time in communicating these latest discoveries to Buffon. He had already written him concerning some enormous hippopotamus-like teeth brought to London by a Mr. Greenwood, who claimed to have found them at a salt lick in America which contained six enormous skeletons with tusks five or more feet long. Collinson now sent two grinders from the Croghan collection with a full account of the circumstances of their discovery and a full confession of his own perplexity concerning them. They were clearly not elephant teeth, yet they were found intermixed with tusks indistinguishable from those of elephants. Could one suppose that there existed an animal with the tusks of an elephant and the teeth of a hippopotamus? In 1778 Buffon printed Collinson's letter in the "Notes" to his *Epochs of Nature* and addressed himself to the queries contained in it. The teeth found by Croghan, he declared, were much larger and more pointed than those brought back from the Ohio country by Longueuil; moreover, they failed to exhibit the clover-leaf pattern typical of hippopotamus teeth. Though found near tusks and elephant bones, they were too different from elephant teeth to warrant the assumption that they really belonged with those remains. Instead, they proved the existence of a third huge animal, differing from the elephant and the hippopotamus as well. This giant *incognitum* must have inhabited both the New and the Old World, for its peculiar teeth had been found in Siberia as well as in North America. Hence naturalists must not only explain the presence of huge elephants and hippopotami in the northern latitudes of the globe at some remote period but must also make room for a hitherto unknown species larger than all the rest and undoubtedly extinct.

The solution to these problems Buffon found in his theory of the earth. Let it be assumed that the earth began its career as a molten mass struck off from the sun and gradually cooled from the poles toward the equator. The northern latitudes would be the first region suited to sustain animal life, and the quadrupeds which came into existence there would be of extraordinary size owing to the intensity of the life-generating forces at this stage of the earth's history. As the globe cooled and these forces diminished in intensity, some species would perish entirely: such had been the fate of the animal with the huge teeth and of the inhabitants of the enormous spiral shells called *cornua Ammonis*. Other species, such as the elephant, the rhinoceros, and the hippopotamus, would move southward with the changing climate, diminishing in stature as the forces of terrestrial nature lost their pristine vigor. Thus, the history of the globe could be read from

99

the bones, tusks, teeth, and shells strewn over its surface or embedded in its crust (14).

The sensation produced in European scientific circles by Croghan's tusks and grinders stimulated American interest in the place where they had been found. Thomas Hutchins marked the site "Big Bone" in his map of the western country published in 1778 and acquainted his readers with Hunter's opinions concerning the animal whose remains had been found there. John Filson, an early historian of the American frontier, expanded on this theme in his *Discovery, Settlement, and Present State of Kentucke* (1784) and echoed Hunter's relief that the monstrous carnivore of North America and Siberia was probably extinct. In 1785 General Samuel H. Parsons, finding himself near the famous Lick, took time off from his explorations to gather some specimens. A few lay at hand on the surface of the ground; the rest had to be dug out of the quagmires bordering the brackish creek which meandered through the valley. He failed to unearth a complete skeleton but carried off four hundred pounds of tusks, bones, and grinders to distribute among the literati of New England. In the following year the *Columbian Magazine* carried an article describing some fossils brought back from the Ohio country by one Major Craig. The anonymous author denounced as "an idea injurious to the Deity" Hunter's opinion that the giant quadruped of the Ohio was extinct. "I believe our globe, and every part and particle thereof, came out of the hand of its creator as perfect as he intended it should be, and will continue in exactly the same state (as to its inhabitants at least) till its final dissolution" (15).

Meanwhile new evidence had come to light in an unexpected direction. In the fall of 1780 the Reverend Robert Annan, a country parson living in Orange County, New York, fifteen miles west of the Hudson River, found four huge teeth while inspecting the progress of swamp-draining operations on his farm. He took them to the parsonage, had them washed, and placing them in what he considered to be their proper order, gazed at them in mute astonishment. He called in a neighbor to behold the miracle, and the two men soon fell to

100

digging at the place where the grinders had been found. They dug up bones in great numbers but too badly decayed to permit handling. Word of the discovery soon reached New York. General Washington came to see the teeth and pronounced them similar to one which had been brought to him from the Ohio country. After the treaty of peace, Dr. Frederick Michaëlis, physician-general to the Hessian troops, came with several companions to dig for bones. The project was hampered by rain, but Michaëlis secured a few specimens to take back to Germany. Another lot was sent to Semittien's Museum in Philadelphia. In 1793 Annan published an account of his discoveries in the *Memoirs* of the American Academy of Arts and Sciences. After describing the grinders in detail, he raised some of the questions stirring in his mind.

> Shall we suppose the species to be extinct over the face of the globe? If so, what could be the cause? It is next to incredible that the remains of this animal could have lain there since the flood. May there not be some of the kind yet surviving, in some of the interior parts of the continent? Comparatively little of it has yet been explored. Some gentlemen...have supposed that their extinction (as it is probable they are extinct) is owing to some amazing convulsion, concussion, or catastrophe, endured by the globe. But I know of none that could produce such an effect, except the flood (16).

News of these discoveries reached Jefferson in Paris through the learned President of Yale College, Dr. Ezra Stiles. It especially interested Jefferson, for he had just published his own views concerning the American mammoth in a small, private edition of his *Notes on Virginia*. In that work he boldly took issue with Buffon and Daubenton. There could be no justification, he declared, for separating the grinders found at Big Bone Lick from the tusks and bones accompanying them, or for regarding them as hippopotamus teeth. They were the only teeth found with the tusks and bones, and none of the bones were those of a hippopotamus. "We must agree then, that these remains belong to each other, that they are of one and the same animal, that this was not a hippopotamus, because the hippopotamus had no tusk nor such a frame, and because the grinders differ in their size as well as in their number and form of their points" (17). Neither

101

could the animal be an elephant. Its teeth were not those of an elephant, and its remains were found in regions much too cold for either the elephant or the hippopotamus. To imagine that the northern regions of the earth once enjoyed a tropical climate was fanciful and unnecessary. Why not suppose, instead, that the mammoth was a creature similar to the elephant in some respects but quite different in others? Why not suppose it a being *sui generis,* a huge carnivore adapted by its Creator to the cold climate of the north? The species need not be very numerous, but there was no reason to suppose it extinct. Quite the contrary: "Such is the economy of nature, that no instance can be produced, of her having permitted any one race of her animals to become extinct; of her having formed any link in her great work, so weak as to be broken" (18).

Jefferson left Paris without having converted the aged Buffon to his view. Back home, he tried again to get some of the bones and teeth which had occasioned the controversy. He got more than he had bargained for when, in 1796, he succeeded in acquiring some remains turned up by workmen excavating for saltpeter in a cave in western Virginia. These remains proved beyond a doubt that still another *incognitum* had existed, a huge, clawed creature quite unlike the monster of the Ohio and the Hudson. Jefferson named it *megalonix,* or "great-claw," and wrote an account of it for the American Philosophical Society. In this memoir he conjectured that the animal was a huge lion more than three times the size of the African lion described and anatomized in Buffon's *Natural History.* He rejoiced at being able to refute Buffon's aspersions on the size and vigor of nature's productions in the New World and declared it likely that the *megalonix,* like the mammoth, would be found still living in some part of the world. "The movements of nature," he asserted, "are in a never-ending circle. The animal species which has once been put into a train of motion, is still probably moving in that train. For if one link in nature's chain might be lost, another and another might be lost, till this whole system of things should vanish piecemeal, a conclusion not warranted by the local disappearance of one or two species of animals, and opposed by the thousands of instances of the renovating power constantly exercised by nature for the reproduction of all her subjects, animal, vegetable, and mineral" (19).

102

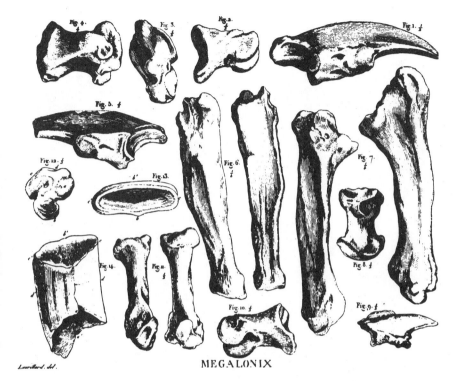

Lacrillard. del.

MEGALONIX

FIG. 4.5 — Bones of an extinct giant sloth **(Megalonix jeffersoni),** found in a cave in western Virginia in 1796 and described by Jefferson and Dr. Caspar Wistar in the **Philosophical Transactions** of the American Philosophical Society in 1799. Jefferson inferred from the animal's claws (upper right, Fig. 1) that it was a giant lion, but Dr. Wistar inferred more correctly that the creature was similar in some respects to the modern sloth and in others to an animal whose bones had recently been found in Paraguay and mounted in Madrid (see Figure 4.6, p. 104). In 1804 Cuvier confirmed Wistar's opinion in his memoirs on the **megalonix** and the **megatherium.**

The description of the same remains by Dr. Caspar Wistar, Professor of Anatomy in the University of Pennsylvania, was less eloquent than Jefferson's but far more accurate and perceptive. Wistar's opinion, offered diffidently because he had nothing but drawings with which to compare the Virginia remains, was that the *megalonix* was a creature similar in some respects to the bradypus, or sloth, drawn by Dau-

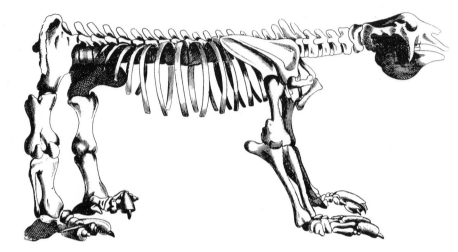

FIG. 4.6 — The first skeleton of a **megatherium** known to science, unearthed in Paraguay and mounted in the Cabinet of Natural History at Madrid. The figure, "drawn by G. Cuvier," appeared in **The Monthly Magazine and British Register** for September, 1796, where it was seen by Dr. Caspar Wistar of Philadelphia and compared with the bones procured from western Virginia in the same year by Thomas Jefferson, who named the creature thus discovered **megalonix**, or "great claw." Jefferson thought that it was a huge lion, but Wistar surmised more correctly that both the **megalonix** and the **megatherium** belonged to the family of edentates, or clawed quadrupeds without cutting teeth, such as the sloths and anteaters.

benton for Buffon's *Natural History* and in other respects to the Paraguayan creature whose skeleton, mounted in the Royal Cabinet at Madrid, had been depicted in a recent issue of the *Monthly Magazine* in London, yet not of the same species with either. This conjecture was verified in 1804 when Cuvier published his studies of the *megalonix* and the *megatherium*. "These two animals," said Cuvier, "will then have formed two species of a single genus, belonging to the family of edentates and intermediate between the sloths and the anteaters, but closer to the first than to the second" (20). For the evidence on which his conclusions concerning the *megalonix* were based Cuvier acknowledged his debt to Jefferson.

Within a few months of receiving Jefferson's memoir on the *megalonix* the American Philosophical Society heard his views on the mammoth attacked by George Turner. While serving as a federal judge in Cincinnati, Ohio, Turner had visited Big Bone Lick and collected some specimens. Upon comparing these with the remains described in various scientific journals, he became convinced that the remains of the carnivorous "mammoth" had been confused with those of a herbivorous creature – "some link in the chain of animal creation, which, like that of the Mammoth, has long been lost." With apologies to Jefferson he declared it unlikely that either of these species was still living. Such huge creatures could not exist unnoticed by man. But although the notion of an inviolable chain of being must be abandoned, it must not be supposed that the extinct species had perished from a defect in their original constitution. "The Author of existence is wise and just in all his works. He never confers an appetite without the power to gratify it." Not nature, but man himself, was undoubtedly responsible for the extermination of these monsters, particularly the carnivorous mammoth (21).

An argument similar to Turner's was presented to the Society by the Reverend Nicholas Collin, Rector of the Swedish Churches in Pennsylvania. Collin suggested that the primitive earth had been in-

FIG. 4.7 – A modern mounting of a **megatherium** skeleton.

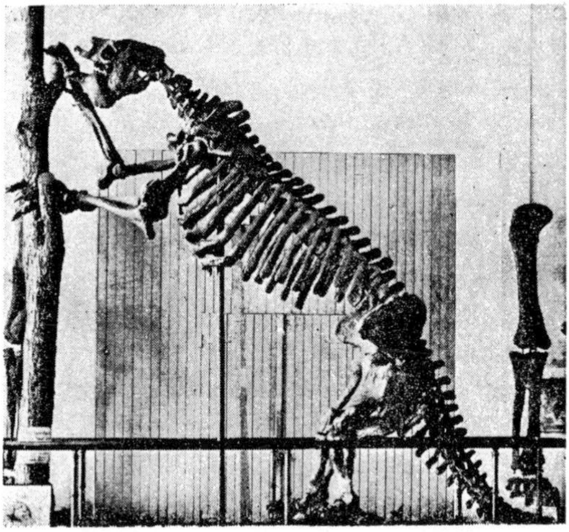

Courtesy of the American Museum of Natural History

habited by barbarous men and monstrous animals and that some of these animals became extinct in the course of man's moral and physical improvement. As mankind increased in numbers and in mastery over nature, large land quadrupeds must have been deprived of their food supply and their places of refuge. In new countries like America their extinction must have been recent, since their remains were often found well preserved in exposed places (22).

Abraham Bradley took issue on this point in his *New Theory of the Earth,* published at Wilkes-Barre, Pennsylvania, in 1801. He argued that the mammoth must have lived before the Deluge because God would not let any of the present creation perish until the extermination of the whole. Governor John Drayton, in his *View of South Carolina* (1802), cited certain teeth and bones found during excavations for a canal between the Santee and Cooper Rivers to refute Jefferson's idea that the mammoth was confined to northern latitudes, but he was equally sceptical of Buffon's theory of the gradual cooling of the globe from the poles toward the equator. That theory, said Drayton, created "a blank in nature" by supposing the earth to have been uninhabited for long periods of time. The Deluge was another possible explanation, but Drayton refused to comment further. Retreating from the "deep labyrinths of philosophical surmise," he declared himself content to adore "that superintending Providence, which, from a mouse, to this object of our admiration, has continued a complete chain of animated nature" (23).

In Europe the mystery of the "elephant bones" was gradually being unraveled. In the last decade of the eighteenth century three great comparative anatomists — Petrus Camper, Georges Cuvier, and Johann Friedrich Blumenbach — arrived independently at the conclusion that there were two species of living elephants — the African and the Asian — and that the Siberian mammoth constituted an extinct species of the same genus. Cuvier was led in this direction when he came across Messerschmidt's drawing of the cranium of a Siberian mammoth in the *Philosophical Transactions* of the Royal Society for 1737. On comparing the figure with his own drawings of the Indian elephant, he was struck by the many differences in the conformation of the skull. When he added these differences to those he had already observed in the lower jaw and some isolated teeth, he could no longer

FIG. 4.8 — The **Megalonix jeffersoni** reconstructed by modern paleontologists.

doubt that the Siberian creature was a separate species of elephant hitherto unknown. This discovery opened an exciting new vista to Cuvier's imagination. Perhaps the mammoth was but one of many extinct species which had once roamed the earth and whose bones were now buried in its crust. Perhaps the whole array of these lost species could be reconstructed by the techniques of comparative anatomy. Thus inspired, Cuvier began the long series of arduous researches which culminated in his great work on fossil bones.*

One of the first fruits of these researches was his memoir "On the Species of Elephants Living and Fossil," read before the National Institute of Arts and Sciences in 1796 and published in its *Memoirs* in 1799. In this essay Cuvier described the African and Asian elephants and presented his reasons for thinking that the Siberian mammoth was specifically different from both. Turning then to the North American remains, he reviewed the long controversy concerning the "hippopotamus teeth" found repeatedly in association with tusks and bones similar to those of the elephant. There could no longer be any question, said Cuvier, that the teeth belonged with the tusks and bones. Yet these teeth were quite unlike those of any known elephant. The crown had only three or four pairs of blunted points, and the laminations were fewer in number and thicker than those of elephant teeth. The huge "clover-leaf" teeth which had puzzled Buffon differed from these only in being worn down by long use. The creature who chewed with them was probably no larger than the Indian elephant, though heavier and thicker of bone. It was probably equipped with a trunk to enable it to reach its food. In short, it was a species of elephant, but a species distinct both from the African and the Asian elephant and from its extinct Siberian cousin, the *Elephas mammonteus*. For its name Kerr's *Elephas americanus* would do, although its remains had been found in both the Old World and the New (24).

Having finished with the subject of elephants, Cuvier gave his listeners a glimpse of the vanished world soon to be revealed by paleontological research.

> Let one add to these two examples of animals whose originals are not known the crocodile of the mountain of St. Pierre at Maestricht, concerning which citizen Faujas will prove...that it differs from the crocodile of the Nile and from that of the Ganges; the quadrupeds of the caverns of Gaylenreuth in the Anspach country, which had been thought related to the polar bear, and which I shall prove to differ from it considerably; the fossil rhinoceri of Siberia, which I

---

* *Recherches sur les ossemens fossiles de quadrupèdes* 4 v. (Paris: 1812) and later editions.

shall prove not to resemble either those of Africa or those of the Indies . . .; the petrified horns of a species of the deer genus which are not those of the elk, nor the reindeer, nor of any known species and which are also found in the mountain of St. Pierre. Let one ask why so many remains of unknown animals are found, whereas almost none is found which can be said to belong to known species, and one will see how probable it is that they have belonged to creatures of a world anterior to ours, to creatures destroyed by some revolutions of our globe; beings whose place those which exist today have filled, perhaps to be themselves destroyed and replaced by others some day (25).

In America the search for fossil remains continued to gain momentum. In the year that Cuvier published his memoir on elephants, the American Philosophical Society sent out a circular letter intended to stimulate discoveries relating to the natural history of America. Signed by Jefferson, Wistar, Turner, and Charles Willson Peale the letter called particular attention to the importance of finding "one or more skeletons of the Mammoth, so called, and of such other unknown animals as either have been, or hereafter may be discovered in America" and suggested Big Bone Lick as a likely place for accomplishing this object. In that same year workmen digging marl on the farm of John Masten in the town of Shawangunk, New York, turned up a thigh bone three feet nine inches in length and eighteen inches in circumference in its narrowest part. Other bones soon came to light; a crowd gathered from the neighborhood as word of these discoveries spread. The farmer, a thrifty, hardheaded German, could see no value in the bones and would have stopped the search if the local minister and several physicians from the vicinity had not dissuaded him. It then occurred to Masten that the bones might be of some value, and he invited everybody to help dig for them. For three days nearly a hundred men toiled in the wet earth, urged on by the excitement of discovery, the hope of profit, the plaudits of the attending gentry, and the ever-flowing bottle. Plenty of bones were dug up, but many were broken or left behind in the unruly enthusiasm of the search. The constant flow of spring water into the pit made digging difficult. Finally the neighborhood was ransacked for bailing implements, but by the fourth day the water had risen so high the enterprise was abandoned.

Word of these events soon reached the learned world. In a letter to the *Medical Repository*, Dr. James G. Graham, one of the physicians who had witnessed the digging, linked the bones and teeth discovered on the Masten farm to others found in the same vicinity.

> Some time in 1782, several of them were discovered in a meadow or swamp about three miles south of Ward's Bridge, in the town of Montgomery, now in the county of Orange, three or four feet below the surface, most of them much decayed. The next discovery of them was made about one mile east of said bridge. In this place three or four ribs were found, about eight feet below the surface, in a very sound state. The swamp here does not contain more than three or four acres, and the remaining bones of the skeleton probably yet remain at its bottom. About three miles east of said bridge some other bones have been found; and about seven miles east of said bridge, a tooth (one of the grinders), and some hair, about three inches long, of a dark dun color, were found by Mr. Alexander Colden, four or five feet below the surface. These I procured, and sent them to Dr. Bayley, of New-York, who has, I am informed, deposited them in Columbia College. And last week another skeleton has been discovered, about three miles east of my house, in the town of Shawangunk, about ten miles north-east of said bridge. These last discovered bones lie about ten feet from the surface and are in a very sound state...These large bones are uniformly found in deep wet swamps only, by farmers, in digging up black mould and *marl* for the purpose of manuring their lands (26).

This variety of discoveries within a small area indicated, added Graham, that the unknown animal to whom the bones belonged had once flourished in great numbers along the Hudson River. "And why Providence should have destroyed an animal or species it once thought proper to create, is a matter of curious inquiry and difficult solution."

A further description of the bones found at Shawangunk came from one Sylvanus Miller, who heard of the discovery as he was passing through Newburgh and turned aside to investigate. He was rewarded by the sight of several leg bones, ribs, and vertebrae, the upper part of a head, a thigh bone five feet long and more than forty inches around the joint, some teeth nearly seven inches long and four inches broad, and some foot bones suggesting a clawed animal. Like Graham,

Miller felt that the occurrence of so many huge bones and teeth in the same district was a phenomenon demanding an explanation "to satisfy the public mind." The editors of the *Repository* responded by publishing a brief account of Cuvier's researches on fossil quadrupeds. The list of extinct species, they said, had reached twenty-three, including the mammoth, the long-headed rhinoceros, the great fossil tortoise, and "a sort of dragon." "But these are not all which the earth contains: there are *parts* of skeletons of which M. Cuvier cannot speak with equal assurance; but of which, however, enough is known to encourage a hope that the list of zoological antiquities will soon be lengthened" (27).

To many naturalists the Shawangunk discoveries seemed to present the long-awaited opportunity to secure a complete skeleton of the American *incognitum*. Dr. Caspar Wistar solicited President Jefferson's aid in this project. Jefferson forwarded the request to Chancellor Livingston in New York, only to learn that Livingston had already tried to get some specimens but had lost out in the general rush, "the whole town having joined in digging for them till they were stopped by the autumnal rains" (28). Charles Willson Peale, creator of Peale's Museum, now took up the search. He had been interested in finding a skeleton of the mammoth ever since the day in 1784 when Dr. Samuel Brown of Lexington, Kentucky, brought him some bones from Big Bone Lick to have drawings made of them for the German savant Frederick Michaëlis. The popular interest aroused by these bones when Peale placed them on temporary exhibit in his picture gallery gave him the idea of adding a museum of natural history to his artistic exhibits. Thus, when news of the discoveries on the Masten farm reached him in Philadelphia in the spring of 1801, he set out at once for Shawangunk. On arriving there, he found the bones spread out in Masten's granary, where they could be inspected for a slight fee. Peale obtained permission to sketch them, then entered into negotiations for their purchase. His offer of two hundred dollars for the bones in the granary and one hundred for the right to dig out the rest was finally accepted, with a gun for the farmer's son and some gowns for his daughters thrown in for good measure. Peale returned to Philadelphia with the bones and, putting them together as best he could, found that they formed an incomplete skeleton of huge size. Delighted with this result, he set out for Shawangunk again, accom-

111

panied by his son Rembrandt, Professor James Woodhouse, and several others. Supplies were purchased in New York with money lent by the American Philosophical Society, and the expedition arrived at the bone pit early in September. In spite of the unusually light rainfall that year, the pit was filled with water. A whole week was consumed in constructing a bailing machine operated by three or four men walking side by side in a wheel twenty feet in diameter. The machine worked efficiently, but the pit was never entirely free of water, and the soft marl oozed out constantly from the sides of the pit. For several weeks twenty-five men labored in the cold muck, vying to entertain and astonish the great crowd of spectators with their discoveries. Travellers on the highway stopped to see what could cause vehicles, horses, men, women, and children to gather around such a strange-looking apparatus. Some stayed to lend a hand at the wheel. Many of the missing parts of the skeleton were turned up, but several of the tail and toe bones, the top of the head, and — most disappointing of all — the lower jaw could not be found. A man less resolute than Peale would have been satisfied with this result, but Peale was determined to have a complete skeleton and, if possible, one found *in situ.* He tried again at a small morass eleven miles distant from the Masten farm. Failing in his object there, he crossed the Walkill and tried a third time in a marshy woodland on the farm of one Peter Millspaw. The digging proceeded for several days with little success. Meanwhile, the nearby area was sounded with pointed iron rods for evidence of buried bones. At last, when both Peale and his workers were ready to admit defeat, Rembrandt struck a collection of solid objects with his sounding iron. Once again the shovels bit into the muck, and this time the weary workers met with success. In rapid succession they dug up a thigh bone, two arm bones, a scapula, and several toe bones, and then, wonder of wonders, a complete lower jaw. Delight and astonishment were on every face — "the unconscious woods echoed with repeated huzzas, which could not have been more animated if every tree had participated in the joy. 'Gracious God, what a jaw! how many animals have been crushed between it!' was the exclamation of all: a fresh supply of grog went round, and the hearty fellows, covered with mud, continued the search with encreasing vigor" (29).

Peale now had two fairly complete skeletons and the remnants of a third. The next three months were occupied in mending broken pieces and assembling the two best skeletons. The missing parts in each were carved out of wood from their counterparts in the other. Only the top of the head and the end of the tail had to be omitted for lack of an original. One of the skeletons was mounted in the southeast chamber of Philosophical Hall, and on Christmas Eve the members of the American Philosophical Society and several foreign dignitaries were invited to inspect it. Soon afterward the doors were thrown open to the public. Now, for the moderate price of fifty cents, the man on the street might see for himself the skeleton of the animal whose tusks and grinders had set the scientific world agog and whose disappearance from the earth, if it should be proved, left a yawning gap in the seamless fabric of creation. The public was quick to take advantage of the opportunity; Peale's mammoth was a popular, as well as a scientific, sensation.

Encouraged by the success of his exhibit, Peale resolved to send two of his sons, Rembrandt and Rubens, to Europe with the second skeleton. A preliminary exhibition in New York garnered two thousand dollars for the trip, and by the end of June, 1802, the young Peales were on their way to England with their precious cargo. The mammoth was placed on display in Pall Mall, but despite Rembrandt's energetic publicity campaign, it was not the success it had been in New York. Efforts to sell the skeleton were unavailing. Financial difficulties and the troubled international situation led to the abandonment of the proposed tour of Europe, and the Peales returned to America late in 1803, hoping to recoup their fortunes by exhibiting the mammoth in American cities. The second skeleton eventually became a prize exhibit in Rembrandt Peale's museum in Baltimore. It was later purchased by Barnum (30) .

Although the London exhibition was not a financial success, it had important scientific consequences. During its course, Rembrandt Peale published two accounts of the skeleton, one a brief brochure which was reprinted in the *Philosophical Magazine,* the other an extended work entitled *An Historical Disquisition on the Mammoth, or Great American Incognitum, an Extinct, Immense, Carnivorous Animal, Whose Fossil Remains Have Been Found in North America.* In

113

this book he attempted to settle the long-standing controversy concerning the tusks, bones, and grinders found in the Ohio country. Disclaiming any further acquaintance with comparative anatomy than could be obtained from perusing the works of Buffon, Hunter, Camper, Cuvier, and others, he nevertheless declared his competence as an artist to detect differences of anatomical form. He then launched into a comparative description of the skeleton which he was exhibiting in London and reached the conclusion that Hunter had been right in regarding the American *incognitum* as an extinct carnivore habituated to northern climates. The complete skeleton proved beyond doubt, said Peale, that the grinders, tusks, and bones found lying near each other by Croghan and Longueuil really belonged together. The conformation of the grinders proved their adaptation for champing animal food — probably shell-fish — and precluded the idea of a herbivorous creature. The tusks differed from those of the elephant both in curvature and in composition. It was plain, then, that the mammoth was a creature *sui generis*. Its disappearance from the face of the earth should occasion no surprise now that Cuvier had described and classified the remains of twenty-three kinds of extinct animals. In America alone the vestiges of at least four extinct species had been found. Besides the tusks and grinders of the American mammoth, the Ohio country had yielded a number of teeth similar to those of the Siberian mammoth and a skull fragment from a giant ox or buffalo.* From western Virginia had come the bones and claws of a huge creature similar in many respects to the extinct sloth, or *megatherium*, of Paraguay. These undeniable facts demonstrated that the surface of the earth had undergone profound revolutions and that the American continent had been linked to the Old World at the time when these races of giant animals lived.

How long these animals have existed, we shall perhaps ever remain in ignorance, as no judgment can be formed

---

*The skull referred to was found in a creek bed a few miles north of Big Bone Lick by the same Dr. Samuel Brown who brought the giant bones to Peale's father in 1784. C. W. Peale and Jefferson had asked Brown to search for the missing head bone of the mammoth. Rembrandt Peale's account of the skull fragment which Brown found appeared in *The Philosophical Magazine* in 1803 titled "Account of Some Remains of a Species of Gigantic Oxen Found in America and Other Parts of the World."

from the quantity of vegetable soil which has accumulated over their bones. Certain we are that they existed in great abundance, from the number of their remains which are found in America. We are likewise sure that they must have been destroyed by some sudden and powerful cause; and nothing appears more probable than one of those deluges, or sudden irruptions of the sea, which have left their traces (such as shells, corals, &c.) in every part of the globe. It is, therefore, extremely probable that whenever and by whatever means the extirpation of these tremendous animals was effected, the same cause must have operated in the destruction of all those inhabitants from whom there might have been transmitted some satisfactory account of these stupendous beings, which at all times must have filled the human mind with surprise and wonder (31).

Peale's performance as a comparative anatomist was inevitably deficient, but the recovery, exhibition, and description of an entire skeleton of the American *incognitum* was of invaluable aid to Cuvier in untangling the mystery of its identity. Before this skeleton became known, the anatomists of Europe had been working with such fragmentary remains as the tusks and grinders collected by Longueuil and Croghan and a few isolated teeth found in Russia and Siberia. In 1785 Michaëlis had returned to Europe with the drawings which Charles Willson Peale had made of a piece of skull and some other bones found at Big Bone Lick by Dr. Samuel Brown and had showed

FIG. 4.9 — The **Mastodon americanus,** which once roamed the forests of North America and which Jefferson thought must still exist: "Such is the economy of nature, that no instance can be produced, of her having permitted any one race of her animals to become extinct. . . . ."

Courtesy of the American Museum of Natural History

them to the famous Dutch anatomist Petrus Camper. Camper, erroneously supposing that the skull fragment belonged to the anterior part of the skull and finding in it no place for the insertion of tusks, reversed his earlier opinion that the Ohio *incognitum* was a kind of elephant. Copies of the drawings were sent to Cuvier by Professor Autenrieth of Tübingen with the suggestion that the fragment really came from the posterior part of the skull. Camper having died in 1799, Cuvier wrote to his son, who by this time had obtained the original bones from which Peale had made the drawings. The son upheld his father at first, but finally conceded that Autenrieth's view was correct. At this juncture the Peales excavated the Hudson River skeletons and sent casts of the bones to Europe. Cuvier received a copy of the *Historical Disquisition* from Rembrandt Peale, and the British anatomist Everard Home sent him a drawing of the skeleton which Peale was exhibiting in London. Translations of the *Disquisition* and figures of the skeleton soon appeared in various learned journals throughout Europe. Thanks to the Peales, wrote Cuvier, the osteology of the American *incognitum* was entirely known excepting only the top part of the cranium.

In 1806 Cuvier undertook to describe, name, and classify all of the elephant-like creatures whose bones and teeth had been found in various parts of the globe. He was able to show that the huge teeth which had puzzled Collinson and Daubenton were similar to those of the hippopotamus in having no cement, or cortex, and in being adapted for feeding on roots, herbs, and aquatic plants. But they were like those of the elephant in their mode of succession: the new teeth pushed the older ones forward in the jaw, the size and number of points of each tooth depending on its position in this succession. From these and other comparisons Cuvier concluded that the American mammoth was not a true elephant but a member of a new genus, which he proposed to call the mastodonts. This genus would include not only Peale's animal — *le grand mastodonte* — but four other species whose teeth and bones had come to light: the mastodont with narrow teeth found at Simorre in southern France, the *petit mastodonte* of Saxony and Montabusard, and the two species whose remains Alexander von Humboldt had collected in the New World — the mastodont of the Cordilleras and the somewhat smaller *mastodonte humboldien*.

Among the true elephants Cuvier distinguished three species: the modern species of Asia and of Africa and the extinct species whose remains had been found in both hemispheres but most commonly in Siberia. The differences in the teeth of these three species, noted earlier by Camper and Blumenbach as well as by Cuvier, were matched by differences in the cranium and the lower jaw. Messerschmidt's drawing of 1737 was still the best representation of the cranium, but it could now be compared with a life-size drawing sent to Cuvier by the St. Petersburg Academy of Sciences of another fossil cranium in their possession, and these two drawings could be matched against the skull fragments which the French astronomer De Lisle had brought back from Siberia a half century earlier. Had Cuvier but known it, much ampler evidence concerning the Siberian mammoth was being unearthed on the frozen delta of the Lena River in the very year he

FIG. 4.10 — Skeleton of an American mastodon, exhumed by Charles Wilson Peale in 1801 in Orange County, New York, and exhibited by his son Rembrandt in London in 1802. This drawing, made by the British anatomist Everard Home, was reproduced by Georges Cuvier in a famous memoir of 1806, in which Cuvier distinguished the mastodon from the woolly mammoth and from modern elephants. Thanks to the Peales, Cuvier wrote, the bone structure of the mastodon was completely known, except for the top of the cranium (sketched in by Cuvier in this figure).

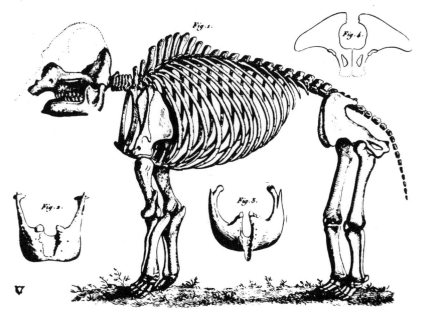

published his memoir. There Michael F. Adams, a member of the St. Petersburg Academy of Sciences, was engaged in collecting what remained of the carcass of a woolly mammoth which had tumbled down from an icy hillock. The trunk, tusks, tail, and one forefoot were missing, and dogs and wild animals had consumed most of the flesh, but Adams was able to secure a nearly complete skeleton by boiling away the meat and ligaments. This skeleton, first figured in 1815, was eventually to find its place in Cuvier's great *Fossil Bones,* but it would serve only to confirm conclusions he had drawn in 1806 from much scantier evidence (32).

Having distinguished the fossil elephant from living elephants of the same genus and from the mastodonts, Cuvier went on to list five other species of extinct pachyderm: one of the rhinoceros, two of the hippopotamus, and two of the tapir kind. It was significant, he noted, that many of these extinct species had lived outside the torrid zone and had been distributed between the eastern and western hemispheres in a much different manner from the present species of the same genera.

What theory of the earth could account for these facts? Buffon had inferred from the situations in which the giant bones and teeth were found — in gravel or marl pits near the surface of the earth — that the animals to which they belonged had lived at a relatively late period

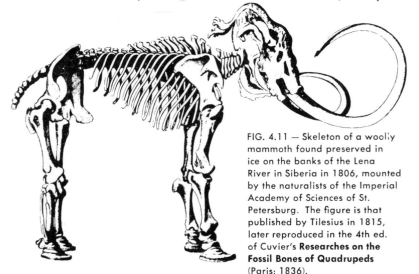

FIG. 4.11 — Skeleton of a woolly mammoth found preserved in ice on the banks of the Lena River in Siberia in 1806, mounted by the naturalists of the Imperial Academy of Sciences of St. Petersburg. The figure is that published by Tilesius in 1815, later reproduced in the 4th ed. of Cuvier's **Researches on the Fossil Bones of Quadrupeds** (Paris: 1836).

in the globe's history, sometime subsequent to the formation of thick beds of fossiliferous stone by marine deposition. Cuvier agreed. But he had no use for Buffon's further supposition that the climate of the globe had undergone a cooling process in the course of which various races of pachyderms had been driven southward to their present habitats and molded to their present form. There was no evidence for these supposed transformations, said Cuvier. The mummies recently unearthed in the tombs of Egypt indicated that the men and animals of the Nile Valley had substantially the same anatomical peculiarities today that their ancestors had had several thousand years ago. Moreover, the range of variation among fossil pachyderms was much the same as that found among northern animals at the present time. The extinction of these pachyderms had probably been caused by some great but transient inundation of the sea — "the last, or one of the last, of the catastrophes of the globe" (33).

While Cuvier was laying the foundations of vertebrate paleontology, his colleague Lamarck was inaugurating the systematic study of invertebrate fossils with a series of memoirs on invertebrate remains found in the vicinity of Paris. In the "Introduction" to the first of these, published in the *Annals* of the Museum of Natural History in 1802, he took issue with the view of earth history which Cuvier was developing to explain the phenomena of the fossil record. At first glance, said Lamarck, the well-marked strata of the globe with their well-defined contents suggest the idea of successive creations separated by geological upheavals.

> For man, who observes and who, in that regard, judges only according to the changes which he himself perceives, the intervals of these mutations are *stationary states* which appear limitless to him because of the brevity of existence of the individuals of his species. Moreover, since the record of his observations and the notes which he could jot down in his annals reach back only a few thousand years (three to five thousand years), a duration infinitely small relative to those in which the great changes which the earth undergoes are effected, everything seems *stable* on the planet which he inhabits, and he is led to ignore the hints which are presented to him on every hand by the *monuments* buried around him or intermixed with the soil which he tramples underfoot (34).

Close attention to these neglected hints, Lamarck went on, must eventually disclose that the processes which produced the fossiliferous strata of the earth's crust were still operating, embedding the remains of the present flora and fauna in layers of sand and dirt which would gradually harden into rock. In due time these slow changes must alter the relative positions of land and sea, modify the climate of the various parts of the globe, and transform the creatures which inhabited it. Thus, Lamarck added to the geologic uniformitarianism of Buffon and Hutton his own revolutionary theory of organic transformation and applied the combined force of these ideas to the interpretation of the fossil record.

In England the exploration of the fossil record and its implications for the traditional view of nature was undertaken by James Parkinson, a London surgeon who devoted his spare time to paleontological researches. Parkinson's three volumes entitled *Organic Remains of a Former World* appeared successively in 1804, 1808, and 1811. In the first volume, subtitled "An Examination of the Mineralogical Remains of the Vegetables and Animals of the Antediluvian World," the author proposed two directions of inquiry into the problems presented by the fossil record. The main reliance must be placed, he declared, on the examination of the chemical and mineralogical processes appointed by God to operate constantly on every mass of matter for the preservation of the earth and its inhabitants. But there was also the testimony of Scripture and ancient tradition that the earth had been devastated by a universal Deluge sometime after its formation and peopling. Therefore, when facts were encountered which could not be explained with reference to processes still in operation, they could perhaps be shown to result from the Deluge.

In effect Parkinson was proposing to combine uniformitarianism with catastrophism. His discussion of the formation of coal deposits illustrates the hopeless character of the undertaking. He began by supposing that coal was produced by "bituminous fermentation" of inundated vegetable materials. He then argued that a Deluge like that described in the Bible would be required to produce and bury the vast measures of coal known to exist under the earth's surface. He was vague as to the cause and effects of this Deluge, but he concluded from the distribution and location of known coal deposits that it must have been universal and violent, destroying completely the original surface of the earth, making deep excavations, and dissolving the most intractable substances. There were in some cases, however, alternating beds of limestone and coal, a phenomenon suggesting the alternate filling and emptying of great water-basins over long periods of time. He explained this by supposing that the Deluge had been followed by a series of lesser floods incident to the readjustment of the globe to its new condition.

This conception of the Deluge appeared to account for the coal measures, but how could it be reconciled with the wisdom and goodness of God? The production of vast stores of coal for human use seemed to provide a partial answer to this awful question, and Parkinson made the most of it. But must God destroy a world to provide man with coal? Perhaps, Parkinson answered, God had a reason for placing man in a ruined world in which the presence and uses of coal must be discovered by human effort. Perhaps He sought by this means to provoke man to the exercise of his innate powers "and even to urge him to a change from the savage to the civilized state" (35). Surely it was not incredible that God should assign man a part in finishing the work of Creation. As for the destruction of the former state of things, might it not have been one of a series of divine reconstructions of the globe, in which the ruins of each world contributed materials for the fashioning of the next? Might not the present world be destined to undergo a like remodelling to fit it for the residence of beings still nobler than man? If so, did not the fossil record open up new and larger views of God's wisdom and providence?

The second and third volumes of Parkinson's work were devoted to the classification and description of fossil plants and animals. In the second, he drew heavily on the works of Wallerius and Linnaeus, but by 1811 he had adopted the classificatory schemes of Cuvier and Lamarck and was prepared to acknowledge their leadership in the field of paleontology. Throughout the work he continued to wrestle with the problem of reconciling the findings of the new science with the traditional view of nature. It could not be denied, he declared: first, that many fossil species and genera had no living counterparts, second, that most living species and genera were not represented in the fossil record, third, that no sharp chronological line could be

121

drawn separating fossil from existing types (since the uppermost layers of the earth's surface contained the remains of both), and fourth, that the tremendous accumulations of fossil zoophytes could not be accounted for by the Deluge but must be regarded as the remains of animals which had lived and died in some primeval ocean. All these facts and inferences could be reconciled with the doctrines of Creation and Providence, said Parkinson, if only men would give up the antiquated notion of an inviolable chain of being. Surely the contrivance by which the disappearance of whole genera of plants and animals was prevented from upsetting the economy of nature was a nobler testimony to the wisdom of the Creator than any supposed scale of being could be. Rightly interpreted, the fossil record refuted both the ancient doctrine of the eternity of the world and the modern theory of geologic uniformitarianism. "Does it not appear from this repeated occurrence of new beings, from the late appearance of the remains of land animals, and from the total absence of the fossil remains of man, that the creative power, as far as respects this planet, has been exercised, continually, or at distant periods, and with increasing excellence, in its objects, to a comparatively late period: the last and highest work appearing to be *man*, whose remains have not yet been numbered among the subjects of the mineral kingdom" (36).

The trend of Parkinson's thought was plain. He was applying the Christian concept of human history to earth history. Formerly the earth had been regarded simply as a stage for human history, the changes on its surface as evidence of the tendency of all things to decline from their original perfection. Now that it appeared undeniable that the earth had a long, eventful history of its own, there was nothing to do but to suppose that terrestrial revolutions, like those in human affairs, had their place in God's plans. If the fossil record showed a progress from simple to complex forms of life, this progress must have been intended by God, and He must have arranged a series of appropriate stage-settings for each act of the drama of perpetual improvement. Thus, by substituting the notion of cosmic progress for the doctrine of original perfection and subsequent decline, Parkinson was able to admit the mutability of the structures of nature without giving up the idea that the material world had been created as a stage for the activities of intelligent beings.

It remained only to reconcile this altered conception of things with Scripture. Like many before him, Parkinson pointed out that the sequence of events described in the first chapter of Genesis was similar in many respects to that suggested by the fossil record. That there were exceptions to the general concordance of Genesis and paleontology he was willing to admit. For example, the alternation of coal and limestone beds in the earth's surface created a problem concerning the second day of creation, and only a few species of birds could be assigned to the third day. The chief difficulty, however, concerned the time period allowed for the accomplishment of the work of creation. This problem could be solved, said Parkinson, by taking the Hebrew word for *day* in a figurative sense, as indicating a period of indefinite length. The resulting vindication of Scripture by science must, he declared, "satisfy or surprise everyone."

In 1812 Cuvier threw the enormous weight of his authority and prestige behind the theory of successive creations in the brilliant "Preliminary Discourse" to his *Ossemens fossiles,* subsequently entitled "Discourse on the Revolutions of the Surface of the Globe." He began by recalling the achievements of physical science in extending the limits of space and discovering the mechanism of the universe. "Would it not also be glorious," he asked, "for man to burst the limits of time, and, by a few observations, to ascertain the history of this world, and the series of events which preceded the birth of the human race?" That the earth had undergone a series of revolutions not only in its topography but also in its flora and fauna could no longer be doubted, he declared. The strata of the earth's crust displayed unmistakable proofs of a series of geological epochs each with its characteristic flora and fauna, only a few species lingering over from previous epochs. Thus, there had been *several* creations, separated by tremendous revolutions involving the wholesale extinction of species. The causes of these revolutions and the time periods separating them could not be definitely assigned. The Wernerian hypothesis of the gradual subsidence of a primitive ocean explained many phenomena, but it could not account for the superposition of marine fossils on deposits of fresh-water fossils, unless numerous resurgences and retreats of the sea were assumed. Much additional study of the fossil record would be necessary before the question of causes and time periods could be raised profitably.

Cuvier had reasons other than scientific caution for putting aside these questions. He was well aware that, if the explanation of the fossil record was to be sought in the everyday processes observable on the earth's surface, the traditional view of nature and the traditional time-perspective would have to be thrown overboard. Hence he sought by every means to prove that the revolutions which had transformed the earth had been sudden upheavals produced by causes no longer in operation and decreasing in violence and scope as the globe assumed its present form. Processes operating slowly and steadily could not produce sudden effects, he declared. The frozen carcasses discovered in the wastes of Siberia proved that the events which had wiped out whole races of living creatures were cataclysmic. The earth had received its destined form: "the thread of operation is here broken, the march of nature is changed, and none of the agents that she now employs were sufficient for the production of her ancient works" (37).

On the question of the Deluge, Cuvier was equally conservative. The evidence indicated, he declared, that the last geologic revolution had taken place recently, not more than five or six thousand years ago. The absence of human fossils suggested that men had begun to spread over the globe at that period and to establish themselves "in places fitted by nature for their reception." It was possible, however, that fossil bones of human beings might eventually be found. Such a discovery would presumably indicate that some human beings had lived before the last great catastrophe and that a few had survived to populate the world thereafter. Perhaps great numbers of mankind had been buried beneath the ocean when the present seas were formed by the submergence of land masses and the concomitant elevation of continents. The fossil record of these continents seemed to show that they had undergone several inundations, each marked by the extinction of many species and genera. "It appears to me," Cuvier concluded, "that a consecutive history of such singular deposits would be infinitely more valuable than so many contradictory conjectures respecting the first origin of the world and other planets, and respecting phenomena which have confessedly no resemblance whatever to those of the present physical state of the world; such conjectures finding, in these hypothetical facts, neither material to build upon, nor any means of verification whatever" (38).

Although Cuvier did not attempt to reconcile paleontology with the Bible, his support of catastrophism gave the traditional view of nature a new lease on life. Jameson praised him for having refuted Lamarck's doctrine that so-called extinct species are the ancestors of existing species and for establishing the historicity of the Deluge and the comparatively recent peopling of the earth's surface. Cuvier might talk of bursting the limits of time, but he could not accept the infinite time-perspective of Lamarck and Hutton without admitting the power of ordinary natural processes to transform the face of the globe by their long-continued action. The doctrine of catastrophism was a compromise between the requirements of scientific integrity and the demands of traditionalism. It depicted a series of creations separated by geologic upheavals divinely appointed to prepare the earth for new forms of life, which were then divinely "called into existence." *It recognized change but not development.* In brief, catastrophism was a means whereby the instability of terrestrial structures could be acknowledged without admitting change to be fundamental and structure superficial. The same means made it possible to continue to believe in the subordination of the inorganic world to the needs of living creatures. The modification of the earth's flora and fauna could be attributed to the rational agency of an all-wise Creator rather than to the merciless requirements of the struggle for survival. Science could remain preoccupied with description and classification. The time-perspective remained vague, prediction was impossible; God was his own timekeeper and scriptwriter. There was one new element in the picture, however. Change could no longer be regarded as a decline from original perfection or as a cycle of processes serving to maintain the status quo. Progress seemed written into the history of the earth. Such was the compromise of catastrophism. It held sway until Sir Charles Lyell reaffirmed Hutton's principles in 1830, without, however, admitting the consequences which Lamarck had deduced from them in regard to the stability of organic forms.

American Mastodon (*Mastodon americanus*)	Woolly Mammoth (*Elephas primigenius*)	Other Finds
	1695 — Traveler's account of *Mammotovoi Kost*, known from tusks, bones, etc., found in Siberia.	
1714 — Cotton Mather's account of huge bones and teeth found near Albany.		
	1728 — Sir Hans Sloane describes tusks, bones, etc., found in Siberia.	
	1737 — Breyne's account of Siberian "elephant bones," with Messerschmidt's drawings.	
1739 — Longueuil collects bones and teeth near the Ohio River.		

1761–64 — Buffon and Daubenton compare specimens from Siberia and from the Ohio with bones, teeth, etc., of modern elephant, conclude all are of one species.

1767 — Croghan's teeth and grinders from Big Bone Lick described by Collinson.

1768 — Hunter compares bones and teeth of Ohio and Siberian creatures with modern elephant, concludes they represent an extinct, carnivorous species different from the elephant.

1771 — Parts of frozen rhinoceros carcass brought to St. Petersburg Museum.

1778 — Buffon declares Croghan grinders different from both elephant and "hippopotamus" teeth.

1785 — The Rev. Robert Annan describes teeth and bones found in Orange County, N.Y., in 1780.
Drawings of Kentucky bones taken to Europe by Michaëlis.

1792 — Kerr describes *Elephas americanus* in his translation of Linnaeus, *System of Nature.*

1799 — Cuvier and Blumenbach independently distinguish Ohio and Siberian creatures from each other and from the modern elephants, Asiatic and African.

1799 — Jefferson and Wistar describe *megalonix* bones found in Virginia in 1796.

126

American Mastodon (*Mastodon americanus*)	Woolly Mammoth (*Elephas primigenius*)	Other Finds
1801 — Peale exhumes two mastodon skeletons near the Hudson River.		
1802–03 — Peale's second skeleton exhibited in London.		
		1804 — Cuvier distinguishes the *megalonix* from the *megatherium* (S. American).
1806 — Cuvier establishes new genus called *mastodon*, including *le grand mastodonte*, figured from Peale's London mammoth.	1806 — Cuvier adopts Blumenbach's name *Elephas primigenius* for woolly mammoth.	
1809 — Capt. Clark collects specimens from Big Bone Lick for President Jefferson.	M. F. Adams collects remains of a mammoth carcass at the delta of the Lena River, (figured by Tilesius in 1815).	
1812 — Cuvier publishes his *Ossemens fossiles de quadrupèdes*, with its "Preliminary Discourse" concerning the revolutions which have happened on the surface of the globe.		

...the number of true species in nature is fixed and limited and, as we may reasonably believe, constant and unchangeable from the first creation to the present day.

JOHN RAY, 1688

...would it be too bold to imagine, that in the great length of time, since the earth began to exist, perhaps millions of ages before the commencement of the history of mankind,... all warm-blooded animals have arisen from one living filament, which THE GREAT FIRST CAUSE endued with animality, with the power of acquiring new parts attended with new propensities, directed by irritations, sensations, volitions, and associations; and thus possessing the faculty of continuing to improve by its own inherent activity, and of delivering down those improvements by generation to its posterity, world without end?

ERASMUS DARWIN, 1794

From Monad to Man

T HE IDEA THAT SPECIES were absolutely fixed sprang alike from a need for scientific order and from age-old philosophical and theological presuppositions. Before the seventeenth century no pressing need had been felt to group plants and animals into species, genera, orders, classes, and the like. In botany, for example, plants had been studied chiefly for their medical virtues and were so listed in the herbals which came down from classical antiquity. There was no clear concept of a species and certainly no idea that one form of plant or animal life might not be transmuted into another form. Stories of such transformations were all too common in botanical and zoological literature.

With the revival of observation and the exploration of vast new continents during the Renaissance, however, biologists were hard pressed to keep up with the rapidly increasing number of new specimens. Otto Brunfels, one of the German "fathers of botany," enumerated only 258 species and varieties in his *Living Pictures of Plants*, published in 1530, and four-fifths of these had been known to classical and medieval writers. A century later the Swiss naturalist Kaspar Bauhin dealt with approximately 6,000 kinds of plants in his *Chart of the Botanical Theater* (1627) (1). Clearly it was time for someone to distinguish essential forms from accidental variations, to define · a stable unit upon which botanical systems could be erected.

This was precisely the task undertaken by the English naturalist John Ray in the second half of the seventeenth century. In a paper communicated to the Royal Society of London in 1674, Ray noted that former botanists and herbalists "mistaking many accidents for notes of specific distinction, which indeed they are not, have unnecessarily multiplied beings." Among these "accidents" he enumerated variations in the size of the plant, the color and number of the flowers, the taste and shape of the fruit, the number of leaves, and the like. Differences of this sort should not be taken as marks distinguishing separate species, he declared. "Diversity of colour in the flower, or taste in the fruit, is no better note of specific difference in plants, than the like varieties of hair or skin, or taste of flesh in animals; so that one may, with as good reason admit a blackmore [Negro] and European to be two species of men, or a black cow and a white to be two sorts of kine, as two plants, differing only in colour of flower, to be specifically distinct; such varieties, both in animals and plants, being occasioned either by diversity of climate, and temperature of the air, or of nourishment and manner of living." This could be shown, he added, by sowing the seeds of plants in different soils and noting the resulting differences in plants springing from the same seed. "For, if you sow the seed, for example, of a single julyflower in good ground, among many that bear single flowers, it shall give you some roots, that yield double, and some of different colours, from the mother-plant, which you may afterward propagate by the slip. The plants, that are most apt to be thus diversified by sowing, are julyflowers, anemonies, larkspurs, columbines, bears ears, stocks, and wall-flowers, primroses and cowslips, tulips, crocuses, blue-bottles, daisies, hepaticas, and violets" (2).

But if the various species of plants could not be distinguished by differences of the sort mentioned, how could they be distinguished? The true criterion, implied in Ray's paper of 1672 and stated explicitly in his *General History of Plants* (1686–1704), was one of common descent, presumed or actually observed. "A species is never born from the seed of another species and reciprocally," he declared. Thus, regardless of differences, individuals belonged to the same species if their descent from the same ancestors could be proved or reasonably inferred. However they might vary in nonessential features, they agreed in the arrangement of their essential parts, preserving that arrangement from generation to generation in an unbroken line of descent from the first examples of the species. The conclusion was inescapable that the forms of the species came ultimately from God

— or so it seemed to one of Ray's training. In the philosophical tradition stemming from Aristotle, whatever was permanent and unchanging, whatever gave form to substance, was divine in origin. In the theological tradition derived from the Bible, it was God who gave form to the earth when it was without form and void. Science and religion were in perfect accord, "the number of species being in nature certain and determinate, as is generally acknowledged by philosophers, and might be proved also by divine authority, God having finished his works of creation, that is, consummated the number of species in six days" (3).

► THE SEARCH FOR NATURE'S METHOD

On the scientific and philosophical foundations laid by John Ray, the Swedish naturalist Carolus Linnaeus built an imposing edifice of systematic natural history. His lifelong ambition was to reduce the earth to order as Newton had the heavens. "God created; Linnaeus arranged" went an eighteenth century saying, and Linnaeus himself noted in his diary that his *System of Nature* was unique, "a work to which natural history has never had a fellow" (4). In it, for the first time, every terrestrial production was assigned its place in one great system of classification. Nothing could have been more contrary to the idea of organic evolution than Linnaeus' aspiration to place every creature in its proper niche, yet the very effort to enumerate the species and genera definitively was to raise problems which could not be solved within the Linnaean framework.

At the heart of Linnaeus' conception of nature was the idea of an *oeconomia,* a rationally ordained system of means and ends. This idea was balanced against the idea of nature's infinite variety, conceived as the manifestation of inexhaustible creative power. Thus the earth, with its delightful variety of climate and topography, was populated with an equally

Carolus Linnaeus,
1707 · 1778

varied assemblage of living beings, yet every creature was perfectly adapted to the region in which it lived. Each species obeyed the command to increase and multiply, yet the different plants and animals were so related to each other and to their environments that the balance of nature was preserved and no species was ever destroyed.

Linnaeus never tired of describing the mechanisms which maintained the adaptation of organism and environment and the equilibrium of the species. "Cats bury their dung. Nothing is so mean, nothing so little, in which the wonderful order, and wise disposition of nature does not shine forth." From this point of view the presence of carnivorous animals in nature was intelligible: death and destruction were but the means of preserving nature in its established course, "so that of all the species originally formed by the Deity, not one is destroyed" (5). But the economy of nature did not exist for its own sake. The mineral kingdom supported the vegetable, the vegetable the animal, and the whole earth existed as a theater for the activities of man, without whom it would have been incomplete, like a body without a head. Man alone was capable of putting nature to its full use. He alone could appreciate the wisdom, power, and goodness displayed in the creation. It was his duty, therefore, to study nature diligently that God might be glorified in His works.

In this view of things, natural history consisted of describing the various productions of the earth, their appearance, their habits, their relations to each other, and their uses (6). The naturalist began, as Adam had, by naming things. The soul of science was method, said Linnaeus, and method consisted in assigning to every natural body its proper name expressive of its relation to other bodies. Names might be artificial and arbitrary, suited only to human convenience, or they might be natural, expressing the real affinities among objects. To discover the natural method of classification was the naturalist's great objective. Linnaeus believed that he had achieved this aim in the case of the genera and species but only fragmentarily for the classes and orders.

Implicit in the idea of a natural method of classification as the goal of science was the belief that nature had been constructed upon a pattern discoverable, at least in part, by human reason. Linnaeus' appeal to botanists to regard the genera and species as natural was based partly upon his conviction that such an assumption was essential to the progress of botany, but even more upon faith that terrestrial phenomena must be intelligible to man because they had been made for human use and instruction. In the fruiting organs of plants Linnaeus

132

believed he could discern characters "written by the hand of God" to aid man in distinguishing the genera. In like manner the species might be discriminated by careful comparison of the number, figure, proportion, and situation of the parts. Throughout his life Linnaeus never gave up the search for the natural method of classification, never ceased to proclaim it as the goal of science. Reviewing the various systems of botany in his *Philosophia botanica,* he declared:

> Besides all the above-mentioned systems or methods of distributing the plants...there is a natural method, or nature's system, which we ought diligently to endeavour to find out....And that this system of nature is no *chimaera,* as some may imagine, will appear, as from other considerations, so in particular from hence, that all plants, of what order soever, show an affinity to some others to which they are nearly allied. In the mean time, till the whole of nature's method is compleatly discovered (which is much to be wished), we must be content to make use of the best artificial systems now in use (7).

It was inevitable that Linnaeus should attribute to the underlying pattern of nature an absolute stability appropriate to its divine origin. In his *Philosophia botanica* he defined species as primordial types created by divine wisdom and perpetuated by generation from the beginning to the end of the world (8). Varieties, on the other hand, he viewed as temporary forms produced by the material environment, by accidents of climate, soil, and culture. The true botanist, he declared, would not concern himself with "the various forms of sporting nature" any further than necessary to distinguish the true species. This done, he would leave varieties to florists and gardeners.

Philosophical preconceptions were equally apparent in Linnaeus' discussion of spontaneous generation. He began by citing the experiments of Francesco Redi which showed that maggots were not generated in putrefied meat if flies were prevented from laying eggs in it. But Linnaeus did not leave the argument there. If spontaneous generation were possible, he continued, the ingenious devices with which plants were equipped for broadcasting seed would be superfluous, but God had not created anything in vain. Moreover, if microscopic animals could be produced by chance, why not the intricate mechanisms of larger plants and animals — a thought too preposterous to be entertained. Finally, spontaneous generation would mean the random appearance of new species of plants and animals, with the result that there could be no inference from the genera to the species. Science would be impossible: "there would be no such thing as certainty, but all confusion" (9). Clearly, then, spontaneous generation was as repugnant to rational philosophy as to observation and experiment.

But although Linnaeus declared flatly that nature was a fixed order of permanent structures, he seems to have had some doubts on this score. In his later works he suggested that new species might be formed by hybrid generation. He described four such hybrid species in his "Dissertation on the Sexes of Plants" and added that the case of certain species of geraniums found in South Africa "would almost induce a botanist to believe that the species of one genus in vegetables are only so many different plants as there have been different associations with the flowers of one species, and consequently a genus is nothing else than a number of plants sprung from the same mother by different fathers. But whether all these species be the offspring of time; whether, in the beginning of all things, the Creator limited the number of future species, I dare not presume to determine" (10). In his *System of Vegetables* Linnaeus suggested that perhaps God created only one plant of each order and then arranged intermarriages so as to form the genera, leaving nature to produce the species within each genus by further crossings (11). Linnaeus returned to this idea in his diary, stating that the impregnations of the original species of each genus had taken place "accidentally" (12). Though he said nothing of the time required for these developments, it is clear that he was looking for a historical explanation of the origin of genera and species and that he conceived the outcome as determined to a considerable degree by chance. In certain passages in the *Species plantarum,* moreover, he speculated that "daughters of time," as he called new species, might be produced by the influence of climate and geography. He suggested, for example, that the species *Achillaea alpina* might be a transmuted form of *Achillaea ptarmica: "An locus potuerat ex praecedenti formasse hanc?"* ("Could this have been formed from the preceding one by the environment?") (13).

Nor was Linnaeus blind to the possibility that species might be-

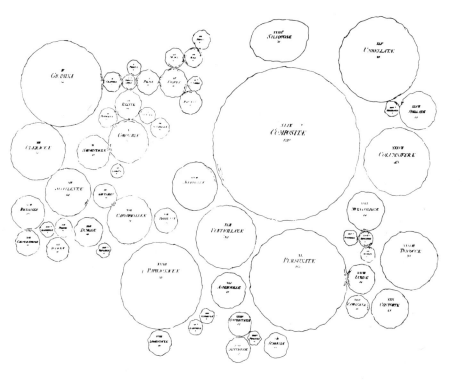

FIG. 5.1 — Linnaeus' conception of the natural affinities of various orders of plants, depicted by his student Paulus D. Giseke in his **Praelectiones in Ordines Naturales** in 1792. According to Giseke: "There are provinces (of circles) of which some are merely neighbors, while others are contiguous. . . . When circles are closely assembled they usually, but not always, indicate neighboring provinces. Thus the first 13 orders are not only as Monocotyledons more closely related to one another than to others, but furthermore even by genera themselves, which are inscribed at the periphery [of the circles representing the orders]." This "genealogico-geographical" conception of plant affinities was reaffirmed by A. P. de Candolle in the early 19th century. Note how different it is both from the "evolutionary tree" of modern biology and from the notion of the continuous scale of nature, or great chain of being, so common in the 18th century. The circles in the diagram are not drawn entirely to scale. A complete list of the orders and transitional genera shown may be found in Note 6 to this chapter.

come extinct. During his travels in Scandinavia and his labors in Count Tessin's museum he found innumerable fossils of marine species for which no living counterparts could be discovered in all the collections of Europe. In the *System of Nature* these were described as *"deperditus,"* sometimes with the suggestion that they had retreated to remote parts of the earth or sea. Of the fossil *Anomiae* Linnaeus wrote:

> The animals which inhabited all these 'wild mussels,' as well as unaltered shells, are nowadays unknown to us..., nor do we know what in the world may have become of them. Still, we shall never believe that a species has entirely perished from the earth (14).

Admit their extinction or not, here were the remains of thousands of otherwise unknown sea creatures embedded high in the mountains. How had they gotten there? Linnaeus did not think for a moment that their presence could be accounted for by the Deluge (15). Ages had been required to accumulate and solidify these deposits, other ages to cover them with strata of a different character. Nowhere, he confessed, had he discovered traces of a universal Deluge, nowhere had he been able to penetrate the accumulations of time to discover the original earth. His mind reeled when he thought of the time which must have elapsed while millions of creatures lived and died, leaving only their shells as a witness to posterity. He could readily believe the world older than the Chinese asserted if Scripture did but permit. Perhaps the collection of fossils was sheer vanity and idle curiosity; he did not presume to say. The fact remained that they were "the only remaining fragments of the ancient world,...far beyond the memory of any history whatsoever" (16).

Linnaeus treated the question of the earth's antiquity gingerly lest he offend the clergy. His botanical works had been excluded from the Papal domains for a brief period as being incompatible with Scriptural botany, and he had been sharply criticized for including man among the *anthropomorphi* in his *System of Nature* (17). In an oration on "The Rise of the Habitable Earth," and again in the *Philosophia botanica,* he proposed a theory designed to reconcile Genesis with the undeniable facts of geology and zoology. He suggested that the habitable earth had been originally a small, ocean-girt island stocked with every species of plant and animal (18). Only on this supposition would it have been possible for Adam to name all the animals. Though it was not expressly indicated in Scripture, Linnaeus deduced the presence of a great variety of plants from the fact that certain animals, especially insects, depended for food upon certain species of plants. Moreover, since living beings required an environment suited to their natures, there must have been a high mountain on the island to provide the requisite diversity of living conditions. As the land had gradually encroached on the sea (a process still going on), the plants and animals had spread to those regions where they were now found. During this same process, presumably, various strata and their fossil contents had been deposited and solidified. As for the Deluge, it might be assumed to have occurred before the present continents were formed. In this manner Linnaeus tried to combine the traditional notion of a stable earth complete in every detail with a new concept which viewed the crust of the globe as a system of matter in motion and found it a graveyard filled with the accumulated remains of species long since vanished. Eden itself, though retained with its full complement of animals, was reconstructed to accord with the laws of nature. Instead of an arbitrary arrangement of things established by divine fiat there was an interdependent system of nature, involving a delicate adjustment of organism and environment.

It seems unlikely that Linnaeus could have been more than temporarily convinced by his own theory of the earth. Despite his great piety and respect for Scripture, his ultimate reverence was for nature. Her laws, he declared, were immutable and perfect. To discover them the naturalist must confess his ignorance and patiently observe her ways. "In every branch of natural knowledge, the first principles should be established on, and confirmed by, repeated observations and experience" (19). From the testimony of nature there was no appeal.

Whatever Linnaeus' private convictions, the prestige of his name was clearly on the side of a static view of nature. Like Newton, he was hailed as a believer in revelation who had shown the world to be a rationally ordained system of perfectly contrived and immutably established structures, a glorious theater for the drama of human salvation.

► SPECIES AND TIME

Among the critics and rivals of Linnaeus none was more widely read than the Comte de Buffon and none offered a more striking contrast in background, temperament, and philosophic outlook. The two men had little in common except a love of science and of scientific reputation. Linnaeus was a naturalist's naturalist;

Buffon, 1707 · 1788

his works provided an index to the book of nature. The son of a country parson, reared in the unquestioning faith of a pious people, he achieved greatness by pure force of character and intellect. Buffon, on the other hand, was a gentleman of independent fortune, attuned from childhood to the temper of the Enlightenment. An accomplished stylist, he wrote to entertain as well as to instruct. A philosopher by nature, he could not be satisfied with mere description but must strive unremittingly to know the causes of things. To him the everyday world of nature, with its appearance of stability, design, and infinite variety, was but a veil of sense perception masking permanent, active, and causative nature — "that system of laws established by the Creator for regulating the existence of bodies, and the succession of beings" (20). This simple system of elements, laws, and forces Buffon declared to be the original nature created by God, the agency by which the variegated world of phenomena was produced, maintained, changed, dissolved, and renewed. "Time, space, and matter are her means; the universe her object; motion and life her end.... The springs she employs are active forces, which time and space can only measure and limit, but never destroy; forces which balance, mix, and oppose, without being able to annihilate each other" (21). Creation and annihilation belonged to God alone, but although nature must act according to the original plan and with the original means made available to her, the variety of effects which she could produce by combination and permutation was inexhaustible. It followed that the visible world was not a stately edifice modeled on a pattern in the Infinite Mind but rather a kaleidoscope of effects produced in the human mind by the operation of the hidden system of nature.

In this view of things the Linnaean concept of natural history as a search for the natural method of classifying terrestrial productions was inadmissible. From the opening essay of the *Natural History* (1749) to the closing chapters of the *Epochs of Nature* (1778) Buffon denounced the systematists unsparingly. The dominant characteristic of visible nature, he argued, was its endless proliferation of forms, each form grading imperceptibly into others. "It seems that everything which can be, is: the Creator's hand seems not to have been opened in order to give existence to a determinate number of species, rather it seems to have thrown forth at one and the same time a world of creatures related and unrelated, an infinity of harmonious and contrary combinations, a perpetuity of destructions and renewals" (22). Nature's ways were not man's ways, and whenever the naturalist, by stressing similarities and ignoring differences, thought himself to have discovered distinct groups of plants and animals, nature confounded him by exhibiting intermediate and ambiguous forms. Thus the bat was half-quadruped and half-bird, and the hog formed a link, in one respect, between the whole-hoofed and cloven-footed animals and, in another, between the cloven-footed and the toed animals. These ambiguities of form were puzzling only to those "who mistake the hypothetical arrangement of their ideas for the common order of Nature, and who only perceive, in the infinite chain of being, some conspicuous points to which they incline to refer every natural phenomenon" (23). At best, systems of classification, if based on a wide variety of characters, might serve as auxiliary means to the first end of natural history — "the complete description and exact history of each particular thing." But this goal was more likely to be achieved when description was undertaken without a predisposition to stress similarities and ignore variations. "Accurate descriptions, without any attempt toward definitions, a more scrupulous examination of the differences than of the similarities, a particular attention to the exceptions, and even to the slightest shades, are the true guides, and I will venture to affirm, the only means we possess of investigating Nature" (24).

Thus far Buffon's discussion of the method of natural history seems to have been inspired more by the ancient notion of a great chain of being than by the example of Newtonian science. But the sequel was to show that Buffon attributed the variety and minute gradation of animate nature to the operations of a system of laws, elements, and forces rather than to the inherent necessity of the ideas in the Divine Mind to manifest themselves in the natural world. To Buffon it was inconceivable that an idea should be the cause of a natural phenomenon. Ideas, he declared, were human constructions derived from sense experience by means of the conceptualizing faculty. They were indispensable in investigating natural phenomena, but they were never to be regarded as the causes or inherent forms of those phenomena. Indeed, the more general and abstract the idea, the less correspondence it had to reality. Nature was to be understood, not by endeavoring to discover the blueprint of creation, but by looking for uniformities in the way natural objects presented themselves to one's senses. "The relations between our senses and the objects which affect them are necessary and invariable" (25). By comparing particular effects and the combinations and circumstances in which they occurred, general effects might be discerned; these, in turn, might eventually be discovered to be particular cases of still more general modes of occurrence. Thus, description was only a preparation for comparison, and the goal of comparison was the discovery of uniformities in nature's operations and the elucidation of particular effects in terms of them. This mode of investigation could never penetrate to first causes since it discovered an order of nature relative to human modes of perception rather than to the real existence of things. Nevertheless, it was man's only key to reliable knowledge and to the control of nature for human convenience.

A second obstacle to the successful study of natural history, Buffon continued, was the belief in final causes, the notion that the uses of things explained their origin and structure. Buffon rejected this idea categorically. "The relation of particular objects to ourselves has no connection with their origin. Moral affinity or fitness can never become a physical reason" (26). Instead of attributing the habits and mechanisms by which birds built their nests to the particular design of the Creator, the naturalist should assume "that they depend, like every other animal operation, on number, figure, motion, organization, and feeling." He should attempt, by comparing the activities of birds with those of bees, ants, and field mice, to discern the general laws regulating behavior of this kind. "Who gives the grandest idea of the supreme Being, he who sees him create the universe, arrange every existence, and found Nature upon invariable and perpetual laws; or he who inquires after him, and discovers him conducting and superintending a republic of bees, and deeply engaged about the manner of folding the wings of a beetle?" What reason was there to assume that every structure or character in nature had a purpose? Was it not more reasonable to believe that all things which could exist did exist regardless of their usefulness to man or to other creatures? Of what use to the hog were the half-formed toes found inside its leg? Of what use were a male's nipples? The stubborn conviction that everything had a use even though not apparent served only to conceal ignorance and thus to obstruct the progress of science (27). The true naturalist would put aside the question of the uses of structures and seek the laws regulating their appearance in nature.

Buffon put his theory of the role of comparison in science into practice by including in his *Natural History* detailed anatomical drawings prepared by his collaborator Jean Daubenton. Centuries of exclusive attention to human anatomy had yielded but meager results, he declared. It was futile to study natural objects singly. Nature must be compared with herself. When this was done, it became apparent that the internal structure of living bodies was far more constant and hence far more important for study than the external features. The striking differences in appearance of the various races of dogs concealed a single internal constitution common to all dogs. When dogs were compared with other creatures, the existence of a uniform plan of organization extending throughout the animate creation was disclosed. "This plan proceeds uniformly from man to the ape, from the ape to the quadrupeds, from quadrupeds to cetaceous animals, to birds, to fishes, and to reptiles..., gradually degenerating from reptiles to insects, from insects to worms, from worms to zoophytes, from zoophytes to plants; and, though changed in all its external parts, still preserving the same character, the principal features of which are nutrition, growth, and reproduction" (28). Thus, life turned out to be a system of physiological processes capable of being performed by many combinations of organs and of sustaining a wonderful variety of external manifestations. The Creator seemed to have employed but one idea, varying it ad infinitum "to give men an opportunity of admiring equally the magnificence of the execution and the simplicity of the design." But the design was a mode of operation, a system of processes, not a pattern of structures. To discover and understand these processes, not to classify their result, was the true aim of natural history.

141

When nature's basic modes of operation had been discerned by comparative studies, the final task remained of devising a theory to account for the general and particular effects observed. Here the naturalist must imagine a state of affairs from which the phenomena observed might be fairly and inevitably derived. "By the nature of the question, then, we are permitted to form hypotheses, and to choose that which appears to have the greatest analogy to the other phenomena of nature" (29). As an example Buffon cited the rival theories concerning the transmission of light, but he qualified it with a warning. Mathematics, he declared, could have only a limited application in natural history since it was adapted only to subjects capable of highly abstract treatment. Natural historians should borrow only the method, not the particular laws and concepts, of mechanics and optics. Descartes had erred when he attempted to explain life on mechanical principles, for what were mechanical principles but uniformities detected in the behavior of inanimate bodies? Why should the properties of matter be limited to those already discovered? It was likely that some properties of bodies were unknowable by human beings. But aside from these, might not animate nature be subject to laws quite different from those operating in the inanimate world? Might not life have its own laws? (30).

The distinguishing characteristic of living bodies, Buffon continued, was that of organization, of interdependence among the parts. "To possess a high degree of sensibility, the animated body must form a whole, not only sensible in all its parts, but so constructed that these parts intimately correspond with each other, in such a manner, that an impression made upon one, must necessarily be communicated to all the rest" (31). Low forms of life exhibited a correspondingly low degree of interdependence in the functioning of their parts. Throughout the whole organic realm, however, there was a pronounced tendency toward the production and proliferation of organized beings, a tendency held in check only by the resistance of brute matter to organization and by the preying of living creatures on each other. "Nature, in general, appears to have a greater bias towards life than death: She seems anxious to organize bodies as much as possible" (32). It followed, then, that the process of generation contained the key to the understanding of organic phenomena.

When Buffon attempted to formulate a theory of generation he took as his model the Newtonian idea of a system of matter in motion. Living bodies, he conjectured, must be systems of organic molecules governed in their motions by "penetrating forces" comparable to the force of attraction. The quantity of organic matter in nature was constant. When an organism died, the molecules composing it returned to the earth, there to be absorbed by plants and supplied by them to animals. The force or principle which organized the molecules into animate bodies and governed their distribution and behavior until the bodily system dissolved Buffon called a *moule intérieur,* an internal mold. Without assuming such a force he could not explain the tendency of matter toward organization.

> In my theory...I first admit the mechanical principle, then the penetrating force of gravity, and, from analogy and experience, I have concluded the existence of other penetrating forces peculiar to organic bodies. I have proved by facts, that matter has a strong tendency towards organization; and that there are in Nature an infinite number of organic particles. I have, therefore, only generalized particular observations, without advancing any thing contrary to mechanical principles, when that term is used in its proper sense, as denoting the general effects of Nature (33).

The notion of a system involved the idea of its equilibrium. The equilibrium of the organic system, Buffon suggested, arose because a fixed proportion existed between the number of internal molds and the quantity of organic matter in the universe. "This number of molds, or individuals, though variable in every species, is, upon the whole, always the same, always proportioned to the quantity of living matter" (34). Apart from such an adjustment, new species would appear, produced by the natural tendency of the organic molecules to organize. In like manner, the balance between fecundity and destruction served to prevent the disappearance of any species. Although every species tended incessantly to increase its numbers, natural causes of destruction preserved the balance among the species, so that the number of individuals in each was relatively constant. Nature cared nothing for particular individuals; her sole concern was to preserve the species. "To her they are all equally dear; for, on each of them, she has be-

stowed the means of subsisting, and of lasting as long as herself" (35).
Likewise nature so adjusted the organic to the inorganic system that
"in each particular species, the climate is formed for the manners, and
the manners for the climate" (36).

To man, Buffon assigned a unique place in the self-sustaining system of nature. Though man closely resembled the higher apes in
figure and internal structure, he differed from them absolutely in
possessing the faculties of thought and speech. Man alone was adapted
to all climates and conditions. Man alone required the civilizing influence of society for the full development of his powers. Whereas
animals were affectable only by present objects, the divine gift of intelligence enabled man to remember the past, to compare it with the
present, and to anticipate the future. Indeed, said Buffon, man's
knowledge of the existence of his own soul was the most reliable
knowledge he possessed. Concerning material bodies he had only the
report of his senses, "and the organs of sensation themselves are only
certain affinities with the objects which affect them" (37). Extension,
impenetrability, and all the other properties attributed to matter,
even matter itself, might have no existence outside the human mind,
but there could be no doubting the existence of the conceiving mind.
In the case of man, therefore, nature's rule of proceeding by slight
gradations in her productions had been broken. If this were not so,
creatures intermediate in intelligence between man and the apes would
be found to exist, but none such was known. "Whatever resemblance,
therefore, takes place between the Hottentot and the ape, the interval
which separates them is immense; because the former is endowed with
the faculties of thought and speech" (38). By virtue of this endowment, man exercised dominion over the brute creation. "It is the
dominion of mind over matter; a right of Nature founded upon unalterable laws, a gift of the Almighty, by which man is enabled at all
times to perceive the dignity of his being" (39).

In some respects the self-sustaining organic system depicted by
Buffon differed little from Linnaeus' *oeconomia naturae.* The notion
of a fundamental simplicity underlying the variety of the world, intended both to astonish and to delight mankind, the idea of a balance
between birth and death, of the permanence of species, the perfect

144

adaptation of the organic to the inorganic, of man's unique character and legitimate dominion over the brute creation — all these were axioms of eighteenth century natural history. Buffon's "internal molds" seemed at times little more than the forms of the species in a new guise, and the organic particles little more than Newton's atoms endowed with the very property which they were intended to explain, a tendency towards organization.

Nevertheless, despite these inadequacies and traditional elements, Buffon's theory of life involved important departures from the prevailing view of nature. For one thing, it tried to conceive organic phenomena as the outcome of temporal process rather than as a static expression of a pattern of creation. The idea of a species, Buffon argued, could be rendered intelligible only "by considering Nature in the succession of time, and in the constant destruction and renovation of beings" (40). If there were no death and regeneration, every individual would be a species, a unique creature forming one link in the great chain of being. It was the maintenance of a uniform pattern in the series of time which constituted and distinguished a species. Hence the only sure method of deciding whether two animals belonged to the same species was to discover whether they could unite and produce fertile offspring. "This is the most fixed and determined point in the history of nature. All other similarities and differences which can be found in the comparison of beings, are neither so real nor so constant" (41). Species alone had a real existence in nature; genera, families, classes, and orders were human inventions and quite arbitrary.

Moreover, as Buffon conceived it, the equilibrium of organic nature was a purely quantitative affair, from which no moral lessons could be drawn. The terrestrial system was stable only because it was a system, not because man required a stage for his activities. There was no moral significance in the fact that cats buried their dung. Moral ideas were human inventions, applicable only to human society. But even human affairs were subject to natural law. The number of the human species was fixed within limits as surely as that of any other. "All physical and moral causes, and the effects that result from them, are balanced, and comprehended within certain limits, which are more or less extended, but never to such a degree as to destroy the equilibrium" (42). If it should be asked for what purpose the generations succeeded each other in endless sequence, the species being maintained at the expense of a continual slaughter of individuals, the answer was that these phenomena "are derived from the very essence of Nature, and depend on the first establishment of the universal machine."

But although Buffon declared the organic system to be in perfect equilibrium, he could not close his eyes to certain phenemona difficult to reconcile with that view of things. In his geological writings he had declared the earth to be a system of matter in motion, every structure on its surface undergoing perpetual alteration by the relentless action of geologic processes. But if the climate and the surface of the globe underwent constant slow change, must not living forms change too? How else could the adaptation of organic to inorganic be maintained? Was not the apparent stability of organic forms as much an illusion as the permanence of the "everlasting hills?" Did not the very notion of a system of separate elements combining and recombining imply brief duration of the structures produced? Why should the organic system differ from the inorganic system in this respect? Did not the natural tendency of the organic molecules to organize open the door to spontaneous generation of living creatures outside the influence of the internal molds? (43). These were some of the questions which must have arisen in Buffon's mind as he pursued the problem of species through thirty-six volumes of the *Natural History*, bringing to the task all the resources of his knowledge, intellect, imagination, and art.

Buffon first raised the problem of the permanence of species in his essay on the ass. Was it possible, he asked, that the horse and the ass were descended from a common original stock? Were there among plants and animals "families conceived by Nature, and produced by time?" The question was of crucial importance: "For, if it were proved, that animals and vegetables were really distributed into families, or even that a single species was ever produced by the degeneration of another,...no bounds could be fixed to the powers of Nature: She might, with equal reason, be supposed to have been able, in the course of time, to produce, from a single individual, all the organized bodies in the universe" (44). Against this subversive possibility Buf-

146

fon marshalled all the available arguments. He declared it contrary to Scripture, to reason, and to experience. The species known in the eighteenth century were no different from those described by Aristotle. Hybrid species, such as the mule, were notoriously infertile. How could such vitiated and defective creatures give rise to a new race of beings? Moreover, if the ass issued from the horse it could only have done so by slow and imperceptible changes, passing through innumerable intermediate forms — but no such intermediate forms were known. Finally, the "families" which naturalists pretended to find in nature were only human inventions, useful aids to the memory but having no real existence in nature. "We may, therefore, without hesitation, pronounce the ass to be an *Ass,* and not a degenerated horse, a horse with a naked tail" (45) .

But although Buffon rejected the idea of the mutability of species with apparent finality he could not dismiss it from his thoughts. Though he scoffed at the "natural families" listed by systematizers, he was convinced from his own researches and experiments on domestic plants and animals that they exhibited real families produced from wild stocks by human cultivation and breeding. The various cereals and the different types of sheep, oxen, and dogs were quite obviously products of human selection operating over long periods of time. "These physical genera are, in reality, composed of all the species, which, by our management, have been greatly variegated and changed; and, as all those species, so differently modified by the hand of man, have but one common origin in Nature, the whole genus ought to constitute but a single species" (46) .

Buffon's attitude toward human modification of natural stocks was curiously inconsistent. Ordinarily he branded it a deforming influence producing degenerate types of life. He contrasted nature in the wild: "dictating her simple but immutable laws, impressing upon every species indelible characters, dispensing her bounty with equity, compensating evil with good," with domesticated nature: "seldom perfect, often changed and deformed" (47) . In these passages the traditional idea of the original perfection of nature, marred only by human imperfection, was uppermost, hence the choice of the word *degeneration* to describe what modern writers would call simply change or variation. At other times, however, Buffon described man

147

as a co-worker with nature in producing new effects. It might, he declared, be more reasonable to regard the horse as an improved ass than to think of the ass as a degenerated horse. Perhaps domestic animals were improved varieties of wild stocks, the essence of man's control over nature lying in his ability to modify her productions for his own convenience and use. "The powers of nature, when united to those of man, are greatly augmented" (48).

The basis of man's power to alter nature Buffon found in natural variability. Man simply reinforced the agency of natural causes. Although nature had provided every species with an ancestral model, the individuals of the species departed from the model in various unaccountable ways, so that it might be said to be "depraved or improved by circumstances." By "circumstances" Buffon meant chiefly changes in environment. Every animal, he declared, was adapted to a particular region with a particular climate and food supply. When animals were forced to abandon their natural habitat, either by human intervention or by "any revolution on the globe," they underwent changes in physique and appearance which, in the course of time, became hereditary and were transmitted to their posterity. Changes in one part of the body produced modifications in other parts, so that the whole appearance was materially altered. "These changes are produced in a slow and imperceptible manner. Time is the great workman of Nature. He moves with regular and uniform steps. He performs no operation suddenly; but, by degrees, or successive impressions, nothing can resist his power; and these changes which at first are imperceptible, become gradually sensible, and at last are marked by results too conspicuous to be misapprehended" (49). The rate of change depended on two factors: the degree of alteration of the environment and the number of progeny produced in a given time period. Since wild animals were much freer from human interference than domestic species and could choose their climate and food supply, they changed more slowly. Among them the amount of variation was determined chiefly by the number of young they produced and the frequency of reproduction, hence the higher forms of life were more permanent than the lower. Change in domestic animals might be spoken of as improvement or degeneration, according as it was useful or deleterious to man, but "with respect to Nature,

148

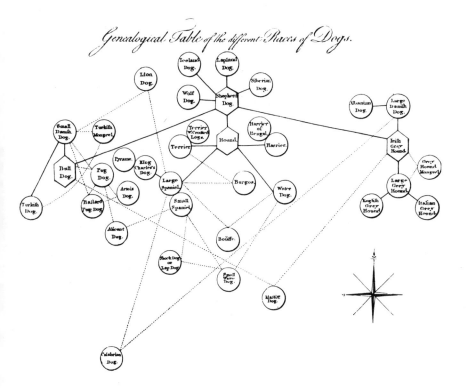

FIG. 5.2 — The derivation of various races and breeds of dogs from the wolf-dog by artificial selection, as portrayed in Buffon's **Natural History** in 1753. A century later Charles Darwin, apparently without having read Buffon, plunged deep into the study of variation under domestication and emerged, like Buffon, with a strong conviction of the importance of artificial selection in producing new types from originally wild stocks.

improvement and degeneration are the same thing; for they both imply an alteration of original constitution" (50).

If change in animal forms was the product of time and circumstance, what limits could be placed upon it? This question became particularly urgent when Buffon came to interpret the results of his comparative enumeration of the quadrupeds of the Old World and of the New World. He was astonished to discover that of two hundred known species of quadrupeds, only 70 were to be found in the New World, 30 of this number being common to both hemispheres. He had declared earlier that true families existed only among domestic animals, but now he asserted that the 30 species of quadrupeds common to both hemispheres and the 130 species peculiar to the Old World were reducible to 15 genera and 9 detached species, "from which all the others have probably derived their origin" (51). Seven of the genera and 2 of the detached species were common to both the Old and the New World, the rest being limited to the Old. The animals peculiar to the New World he found to comprise 10 genera and 4 detached species (see Figure 5.3, pp. 156, 157).

Buffon's first interpretation of these facts was that the quadrupeds of the Old World upon passing into the western hemisphere had degenerated under the influence of a foreign soil and climate. Were not the animal species of the New World fewer in number, less prolific, and smaller in stature than those of the Old? "Even those which, from the kindly influences of another climate, have acquired their complete form and expansion, shrink and diminish under a niggardly sky and an unprolific land, thinly peopled with wandering savages, who, instead of using this territory as a master, had no property or empire; and, having subjected neither the animals nor the elements..., existed as creatures of no consideration in Nature, a kind of weak automatons, incapable of improving or seconding her intentions" (52). Climate and geography might well explain why the reptiles and insects were so large, the quadrupeds so small, and the men so frigid in the New World. The American continent, being a relatively recent production of nature, had probably just begun to feel the beneficial effects of human cultivation.

This theory was not without its difficulties. For example, some of the American quadrupeds were *larger* and *more prolific* than their

Old World counterparts. Buffon suggested that these species might be native to the western hemisphere, then hastily withdrew this heretical suggestion, proposing instead that the climate and soil of the New World had improved rather than degenerated these exceptional species. Other animals, such as the armadillo and sloth, were different from any found in the Old World. Moreover, there was the problem of explaining how the Old World quadrupeds had passed to the western hemisphere. Citing the reports of recent Russian explorations in Siberia, Buffon conjectured that Asia and America were connected in the north. If this were true, it would help explain the fact that all the species common to the two hemispheres were inhabitants of northern climes. But if the other species in the western hemisphere were to be derived from the Old World it was necessary to assume that the two hemispheres had once been connected farther south.

> Hence, to discover the origin of these animals, we must have recourse to the period when the two Continents were united. We must, at the same time, consider the two hundred species of quadrupeds as constituting thirty-eight families: And, though this is by no means the present state of nature, but, on the contrary, a state of much greater antiquity, which we can reach only by inductions and relations almost equally fugitive as time, that seems to have effaced their traces; we shall, however, endeavour to ascend, by facts and monuments still existing to those first ages of Nature, and to exhibit those epochas which shall appear to be most clearly indicated (53).

In attempting to ascend to the first ages of nature Buffon came face to face with the question of the extinction of species. If changes in the environment could modify the forms of living organisms, might not sufficiently great climatic fluctuations cause their extinction? How else could one account for the giant bones and teeth found in various parts of the earth? In his *Epochs of Nature* (1778) Buffon appeared to argue once more for the permanence of species, "especially in the larger species, whose characters are more fixed" (54). He accounted for the mammoth bones by suggesting that the creative powers of nature had been more energetic when the globe was younger and warmer. In his essay on "Animals Common to Both Continents" he had drawn a more radical inference from the fate of the mammoth:

This species was unquestionably the largest and strongest of all quadrupeds; and, since it has disappeared, how many smaller, weaker, and less remarkable species must likewise have perished, without leaving any evidence of their past existence? How many others have undergone such changes, whether from degeneration or improvement, occasioned by the great vicissitudes of the earth and waters, the neglect or cultivation of Nature, the continued influence of favorable or hostile climates, that they are now no longer the same creatures? Yet the quadrupeds, next to man, are beings whose nature and form are the most permanent. Birds and fishes are subject to greater variations: The insect tribes are liable to still greater vicissitudes: And, if we descend to vegetables, which ought not to be excluded from animate Nature, our wonder will be excited by the quickness and facility with which they assume new forms (55).

In his idea that "the least active, and the worst defended, as well as the most delicate and heavy species, have already, or will soon disappear" (56), Buffon was very close to the idea of natural selection. The variability of all natural productions was, for him, a logical consequence of the axiom that all which can exist does exist. The evidence pointing to the extinction of species led him to restate the axiom: "Everything that can happen has been accomplished in time, and either exists or has existed in Nature" (57). From his observation of nature he further concluded that the weak and helpless types would be eliminated by the strong. The sloths, being ill-equipped for survival, would have perished long ago if they had lived in regions occupied by men and larger animals. And they were but one example of many defective creatures — "Those imperfect sketches of Nature, which being hardly endowed with the faculty of existence, could not subsist for any length of time, and have accordingly been struck out of the list of beings" (58). Buffon also recognized that climatic changes resulting from the cooling of the globe or from revolutions on its surface might destroy creatures unable to adjust to the new conditions. In general, however, he stressed the effect of man's increasing dominance on the habits, forms, and survival of plants and animals. Man, he declared, had gained his dominion by conquest, and the losers in the struggle had been subjugated and transformed, or driven to new and less favorable environments, or completely destroyed. In the struggle for survival, man had been the most rapacious and successful of all the animals.

Although Buffon had grown more conservative in his estimate of the mutability of species by the time of the *Epochs of Nature,* his general view of nature was as bold and unorthodox as ever, and more so in some respects. Whereas he had formerly limited the production of organisms by spontaneous generation to the lowest fringes of the animal kingdom, he now implied that the successive flora and fauna of the great epochs of nature had been produced spontaneously. "If," he declared, "the greatest part of the animate creation were suppressed, one would see new species appear because these organic molecules, which are indestructible and always active, would unite to form other organized bodies, but being entirely absorbed by the internal molds of existing beings, new species cannot be formed, at least in the first classes of Nature, such as those of the large animals" (59). This idea of a production *de novo* of the earth's flora and fauna by spontaneous generation apparently appealed to Buffon, for he developed it more fully in the "Supplement" to his chapter on generation. After declaring spontaneous generation to be not only "the most frequent and the most general" but also "the most ancient, that is to say, the first and most universal" mode of generation, he invited his readers to imagine that God had decreed the sudden extinction of all living things. The organic molecules, finding themselves liberated from the internal molds, would proceed to produce an infinite variety of organized beings, some of them well adapted to survive and reproduce, others doomed to early extinction. In due time a flora and fauna not unlike that which had been extinguished would eventuate by a process of natural selection, but the productions of the new epoch of nature would be smaller and less powerful than those of the earlier one, since the quickening heat of the globe would be less intense at the time of the second generation than at the time of the first. In the long run, moreover, the total of organic molecules must decline as the globe lost the heat essential to their production (60).

These speculations indicate how far Buffon had drifted from the traditional concept of nature with its unchanging species, its wise design, its perfect balance, its subordination of structure to function. They also show how far he was from a truly evolutionary concept of nature. In many ways his ideas were more akin to the speculations of pagan antiquity than to those of modern evolutionists. Although he stressed the variability of nature and recognized that some variants were more likely to endure than others, he never pushed the idea of natural selection to its logical conclusion. He used it to explain how species disappeared but not how they were modified. He saw the elimination of the weak and maladapted as an occasional extraordinary event rather than as a process operating relentlessly within each species as well as among species. He viewed change more in terms of degeneration than of improvement. He regarded climate and food supply as positive influences on the form of organisms rather than as negative conditions producing change of form by establishing conditions of survival. In all these questions, however, his ultimate appeal was to observation and experiment:

> In general, kindred of species is one of those mysteries of Nature, which man can never unravel, without a long continued and difficult series of experiments. . . . Is the ass more allied to the horse than the zebra? Does the wolf approach nearer to the dog than the fox or jackal? At what distance from man shall we place the large apes, who resemble him so perfectly in conformation of body? Are all the species of animals the same now that they were originally? Has not their number augmented, instead of being diminished? Have not the feeble species been destroyed by the stronger, or by the tyranny of man, the number of whom has become a thousand times greater than that of any other large animal? What relation can be established between kindred species, and another kindred still better known, that of different races in the same species? Does not a race, like the mixed species, proceed from an anomalous individual which forms the original stock? How many questions does this subject admit of; and how few of them are we in a condition to solve? How many facts must be discovered before we can even form probable conjectures? (61) .

These difficulties, Buffon continued, should exhilarate rather than dismay the naturalist; he should rejoice in the prospect of an endless series of researches, discoveries, and new problems. "There is no boundary to the human intellect. It extends in proportion as the universe is displayed. Hence man can and ought to attempt everything: He wants nothing but time to enable him to obtain universal knowledge" (62). Moreover, since knowledge was power, the possibility of controlling nature for human convenience was equally unlimited.

Buffon had come a long way from the Christian concept of the earth as a stage for the drama of man's redemption by divine grace. Burning with the thirst for knowledge and intoxicated with the sense of man's potential control over nature, he proclaimed man's power to be master of his own fate. Hitherto, he declared, man had pursued evil more energetically than good, amusement more diligently than knowledge, but there was reason to hope that he would at last discover peace to be his true happiness and science his true glory.

► THE PATH OF NATURE

To the problem which Buffon had set for natural historians — the derivation of organic phenomena from the operations of a system of matter in motion — Jean Baptiste Pierre Antoine de Monet, Chevalier de Lamarck, provided the first thoroughgoing solution. Like his friend and benefactor Buffon, Lamarck defined nature as the system of laws and forces governing the motions of matter, "an order of things separate from matter, determinable by the observation of bodies, the whole of which constitutes a power unalterable in its essence, determined in all its acts, and constantly acting on every part of the universe" (63). Along with Buffon, he regarded organization as the peculiar characteristic of living bodies, but he rejected the attempt to explain organization by assuming an organic matter endowed with a tendency to organize. Life was not a property of a special kind of matter but a phenomenon accompanying certain states of ordinary matter, Lamarck argued. Whenever a gelatinous body took on a certain structure and chemical composition through the action of attractive and repulsive forces, it became capable of reacting to external stimuli by the movement of fluids within its mass; it became a living body. More highly organized bodies were capable of more complicated forms of behavior, but every organic phenomenon sprang from a material organization essential to its existence. "Every fact or phenomenon observed in a living body is, at one and the same time, a physical fact or phenomenon and a product of organization" (64).

Intelligence was no exception to this rule, Lamarck declared. There was nothing metaphysical, nothing foreign to matter, in the intellectual faculty. It appeared in rudimentary form in the lowest

Lamarck,
1744 · 1829

155

OLD AND NEW WORLDS

1. Ox, bison, etc.
2. Elk, reindeer, etc.
3. Wolf, fox, etc.
4. Marten, ferret, etc.
5. Rabbit, rat, etc.
6. Walrus, seal, etc.
7. Bat, vampire, etc.

8. Bear } Detached
9. Mole } species

NEW WORLD ONLY

1. Sapajous (monkeys)
2. Sagoins (monkeys)
3. Opossum, etc.
4. Cougar, jaguar, etc.
5. Coatis
6. Stinking weasels
7. Agoutis
8. Armadillos
9. Ant-eaters
10. Sloths

11. Tapir
12. Cabiai } Detached
13. Llama } species
14. Peccary

PREHISTORIC LAND BRIDGE

FIG. 5.3 — Like Charles Darwin a century later, Buffon was led to suspect the mutability of species by comparing the similarity-with-a-difference prevailing among animals in different parts of the world. The map shown here was constructed from the data given by Buffon in his chapter on "The Degeneration of Animals."

GEOGRAPHIC
DISTRIBUTION
of QUADRUPEDS

OLD WORLD ONLY

1. Horse, zebra, etc.
2. Sheep, goats, ante-
   lopes, etc.
3. Wild boar, etc.
4. Panther, leopard, etc.
5. Porcupine, hedge-
   hog, etc.
6. Hyena, civet, etc.
7. Apes, baboons,
   true monkeys, etc.
8. Scaly lizards, etc.

9. Elephant
10. Rhinoceros
11. Hippopotamus ⎫ Detached
12. Giraffe   ⎬ species
13. Camel   ⎭
14. Lion
15. Tiger

LAND BRIDGE

(UNEXPLORED)

According to Buffon's *Natural History* — 1766

R. HAUPT '59

mammals and increased in prominence as the nervous system gained in complexity. Though most highly developed in the human species, even there it was dominated by habit and feeling. The "mind" which some naturalists postulated in order to explain mental phenomena was as much a fiction as the universal deluge invoked by geologists (65).

With this argument Lamarck put aside Descartes' dualism of mind and matter and adopted a radical positivistic materialism which made psychology a branch of zoology. The field of observable phenomena was the realm of reality and the only source of positive knowledge. Beyond it lay the realm of imagination, a world of fantasy and fiction providing nothing substantial but hope. "Who has observed a single phenomenon which has not been produced by bodies, the relations between different bodies, the changes of place, state, or form which bodies undergo?" (66). Positive knowledge could not give absolute certainty, but since it was the only knowledge available, "let us collect with care the facts which we can observe, let us consult experience wherever we can, and when this experience is inaccessible to us, let us assemble all the inductions which observation of facts analogous to those which escape us can furnish, and let us assert nothing categorically: in this way, we shall be able little by little to discover the causes of a multitude of natural phenomena, and, perhaps, even of phenomena which seem the most incomprehensible" (67).

From his theory that life was the product of a certain organization of matter by which bodies were enabled to interact with their environment, Lamarck concluded that the stability of organic forms was proportionate to the stability of the conditions of life. His geologic studies had taught him that the earth's surface had undergone constant slow change, and reflection on the nature of the system of matter in motion had convinced him that heavenly bodies had changed imperceptibly, "with the result that no physical body has absolute stability" (68). How unlikely, then, that living bodies, which depended for their very existence on close adaptation to their environment, should have preserved their forms unchanged. The apparent stability of species arose from man's limited time perspective. The fact that the species of animals known to have lived in ancient Egypt corresponded to those found in that region in modern times proved nothing. It would be better to compare the so-called extinct species, whose re-mains were found in the earth's crust, with living species and note their similarity-with-a-difference. It would then be evident that the modern species were descendants of "extinct" species which had undergone modification in the course of time. The surprising thing was not that species, genera, and even orders had been transformed during the earth's long history but that some forms had changed very little, if at all. These exceptional cases, Lamarck suggested, must be relatively recent productions of nature which had not had time to experience radical alterations in a changing environment (69).

Thus the mutability of species, a conception toward which Buffon had slowly groped his way, became Lamarck's starting point, the postulate which determined his approach to natural history. For the naturalist who believed organic forms to be constantly changing, the problem of classification was subordinate to the more difficult task of discovering first, the mechanisms and laws governing change of form, and second, the direction of change. "The object of the study of animals," wrote Lamarck, "is not only to know the different races and to distinguish among them by fixing their particular characters, but it is also to discover the origin of the faculties which they exercise, the causes which give rise to life and sustain it, and finally the causes of the remarkable progression which they exhibit in their organization and in the number as well as in the development of their faculties" (70). Lamarck recognized that the adaptation of organism to environment was a dynamic relationship and that the environment, far from being arranged so as to minister to life, established the constantly changing conditions to which life must adapt if it was to survive. From this he concluded that "the conformation of individuals and of their members, their organs, their faculties, &c. &c. are entirely the result of the circumstances into which the race of each species has been thrown by nature" (71). The problem of the naturalist, then, was to discover the means by which the adaptation of living creatures to their changing surroundings took place.

The degree of development of any particular organ, Lamarck asserted, was determined by its usefulness to the creature which possessed it. Useful organs would be exercised and developed, less useful ones would tend to wither away. As the conditions of life changed,

159

organs which once were very useful and hence highly developed might become less useful and tend to disappear. These effects of use and disuse, Lamarck declared, would be transmitted to succeeding generations. Thus, the way was opened for radical alterations in the structure and appearance of species. But if organs could be greatly developed or entirely lost in virtue of their utility or disutility, might not a new configuration of the conditions of life occasion the genesis of an entirely new organ designed to meet the requirements of the novel situation? The process by which animals developed new organs to meet the challenge of new circumstances Lamarck described as follows:

> They feel certain needs, and each felt need, stirring their inner consciousness [*sentiment intérieur*], immediately causes fluids and forces to be directed toward the point of the body where an action capable of satisfying the need can take place. But, if there exists at this point an organ appropriate to such an action, it is stimulated to act; and if no organ exists and the felt need is pressing and sustained, little by little the organ is produced and develops by reason of its constant, vigorous use (72).

Thus, use and disuse, organic response to felt needs, and the inheritance of acquired characters combined, "with the aid of plenty of time and circumstances," to produce the astonishing variety of animals inhabiting the earth (73).

Lamarck's theory of the causes, mechanisms, and laws of organic change carried with it a revolutionary implication concerning the direction of change. Whereas Buffon had conceived variation as a process of degeneration or random deviation from innumerable ancestral forms, Lamarck viewed it as an evolution from simple beginnings in accordance with the laws governing the response of organized beings to changes in their environment. The "uniform plan of organization" which Buffon had discerned extending from the polyp to man Lamarck declared to be nature's plan of operations, her line of march. The various combinations and systems of organs by which the fundamental life processes were performed and new levels of existence were made possible had been developed, said Lamarck, not for the sake of mere variety but in response to the felt needs of living creatures struggling to survive amid changing circumstances.

If one should compare these different [living] objects with each other and with what is known concerning man; if one should contemplate, from the simplest animal organization to that of man, which is the most complex and perfect, the *progression* exhibited in the composition of the organization as well as the successive acquisition of different special organs and, consequently, of as many new faculties as new organs developed; then one might perceive how *needs*, reduced in the beginning to almost nothing and increasing gradually in number thereafter, have produced a tendency to actions appropriate to satisfy them; how actions become habitual and energetic have occasioned the development of the organs which perform them; how the force which excites organic movements can, in the most imperfect animals, be located outside of them and yet stimulate them; how, later on, this force has been transferred into the animal itself; finally, how it has there become the source of sensibility and eventually, of acts of intelligence (74).

In accordance with this view of things Lamarck classified animals according to the complexity of their internal organization, beginning with the "apathetic" invertebrates and proceeding through the "sensitive" invertebrates to the vertebrates, or *animaux intelligents*. This arrangement, he declared, "seems to me to represent as closely as possible that of the increasing complexity of organization of animals, that which must govern their distribution in a general series, that which indicates, at least in general, the path which nature has followed in giving existence to the different races of beings" (75).

The "development hypothesis," as Lamarck's theory came to be called, plainly implied that man was the latest product of the process of organic transformation. In his *Philosophie zoologique*, published in 1809, Lamarck described how a race of apes, "moved by the need to dominate and to see far and wide," might slowly undergo structural changes so as to become adapted to an upright posture and bipedal locomotion. Thereafter a felt need for improved means of communication might result in the development of organs of speech and a language. "Such would be the reflections which one might indulge if man, considered here as the preëminent race, were distinguished from animals only by organizational characteristics and if his origin were not different from theirs" (76).

Lamarck made little appeal to the fossil record to support his daring and highly subversive hypothesis. He seems to have thought that it followed plainly and inevitably from the laws of organic transformation and from the progression of forms disclosed by comparative anatomy. Conceding that the development of new organs in response to felt needs was difficult to prove, he argued that it was implied by the law of use and disuse. The inheritance of acquired characteristics he took for granted, as did many others in his day (77). He admitted that the progression of forms from simple to complex was discernible "in the principal masses" only and was often obscured by variations of gradation and by lines of development which seemed to diverge from the main path and even to terminate in blind alleys. These irregularities, he declared, were distortions of the normal development of animal organization; they were produced by accidental circumstances which acted as tangential forces deflecting the main evolution of life. In classifying animals, therefore, those characteristics and organs which resulted from "the direct operations of nature" must be distinguished from, and invariably given the preference over, those which were produced by "the influence of the circumstances of habitation, as well as of the habits which the different races have been forced to contract" (78). For example, all "insects" should have heads according to the plan of nature, but many annelids, cirripeds, and mollusks had only a very rudimentary head or none at all because of the action of causes "foreign to nature."

This notion of a plan of nature distorted but never destroyed by accidental alterations recalls Linnaeus' conception of varieties as random deviations from the norms of the species. For Lamarck, as for Linnaeus, the plan of nature provided the basis of a "natural" method of classification. Although Lamarck substituted a dynamic for a static concept of the realization of the plan, he, no less than Linnaeus, regarded those structures which expressed the plan as being more fundamental and important than those which stemmed from causes "accidental and therefore variable." Apart from such fortuitous causes, Lamarck declared, the progression of life from simplicity to complexity would be entirely regular. Actually, there were innumerable deviations from the main line of development, but the balance of nature was so contrived that "the progress achieved in perfecting organization is never lost" (79).

162

In this concept of an interplay between "the cause which tends incessantly to complicate organization" and "the influence of circumstances," Lamarck seemed to abandon mechanism in favor of teleology. That is, he seemed to suggest that the evolution of life was not simply the resultant of material causes but instead was directed, as if by some divine purpose, toward the realization of a preordained pattern. In other passages, however, he asserted that organic transformation occurred only in response to changes in the conditions of life; without

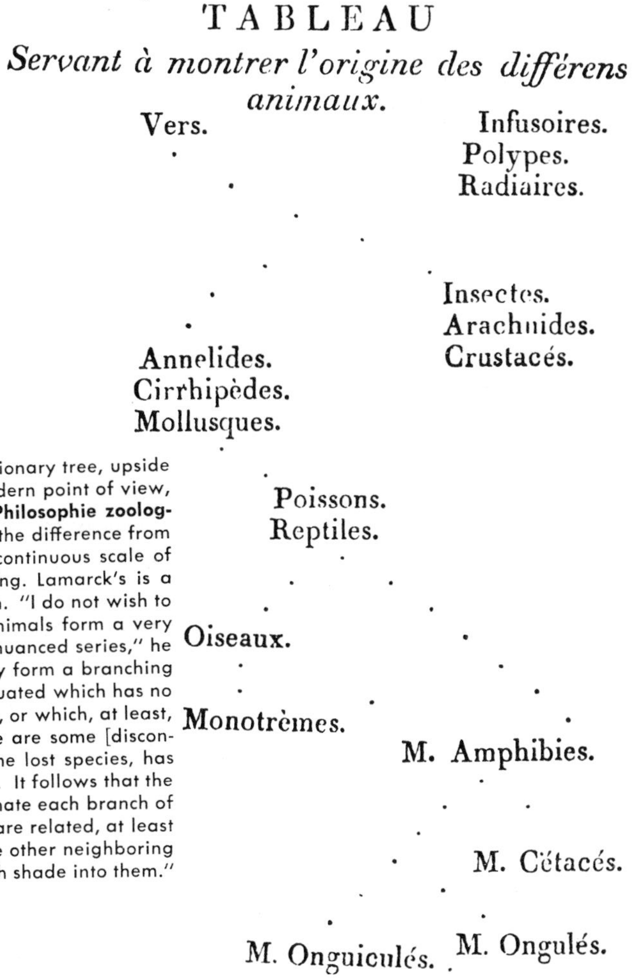

TABLEAU
Servant à montrer l'origine des différens animaux.

Vers.   Infusoires.
                                         Polypes.
                                         Radiaires.

                                         Insectes.
                                         Arachnides.
Annelides.   Crustacés.
Cirrhipèdes.
Mollusques.

Poissons.
Reptiles.

Oiseaux.

Monotrèmes.

M. Amphibies.

M. Cétacés.

M. Onguiculés.   M. Ongulés.

FIG. 5.4 — The first evolutionary tree, upside down from the modern point of view, published in Lamarck's **Philosophie zoologique** in 1809. Note the difference from the old notion of the continuous scale of nature, or chain of being. Lamarck's is a truly branching evolution. "I do not wish to say . . . that existing animals form a very simple and evenly nuanced series," he wrote, "but I say that they form a branching series irregularly graduated which has no discontinuity in its parts, or which, at least, if it is true that there are some [discontinuities] because of some lost species, has not always had such. It follows that the species which terminate each branch of the general series are related, at least on one side, to the other neighboring species which shade into them."

such changes the laws governing the genesis of organs and their development or attrition would have no occasion to operate (80). Nature knew no intent or purpose; she was bound by necessity to produce exactly what she did produce and nothing else. If all her productions were destroyed, she would eventually reproduce them, hence "evil," "chance," and "disorder" had no real existence in nature. "Each part, having necessarily to change and to cease its existence in order to constitute another, has an interest contrary to that of the whole; and if it reasons, it finds that whole badly constructed. In reality, however, the whole is perfect and fulfills completely the end for which it is destined" (81).

Lamarck was trying to have his cake and eat it. On the one hand, he was determined to regard all organic structures and their modifications as the outcome of an interplay of purely physical forces governed by law. On the other hand, he sought to discern a direction, a line of march, in the successive organic productions of nature. But if there were no inherent tendency in living matter toward complication, if nature had no intention to produce such a trend, if organic forms changed only in response to changes in the conditions of life, how could one distinguish between "natural" and "accidental" transformations? Were not all changes equally "natural" and "accidental," all necessary products of the operation of the system of nature? Were not the notions of a "plan of operations" and a "line of march" as relative to the human point of view, as little grounded in nature, as ideas of "disorder" and "evil"? So Buffon would have argued.

Quite apart from the difficulty of distinguishing the "accidental" from the "natural" in organic evolution, there was the even greater difficulty of understanding how a system of matter in motion could operate to produce an irreversible trend in a given direction. Geologic processes acted only to disintegrate, transport, distribute, rearrange, solidify, and elevate the materials of the earth's surface in never-ending cycles. How could these cyclical transformations on the earth's surface give rise to bodies so organized as to interact with their environment and reproduce their kind, and why should the process of interaction produce a steady trend toward complication of organization and the emergence of properties and faculties totally different from those of inorganic matter? Buffon, conceiving organic change as the alteration of ancestral forms by the direct influence of the environment, had concluded that variation of animal form was as random as the combination of forces which produced it. From the human point of view animal stocks might be said to improve or degenerate, but nature knew only change. All possible forms would be produced in the course of time. Those ill-equipped to survive would perish sooner or later, only to be produced again as the ages rolled on. That the conditions of life might constitute a selective factor capable of establishing a direction of change by the process of elimination Buffon did not guess. To him change in organic forms was but another aspect of the universal variability of nature, an inherent property of the system of matter in motion.

Lamarck, however, had to explain how this system of nature could produce not merely a variety of organic forms but an ordered succession of them proceeding from simple to complex, from apathetic to intelligent. He sought the answer to this problem in the principle of adaptation. It seemed obvious to him that the changes which living creatures underwent must adapt them to their altered environments if they were to survive. Moreover, he observed that plants and animals throughout the globe *were* adapted to their surroundings. Since there was no reason to believe that changes produced by the direct action of the environment would be adaptive, the source of adaptive change had to be found in the response of the creatures themselves to the requirements laid upon them by alterations in the conditions of life (82). Animals were not indifferent to their survival or extinction. Confronted with new situations, they made efforts, more or less blind and groping, to obtain sustenance and to protect themselves from their enemies. These efforts, when seconded by "favorable circumstances" and prolonged a great while, resulted in the physiological changes by which the animals became adapted to their new environment. This constant struggle to adapt to slowly changing conditions produced the evolution of more and more complex organisms. *Change was directional because it was adaptive.*

So Lamarck's argument ran, but he seems to have overlooked the possibility that several different responses to new conditions of life might have adaptive value. Why could not any one of a number of organs satisfy the gasteropods' need for a means of exploring their environment? Why should all gasteropods develop in the same direction? Was not a selective factor needed to eliminate all but one or a few of the manifold possibilities of adaptation? Instead of looking for such a factor, Lamarck talked vaguely of "the cause which tends to the complication of organization," thus introducing an idea inconsistent with his materialistic outlook.

In emphasizing felt needs as a positive agency in organic transformation Lamarck stopped short of a completely mechanistic interpretation of life. Was his recognition of a psychological factor in organic development simply a carry-over of the traditional idea of a pre-established harmony between the organic and inorganic realms, insuring that the basic needs of all living creatures should be provided for? Or was it a conscious affirmation that the mechanical concept of nature was inadequate to explain organic phenomena? Whatever the case, it was not until Charles Darwin combined Lamarck's emphasis on the effort of living creatures to survive amid changing conditions of life with Buffon's idea of random variation and the extinction of the least fit that the traditional view of nature felt the full impact of the mechanistic concept (83).

► **PERPETUAL IMPROVEMENT IN NATURE**

Although Lamarck provided the first systematic elaboration of the evolutionary idea, the idea was not original with him, nor were his writings the best vehicle to make it known. An earlier (and from the literary point of view much more effective) champion of the development hypothesis was "the celebrated Dr. Darwin," as Charles Darwin's grandfather, Erasmus, was known to the reading public in Britain and America. Physician, botanist, anatomist, and poet, Dr. Darwin set forth his novel conceptions in both prose and verse, but most forcibly in the chapter on generation in his *Zoonomia: or the Laws of Organic Life,* published in the years 1794–1796. Like Buffon and Lamarck,

Erasmus Darwin, 1731 · 1802

Darwin sought to explain the phenomena of life in terms of the operations of a system of matter in motion (84). The peculiarity of living matter, he observed, was its capacity to undergo progressive transformations resulting in the appearance of new structures, new needs, and new functions. The development of an adult organism from a tiny piece of living matter was a process of this kind. It posed the problem of explaining the dif-

166

ferent forms and qualities of the adult animal in terms of "the different irritabilities and sensibilities, or voluntarities, or associabilities, of this original living filament; and perhaps in some degree from the different forms of the particles of the fluids, by which it has been at first stimulated into activity" (85). Generation was not simply a blowing up of a tiny individual already formed, nor was it a matter of the accretion and distribution of organic molecules under the influence of an internal mold. Instead, it was a process of differentiation and development involving the acquisition of new parts: "with the acquisition of new parts, new sensations, and new desires, as well as new powers, are produced; and this by accretion to the old ones, and not by distention of them" (86). In short, generation was a process of organic transformation produced by the interaction between matter possessing certain propensities and the forces which acted upon it from within and without.

Darwin's theory of organic evolution seems to have been a speculative generalization of his theory of generation, bolstered with facts and ideas drawn from comparative anatomy, geology, botany, and zoology.

> From thus meditating on the great similarity of the structure of the warm-blooded animals, and at the same time of the great changes they undergo both before and after their nativity; and by considering in how minute a proportion of time many of the changes of animals described have been produced; would it be too bold to imagine, that in the great length of time, since the earth began to exist, perhaps millions of ages before the commencement of the history of mankind, would it be too bold to imagine, that all warm-blooded animals have arisen from one living filament, which THE GREAT FIRST CAUSE endued with animality, with the power of acquiring new parts attended with new propensities, directed by irritations, sensations, volitions, and associations; and thus possessing the faculty of continuing to improve by its own inherent activity, and of delivering down those improvements by generation to its posterity, world without end? (87).

Giving his imagination free rein, Darwin went on to speculate that not only warm-blooded animals but all living forms had been derived from a single original filament. Had not the great Linnaeus pictured

167

all plants as arising from a few natural orders? Did not the fossil record show that vegetables were in existence long before the first animals, and the simple kinds of animals long before the more complex? Was not the extinction of various species, attested by the fossil record, a logical and necessary consequence of "this gradual production of the species and genera?" (88).

Darwin's concept of the mechanism of organic transformation anticipated Lamarck's. Throughout their lives, wrote Darwin, "all animals undergo perpetual transformations; which are in part produced by their own exertions in consequence of their desires and aversions, of their pleasure and their pains, or of irritations, or of associations; and many of these acquired forms or propensities are transmitted to their posterity" (89). Food, a mate, and security were the three great needs of animal life. In the effort to satisfy these, animals had developed organs and characteristics perfectly adapted to the functions they performed. That they had been developed in the course of time was shown by the presence of useless appendages in nearly all plants and animals. Man himself bore anatomical traces of his evolution from lower forms of life (90).

This notion, far from dismaying Darwin, suggested to him the intoxicating possibility that "all nature exists in a state of perpetual improvement by laws impressed on the atoms of matter by the great CAUSE OF CAUSES." The very earth itself showed evidences of having been produced gradually by the operation of natural processes. Perhaps, then, the Scottish philosopher David Hume was right in speculating that the world had been generated rather than created.

> What a magnificent idea of the infinite power of THE GREAT ARCHITECT! THE CAUSE OF CAUSES! PARENT OF PARENTS! ENS ENTIUM!
>
> For if we may compare infinities, it would seem to require a greater infinity of power to cause the causes of effects, than to cause the effects themselves. This idea is analogous to the improving excellence observable in every part of the creation; such as in the progressive increase of the solid or habitable parts of the earth from water; and in the progressive increase of the wisdom and happiness of its inhabitants; and is consonant to the idea of our present situation being a state of probation, which by our exertions we may improve, and are consequently responsible for our actions (91).

168

Thus at one stroke Darwin wrested from the new biology a cosmic underwriting of the gospel of human progress and invited Christians to exchange the hope of salvation in the next world for a share in building the increasingly better life on earth which God had prepared from the foundations of the world.

▶ THE CONDITIONS OF EXISTENCE

The most powerful opponent of the development hypothesis was the brilliant and personable Cuvier, Lamarck's colleague in the Museum of Natural History (92). Acclaimed as the Aristotle of the nineteenth century for his work in comparative anatomy, vertebrate zoology, and paleontology, admired for his talents as a minister of state and his felicity as a writer, Cuvier raised himself even higher in the esteem of his contemporaries by placing himself athwart the tide of new ideas which threatened to breach the traditional view of nature. By an intellectual tour de force he held the views of Linnaeus and Buffon in unnatural embrace and banished Lamarck's evolutionism from the circle of scientific respectability.

From Buffon he adopted the conception of an organism as a system of matter in motion, discarding only the notion of a special "organic" matter. A living body, said Cuvier, "...may be considered as a kind of furnace, into which inert substances are successively thrown, which combine among themselves in various manners, maintain a certain place, and perform an action determined by the nature of the combinations they have formed, and at last fly off in order to become again subject to the laws of inanimate nature" (93). In this vortex only the form was relatively constant, changing slowly in the cycle of birth, growth, reproduction, and death. This self-perpetuating vital motion was quite different from the ordinary motions of matter. Since its effect was to counteract and even to reverse the ordinary behavior of material particles, it could scarcely be produced by the usual chemical affinities, "and yet we know of no other power in nature capable of re-uniting previously separated molecules" (94).

Cuvier, 1769 • 1832

Life was thus a system of processes exhibiting and sustaining a pattern of structures. The structures, in turn, were so constructed and related as to carry on the processes essential to the continued operation

169

of the system. There was the basic structure of fibrous walls and circulating fluids common to all living tissue and adapted equally well to the fundamental processes of absorption, assimilation, exhalation, and the like. But the particular arrangement of tissues so as to form a functioning whole was different in every kind of creature and constituted the organization peculiar to that creature. This form, or "internal mold" as Buffon would have called it, Cuvier called a prototype, declaring that it was neither produced nor changed by its own agency. He recognized, however, that organic forms were never transmitted entirely unaltered from one generation to the next, since climate, nutriment, and other influences produced variations from the norm (95).

At this point, the point where Buffon had gone on to study the human and natural causes of variation and to grasp the possibility of organic transformation, Cuvier stopped short. Experience had shown, he asserted, that variation was confined within rigid limits. Historical evidence indicated that these limits had always been the same. Nature had precluded the mixing of species by inculcating an instinctive repugnance to it within each species and by dooming hybrid productions to eventual, if not immediate, sterility. The doctrine that all differences of form among animals were produced by circumstances was an unfounded hypothesis. If modern species were but modified forms of extinct ones, why were not the intermediate forms found in the earth's strata? If living creatures had not changed substantially in the four thousand years since the pyramids were built, what reason was there to think that they changed at all? The evidence was all to the contrary. "Fixed forms that are perpetuated by generation distinguish their species, determine the complication of the secondary functions proper to each of them, and assign to them the parts they are to play on the great stage of the universe" (96).

In comparative anatomy, too, Cuvier began as a disciple of Buffon, echoing his dictum that comparison was the sovereign key to the discovery of general laws in zoology. The very interdependence of bodily functions which precluded discovering the function of a particular organ by extracting it from the whole made it possible, said Cuvier, to achieve the desired result by comparison. "The different classes of animals exhibit almost all the possible combinations of organs...while at the same time it may be said that there is no organ of which some

170

class or some genus is not deprived. A careful examination of the effects which result from these unions and privations is therefore sufficient to enable us to form probable conclusions respecting the nature and use of each organ, or form of organ" (97). Not only must the parts of an animal be so related as to function together harmoniously; they must also be so constructed as to enable the animal to maintain itself in its environment. Thus the laws governing organic form were the laws of the conditions of existence. "As nothing can exist without the re-union of those conditions which render its existence possible, the component parts of each being must be so arranged as to render possible the whole being, not only with regard to itself but to its surrounding relations" (98). It was this great fact which permitted the anatomist to reconstruct an extinct animal from the clues given by a few teeth and bones.

At this juncture, where Lamarck had argued that organic change resulted from the efforts of living creatures to adapt to changing conditions of existence and Buffon had inferred the extinction of poorly adapted forms, Cuvier gave up searching for general laws governing nature's operations and looked instead for the patterns upon which nature's productions had been modeled. The system which he hoped to discover by relating animal structure to conditions of existence was not Buffon's hidden system of laws, elements, and forces, but Linnaeus' *systema naturae,* "or a great catalogue in which all created beings have suitable names, may be recognized by distinctive characters, and be arranged in divisions and subdivisions, themselves named and characterized, in which they may be found" (99). That Cuvier never grasped the dynamic conception of nature is shown by his identifying the conditions of existence with final causes (100). Beginning with the obvious fact that creatures could not survive unless their organs were harmoniously interrelated, he jumped to the conclusion that this very necessity was the cause, or explanation, of the harmonious adjustment. In other words, animals and their environments had been planned to fit each other. *Nommer, classer, et décrire* — to name, to classify, and to describe — was the beginning and the end of science.

In comparing the structures by means of which fundamental life processes were performed in the animal kingdom Cuvier derived four main organic types — vertebrata, mollusca, articulata, and radiata — "on which all animals seem to have been modelled, and whose ul-

171

terior divisions...are merely slight modifications, founded on the development or addition of certain parts, which produce no essential change in the plan itself" (101). The notion of a great chain of being, said Cuvier, was an illusion produced by concentrating attention on single organs or groups of organs. By studying a single organ as it appeared in the various species of a class the naturalist might often discern a graduated series passing from full articulation in one species to mere vestige in another. These vestiges seemed to be left by nature "only to show how strictly she adheres to the law of doing nothing by sudden transitions," but she broke the pattern of gradation abruptly when she passed from one class to another (102).

Cuvier extended to the animals within each class the idea of variation from a primitive norm which Buffon had applied to species only; infinite variety and continuous gradation were to be found *within* the classes but not *between* them. But he was too much a rationalist and believer in the variability of nature to accept the existence of only four major types of animals as an arbitrary, irreducible fact. Why should nature be so frugal with main types and so lavish with lesser combinations? For the explanation of this circumstance Cuvier turned to his favorite theme, the conditions of existence:

> Nature never oversteps the bounds which the necessary conditions of existence prescribe to her; but whenever she is unconfined by these conditions she displays all her fertility and variety. Never departing from the small number of combinations that are possible, between the essential modifications of important organs, she seems to sport with infinite caprice in all the accessory parts. It even frequently happens, that particular forms and dispositions are created without any apparent view to utility. It seems sufficient that they should be possible, that is to say, that they do not destroy the harmony of the whole (103).

In this passage Cuvier seemed to imply that his four primary types of organization were the only types possible within the limits imposed by the conditions of existence. But the sovereign phrase "conditions of existence" contained an ambiguity: there were internal and external conditions of existence, the latter of which Cuvier might better have called conditions of survival. Was nature limited in her produc-

172

tion of creatures by both types of condition? Buffon had argued that nature produced all types which met the requirements of internal harmony, leaving those which were badly adapted to external conditions to perish in due course. But Cuvier made no allowance for the production of species incapable of surviving or for changes in the conditions of existence. He had to admit the extinction of species in order to avoid Lamarck's argument that organic forms changed with the passage of time, but he attributed their extinction to a change in the plan of creation rather than to "accidental" changes in the material environment (104). Cuvier's description was ambiguous in another way. He spoke of nature as actively producing a variety of living creatures, but he was also committed to the doctrine of a fixed creation. Lesser combinations within the four main types of organization could not, like the varieties of species, be regarded as products of time and circumstance, since each species in each class was regarded as a prototype. Hence nature as a system operating in space and time could not properly be said to produce anything except variation within fixed specific limits; for the rest, she simply displayed the diversity with which she had originally been created (105).

In short, Cuvier was attempting the impossible task of reconciling two conflicting views of nature. In one, the system of nature was the original pattern upon which the visible structures of the world had been constructed once and for all; in the other, it was the system of material particles moving in space and time by which the world of phenomena was produced, maintained, and changed. In one, the particular forms exhibited in organic nature were derived from the initial blueprint of creation; in the other, they resulted from the interactions of material bodies. In one, the conditions of existence were final causes, expressing the pre-established harmony of all created things; in the other, they were the changing configurations of the environment to which organic forms must adapt in order to survive. In one, the variety of the world was a necessary expression of divine creativity; in the other, it was the necessary product of the motions of matter in infinite time and space. That Cuvier was able to make these views seem compatible testified both to the greatness of his prestige and to the strength of the contemporary demand that the new science be reconciled with the traditional view of nature (106).

173

How much doth the hideous monkey resemble us!

ENNIUS, 240–169 B.C.

Thus the ape, which philosophers, as well as the vulgar, have regarded as a being difficult to define, and whose nature was at least equivocal, and intermediate between that of man and the animals, is, in fact, nothing but a real brute,...

BUFFON, 1766

6.

Man's Place in Nature

B Y EIGHTEENTH CENTURY STANDARDS the noblest branch of natural history was that which treated the natural history of man. No account of the origins of evolutionary ideas would be complete which failed to reckon with the powerful influence of anthropology on eighteenth century concepts of nature, human nature, and human history.

As various accounts of anthropoid creatures were subjected to scientific scrutiny, the eternal question of man's place in nature was raised anew. As primitive peoples were visited and their appearance, languages, manners, and customs described, the idea that human history had been a progress from savagery rather than a decline from original perfection took firm root. And as familiarity with racial types increased, a noisy controversy arose concerning the unity or diversity of the human species.

Those who believed that all human races were varieties of a single species were hard pressed to explain how nature had managed to produce so great a variety of physical types in the few thousand years traditionally allotted to man's career on earth. The influence of these developments on the progress of evolutionary thinking in biology cannot be calculated exactly, but unquestionably the connection was close and significant.

In 1735 Linnaeus placed the question of man's rank in nature in a new and startling light. In his *System of Nature* man was classified as a quadruped and placed in the same order, *Anthropomorpha,* with the ape and the sloth. The terse entry read as follows:

175

CLASS I – QUADRUPEDS

Body hairy. Feet four. Female viviparous, lactiferous.

ANTHROPOMORPHA Incisors four on either side	*Homo*	Know thyself	
	Simia	Anterior digits 5	Posterior 5
		Posterior similar to anterior	
	Bradypus	Digits 3 or 2	3

The satyr, tailed, hairy, bearded, with a human body, much given to gesticulations, extremely lascivious, is a species of ape, if one has ever been seen. The *tailed men,* also, of whom modern travellers relate so much, are of the same genus (1).

The die was cast. The issue concerning man's relation to the animals had been submitted to the decision of comparative anatomy.

► **N O T   A   M O N K E Y   N O R   Y E T   A   M A N**

Even before Linnaeus threw down the gauntlet, other men had noted the structural similarities between man and the apes. Aristotle and Galen had discussed the subject, and their work had been carried further by the anatomists of the seventeenth century. In England the most brilliant and indefatigable of these was Edward Tyson, graduate

Edward Tyson
1650 • 1708

of Oxford, member of the Royal Society, and a leading physician in London. In his spare time Tyson dissected a great number of animals including an opossum and a rattlesnake from America and an ostrich and lion from Morocco. He published anatomical studies of several of these. In the spring of 1698 he undertook the dissection of an infant chimpanzee which had died a few months after being brought to London from·Angola, Africa. The results were published in 1699 under the captivating title: *Orang-Outang: Or the Anatomy of a Pygmy Compared With That of a Monkey, an Ape, and a Man.* The purpose of the comparison, Tyson explained in the dedicatory epistle to Lord John Somers, was to exhibit the scale of nature concerning which Aristotle had written and in which, said Tyson, the pygmy formed the connecting link between the animal and the rational, "as your Lordship, and those of your High Rank and Order for Knowledge and Wisdom, approaching nearest to that kind of being which is next above us, connect the Visible, and Invisible World" (2). This scale of nature was no mere figure of speech to Tyson. The perfection of natural history, he asserted, would be "to enumerate and remark all the different Species, and their Gradual Perfections from one to another." To this end the various animals must be dissected and their anatomy studied diligently.

Applying comparative techniques to the pygmy, Tyson discovered forty-eight respects in which it resembled man more than monkeys and apes, and only thirty-four in which it approximated more to apes and monkeys than to man. It was, so to speak, "an intermediate link between an Ape and a Man." It was *not* a man, however. The similarities enumerated were all physical, and they in no way destroyed the essential distinction between man and "not-man." In short, the pygmy was "no man, nor yet the Common Ape; but a sort of Animal between both; and tho' a Biped, yet of the Quadrumanus-kind; though some Men too, have been observed to use their Feet like Hands, as I have seen several" (3).

The real distinction between man and the pygmy must be sought in the "Nobler Faculties," such as the power of speech, Tyson concluded. But he could find no physiological reason why the pygmy should not speak. Its larynx was quite like the human larynx, and its brain seemed as large in proportion to its body as in the case of man. Evidently the mental faculties must spring from a higher principle than that of organized matter, and "there is no reason to think, that Agents do perform such and such actions, because they are found with organs proper thereunto" (4).

Although Tyson pretended to find satisfaction in this paradox as providing an irrefutable answer to atheism, the reader senses his reluctance as a natural historian to accept a radical dissociation of structure and function. The science of physiology attempts to understand the human body as an interdependent system of structures performing functions useful or indispensable in the life of the organism. Tyson himself attempted to prove that the pygmy had been designed by nature to walk erect by citing anatomical peculiarities "particularly contrived for the advantage of an erect posture." Why, then, did he welcome an apparent exception to the close correlation of structure and

177

function which in other respects he took for granted? The following passage suggests the answer:

> The *Organs* in *Animal* Bodies are only a regular Compages of Pipes and Vessels, for the *Fluids* to pass through, and are passive. What actuates them are the Humours and Fluids: and Animal Life consists in their due and regular motion in this *Organical* Body. But those *Nobler Faculties* in the *Mind* of *Man,* must certainly have a *higher Principle;* and *Matter organized* could never produce them; for why else, where the *Organ* is the same, should not the Actions be the same too? and if all depended on the *Organ,* not only our *Pygmie,* but other *Brutes* likewise, would be too near akin to us. This Difference I cannot but remark, that the *Ancients* were fond of making *Brutes* to be *Men;* on the contrary now, most unphilosophically, the *Humour* is, to make *Men* but meer *Brutes* and *Matter.* Whereas in truth *Man* is part a *Brute,* part an *Angel;* and is that Link in the Creation, that joins them both together (5).

On the one hand, Tyson sought to explain an apparent disunion of structure and function in terms of the traditional idea that mind is separate from and superior to matter. On the other hand, he was ready to vindicate this traditional idea against any who might argue from the similarity of the organs that the pygmy was a man. Such an argument would not only make mind a function of material organization but would endanger the whole concept of nature as a scale of beings essentially distinct from one another, a scale in which man occupied a special position. If man could not be shown to be structurally different from the apes, he might still be assigned a higher place in nature by virtue of "Nobler Faculties" deriving from a "higher Principle" than material organization (6). But this explanation of the supposed facts was as dangerous to natural theology as it was unconvincing in natural history. The perfect adaptation of structure to function was the chief argument of books like John Ray's *The Wisdom of God Manifested in the Works of the Creation.* Tyson's paradox of corresponding structures without corresponding functions implied an arrangement of nature based on God's arbitrary and inscrutable will rather than on a plan intelligible to human reason.

Tyson's difficulties would have been largely removed if he had been able to dissect several mature chimpanzees and thus discover that the pygmy's brain was not so large as the human brain either absolutely or relative to body size. He would then have had a physiological difference to match the difference in mental capacity and to confirm the pygmy's intermediate position between man and the apes in the great chain of being. In any event Tyson did not intend to let the apparent facts of comparative anatomy disturb his belief in the hierarchy of being or to countenance the heretical idea that quantitative changes in the system of matter in motion were the cause of the qualitative differences observed among nature's productions.

► THE PONGO AND THE JOCKO

When Buffon turned his attention to apes in the fourteenth volume of his *Natural History,* published in 1766, he drew heavily on Tyson for both information and ideas. Tyson had discussed with much quotation and reproduction of plates the works of the early naturalists: Aristotle's and Galen's accounts of the *pithecus;* the compilations of ape-lore by Gesner, Aldrovandi and Jonston; Nicolaas Tulp's description in 1641 of the "Indian Satyr," an Angolese chimpanzee; the account in 1658 by Jakob de Bondt, a Dutch physician in Batavia, of a female orang-outang; and the detailed anatomy of four species of monkeys by Claude Perrault and his colleagues of the French Academy of Sciences (7). From the travel books of his own day Tyson had culled numerous descriptions of manlike creatures bearing such exotic names as *quoias-morrou, drill, barris, engeco,* and *pongo,* the last two of these described in an early seventeenth century work, *Purchas His Pilgrimes,* in a chapter entitled "The Strange Adventures of Andrew Battell, of Leigh in Essex, Sent by the Portugals Prisoner to Angola, Who Lived There and in the Adjoining Regions Neere Eighteene Yeeres." According to Battell, the woods in that region of Africa were inhabited by two kinds of monsters:

Buffon,
1707 . 1788

> The greatest of these two Monsters is called, *Pongo,* in their Language: and the lesser is called, *Engeco.* This *Pongo* is in all proportion like a man, but that he is more like a Giant in stature, then a man: for he is very tall, and hath a mans face, hollow eyed, with long haire upon his browes. His face and eares are without haire, and his hands also. His bodie is full of

179

haire, but not very thicke, and it is of a dunnish colour. He differeth not from a man, but in his legs, for they have no calfe. Hee goeth alwaies upon his legs, and carrieth his hands clasped on the nape of his necke, when he goeth upon the ground. They sleepe in the trees, and build shelters for the raine. They feed upon Fruit that they find in the Woods, and upon Nuts, for they eate no kind of flesh. They cannot speake, and have no understanding more then a beast. The People of the Countrie, when they travaile in the Woods, make fires where they sleepe in the night; and in the morning, when they are gone, the *Pongoes* will come and sit about the fire, till it goeth out: for they have no understanding to lay the wood together. They goe many together, and kill many *Negroes* that travaile in the Woods. Many times they fall upon the Elephants, which come to feed where they be, and so beate them with their clubbed fists, and pieces of wood, that they will runne roaring away from them. Those *Pongoes* are never taken alive, because they are so strong, that ten men cannot hold one of them: but yet they take many of their young ones with poisoned Arrowes. The young *Pongo* hangeth on his mothers bellie, with his hands fast clasped about her: so that, when the Countrie people kill any of the females, they take the young one, which hangeth fast upon his mother. When they die among themselves, they cover the dead with great heapes of boughs and wood, which is commonly found in the Forrests (8).

In addition to these sources and Tyson's own work, Buffon had the naturalists and travellers of the eighteenth century to draw on — nomenclators and classifiers, such as Linnaeus and Brisson, and a host of voyagers to the remote regions of the earth. Those accounts of the great apes which seemed at all reliable or credible Buffon quoted in full, "because every article is important in the history of a brute which has so great a resemblance to man." He noted which animals (the gibbon, the Barbary ape, and the "small orang") he had seen himself, and supplemented his own account with plates executed by his anatomist, Daubenton.

Buffon was not much interested in classification for its own sake; for convenience, however, he proposed a classification based on external characters. An ape, he declared, was "an animal without a tail, whose face is flat, whose teeth, hands, fingers, and nails resemble those of man, and who, like him, walks erect on two feet" (9). Baboons, on the contrary, had a short tail, a long face, a broad muzzle, canine teeth proportionally larger than those of men, and callosities on the buttock. Monkeys constituted a third group possessing tails longer than their bodies. Having established these divisions, Buffon immediately denied that there was anything "natural" about them. Nature knew nothing of perfectly distinct genera. Her productions were always "gradual and marked by minute shades." Thus the Barbary ape was an intermediate form between apes and baboons, as the *maimon* was a link between baboons and monkeys.

Buffon recognized three kinds of apes — the "orang-outang," Aristotle's *pithecus,* and the *gibbon.* The "orang-outangs" he divided into two species, the *jocko* and the *pongo,* differing chiefly in size. As examples of the smaller orang, or *jocko,* he listed the chimpanzees described by Tyson and Tulp and another which he kept in his own house. The *pongo* he knew only from the accounts of travellers:

> ...an ape as tall and strong as man, and equally ardent for woman as for its own females; an ape who knows how to bear arms, to attack his enemies with stones, and to defend himself with clubs...he resembles man still more than the pigmy [Aristotle's *pithecus*]; for, independent of his having no tail, of his flat face, of the resemblance of his arms, hands, toes, and nails to ours, and of his walking constantly on end, he has a kind of visage with features which approach to those of the human countenance, a beard on his chin, and no more hair on his body than men have, when in a state of nature ...This orang-outang or pongo is only a brute, but a brute of a kind so singular, that man cannot behold it without contemplating himself, and without being thoroughly convinced that his body is not the most essential part of his nature (10).

Buffon suspected that the creatures he had labeled *jocko* were perhaps immature pongos which had died before reaching man-sized stature, hence he did not hesitate to apply to the *pongo* information obtained by dissection and observation of the *jocko.* From Tyson he quoted, with corrections, the entire catalogue of characters in which the "orang-outang" approximates to man and to the apes. He concluded that the great orang-outang, even more than Tyson's pygmy, might be considered either the first of the apes or the last of men in its physical structure and appearance. But man was capable of thought and speech, powers which removed him infinitely from the apes. Like Tyson, Buffon held that "matter alone, though perfectly organized, can produce neither language nor thought, unless it be animated by a superior principle" (11). The intelligence and human qualities of the ape had been exaggerated, he declared. The dog was more intelligent than the ape and more akin to man in temperament. The fact that apes were domesticated with great difficulty proved the brutal quality of their nature. Their so-called "human actions" arose from imitation and general structural resemblance, not from a true psychological similarity to man. The crowning proof of their brute nature was their failure to speak even when equipped with organs adequate for that purpose. Consideration of this paradox drew Buffon into a discussion of the interrelations between length of infancy, the formation of society, and the development of speech, a discussion which will be examined later in connection with the speculations of Rousseau and Lord Monboddo. It suffices here to quote Buffon's general conclusion:

> Thus the ape, which philosophers, as well as the vulgar, have regarded as a being difficult to define, and whose nature was at least equivocal, and intermediate between that of man and the animals, is, in fact, nothing but a real brute, endowed with the external mark of humanity, but deprived of thought, and of every faculty which properly constitutes the human species; a brute inferior to many others in his relative powers, and still more essentially different from the human race by his nature, his temperament, and the time necessary to his education, gestation, growth, and duration of life; that is, by all the real habitudes which constitute what is called *Nature* in a particular being (12).

► **THE LINE BETWEEN APES AND MEN**

In the same year that Buffon described and classified the apes and monkeys, 1766, Linnaeus issued the twelfth and last of his own editions of the *System of Nature*. Once again he wrestled with the problem of man's relation to the apes, but he was no nearer a solution than he had been thirty years earlier, when he placed man in the same order with the ape and the sloth. Meanwhile the issue had grown more urgent. From Linnaeus' point of view classification was no matter of mere

FIG. 6.1 — Buffon's **jocko,** or "smaller orang," from his volume on apes and monkeys, published in 1766. The figure, drawn from an immature chimpanzee in Buffon's possession, was made as human-looking as possible, a distortion which drew a sharp protest from the Dutch anatomist Petrus Camper.

convenience but the heart and soul of science. Nature was to be understood not as a continuously graded series of forms but as an aggregate of separate groupings marked off from each other by permanent and distinctive differences in the number, position, figure, and proportion of the parts. Hence, the natural historian's problem with respect to man was to find a generic physical character by which to differentiate him from the apes. Reproached by Gmelin for including man in the same order with the apes, Linnaeus replied:

> ...I demand of you, and of the whole world, that you show
> me a generic character...by which to distinguish between
> Man and Ape. I myself most assuredly know of none. I wish
> somebody would indicate one to me. But, if I had called man
> an ape, or vice-versa, I should have fallen under the ban of
> all the ecclesiastics. It may be that as a naturalist I ought to
> have done so (13).

In his published works Linnaeus was always careful to stress man's mental and spiritual superiority over the apes. But he refused to regard this advantage, however great, as a proper ground for classification in natural history. Speech, though distinctive of man, was "only a sort of power or result, and not a characteristic mark taken from number, figure, proportion, or position" (14).

In working out a classification based on physical characters, Linnaeus was severely handicapped by lack of specimens and reliable descriptions of the subjects to be classified. He seems not to have known Tyson's work. The only ape he mentions as being accessible to him for examination was a young chimpanzee sent to him about 1760 by the English naturalist George Edwards. For the rest he relied on accounts of naturalists and travellers. With respect to both of these sources he was less cautious and sceptical than Tyson and Buffon. The chimpanzees described by Tulp, Edwards, and Scotin he grouped under the generic name *Simia satyrus*. To another group culled from Gesner, Jonston, and Brisson he gave the name *Simia sylvanus*. But when he came to de Bondt's account of the Bornean orang-outang, he was misled by the figure, that of a hairy woman, and by the author's account of the creature's becoming modesty. Tyson had slyly added a fig leaf when he reproduced de Bondt's figure, and Buffon was no less sceptical on this score. Linnaeus, however, was sufficiently impressed by this and similar accounts to place the creature in a separate human genus, *Troglodytes*. The tenth edition of the *System of Nature* (1758) included in this genus the species *Homo nocturnus*, also called *Homo sylvestris orang outang*, described as follows:

Linnaeus, 1707 · 1778

> It lives within the boundaries of Ethiopia (Pliny), in the
> caves of Java, Amboina, Ternate. Body white, walks erect,
> less than half our size. Hair white, frizzled. Eyes orbicular:
> iris and pupils golden. Vision lateral, nocturnal. Life-span

twenty-five years. By day hides; by night it sees, goes out, forages. Speaks in a hiss. Thinks, believes that the earth was made for it, and that sometime it will be master again, if we may believe the travellers (15).

To distinguish creatures of this sort from man was extremely difficult, said Linnaeus. No constant character had been found by which to establish them in a separate genus. But they were certainly not of the same *species* as man, "nor of common descent or blood with us," for their eyes were equipped with a nictating membrane of the kind found in owls and bears. Here Linnaeus, feeling perhaps that his scientific argument had led him onto shaky theological ground, added cryptically: "Nor do I say that the troglodytes of Pliny are Prae-Adamites, although we are the final handiwork of the Creator" (16).

Linnaeus also attempted to find a place in his classification for the various "tailed men" described by travellers. In 1758 he declared himself uncertain whether these tailed creatures were men or apes, but in 1760 his pupil Hoppius gave them the generic name *Lucifer* and assigned one species, *Homo caudatus vulgo dictus,* to the East Indies and the Nicobar islands. As evidence of this creature's existence, Hoppius reproduced a figure from Aldrovandi and an account by a Swedish traveller of an encounter with a horde of tailed men. The genus *Lucifer* he ranked second only to *Troglodytes* in resemblance to man. After them he placed the *Satyr* and *Pygmy* kinds, described from accounts of chimpanzees. Even with respect to these two ape-kinds he could discover no constant physical character, except possibly the "corner teeth," in which man differed from every species of ape. Moreover, apes resembled men in many things other than physical appearance.

> They often go erect, and on their hinder feet alone; they pick their food, and carry it to their mouths with their hands; they drink liquids from cocoa-nuts scooped out, and when short of water, dig wells with their feet. They are omnivorous like us... They are always hunting after lice; they remove dirt from their bodies; they are fond of games as boys, they are capital rope-dancers, always clever gesticulators, at whom you can never laugh enough. They are malicious by nature, ready for every mischief, given to theft, very salacious even

FIG. 6.2 — Four representations of anthropoid creatures, culled from the works of early naturalists by Linnaeus' pupil Hoppius and published in the Swedish **Amoenitates Academicae** in 1760. The fact that Linnaeus attempted to find a place for all of these creatures in his classification of mankind in his **System of Nature** reveals strikingly the limitations of mid-eighteenth century anthropology.

when pregnant. Very mindful of injuries, and difficult to be appeased; anxious, but at the same time timid hunters; imitators of every folly; very difficult to castrate; both fathers and mothers are very fond of their children, even after having had as many as nine. They run away from crocodiles and serpents, and what you would be surprised at, even from those who are ill of contagious fevers (17).

Hoppius' account concluded with an appeal to kings and princes to procure living specimens of apes and troglodytes, so that philosophers might investigate at first hand "how far the power of the human mind surpasses theirs, and what is the real difference between the brute and the rational being."

Linnaeus' last classification of men and apes in the twelfth edition of the *System of Nature* contained few new facts or ideas. Quoting the

Latin poet Ennius once more on the disgusting similarity of apes to men, *Simia quam similis turpissima bestia nobis,* he added his own final comment: "It is remarkable that the stupidest ape differs so little from the wisest man, that the surveyor of nature has yet to be found who can draw the line between them" (18). But he qualified this negative conclusion from natural history by an affirmation of human dignity based on man's unique mental and moral life. Man knows himself to be the ultimate end of creation, a rational being introduced onto the earth to contemplate and admire the Creator in His works. *"What else has been revealed must be explained by theologians."*

► THE ORANG-OUTANG DESCRIBED

In the years after 1766 the chief additions to the growing stock of information concerning anthropoid apes came from Holland, where the publication of Buffon's fourteenth volume excited a lively competition to obtain and describe the Bornean orang-outang. The celebrated anatomist Petrus Camper procured his first in 1770, about the same time that Professor Jean Allamand obtained one for the museum at the University of Leyden. In 1776 Arnout Vosmaer, Director of the Natural History Collection of the Prince of Orange, received a live female orang-outang from Borneo. He kept it at his house a month for observation, until public curiosity forced him to place it in the menagerie. In 1778 he published a detailed description of its appearance and behavior, comparing it to the various apes described by previous authors. The Bornean orang-outang, Vosmaer reported, was considerably different from the apes described by de Bondt, Tyson, and Buffon. It had few of the human characteristics commonly ascribed to it. In height it was but half of the five feet claimed by Buffon for the *pongo.* Was it, then, of a different species from the great orang-outang, or was Buffon's man-sized ape a figment of the imagination? Vosmaer inclined to the latter view. His correspondents in Batavia had assured him that the oldest inhabitants there had never seen an orang of human size, and their testimony had been confirmed by a Swiss officer recently returned from Borneo. Moreover, of ten apes received from

Arnout Vosmaer, 1720 · 1799

187

FIG. 6.3 — A young female orang-outang, procured alive from Borneo in 1776 and described by Arnout Vosmaer, Director of the Natural History Collection of the Prince of Orange, in 1778. After observing the animal for several months, Vosmaer concluded that it had few of the human characteristics attributed to the "orang" by Buffon and other writers (who were actually describing young chimpanzees).

the East Indies, none had measured more than two and a half feet in height. If twenty years of research had failed to discover the great orang-outang, the explanation must be that the creature did not exist (19).

Vosmaer's ape was sent to Petrus Camper for dissection when it died. Camper had begun dissecting monkeys in 1754 to improve his

Petrus Camper,
1722 • 1789

understanding of Galen's works. Aroused by Galen's account of the *pithecus* and Tyson's description of the pygmy, Camper determined to get a tailless ape from Borneo. From correspondents he obtained two orang-outangs, and

Vosmaer, Allamand, and others gave him access to their specimens. In the end he was able to dissect five orang-outangs and to examine several others. In 1779 he published an "Account of the Organs of Speech in the Orang Outang" in the *Philosophical Transactions* of the Royal Society. His *Natural History of the Orang Outang and Other Kinds of Apes* appeared in 1782.

In these works Camper reinforced with anatomical evidence Vosmaer's conjecture that the Bornean orang-outang was a species hitherto undescribed. It was neither the *pongo* nor the *jocko* of Buffon, nor Tyson's *pygmy*, nor Tulp's *Indian Satyr*, nor Galen's *pithecus*. Moreover, it possessed none of the human traits frequently attributed to apes. Its speechlessness, said Camper, was not a ruse adopted to avoid enslavement but a necessary effect of the structure of its vocal organs.

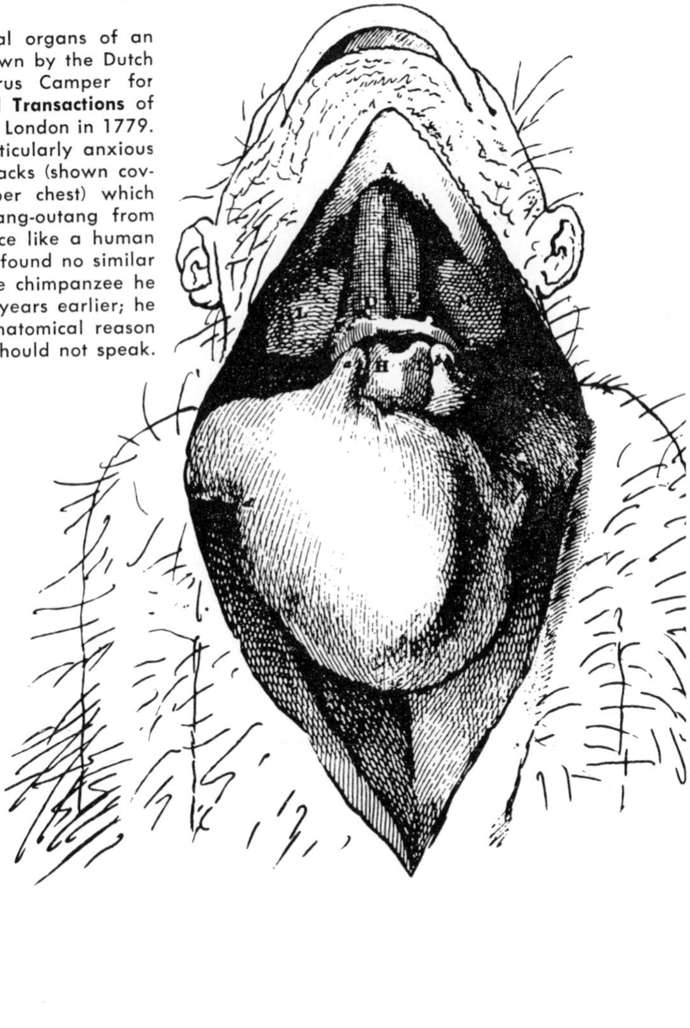

FIG. 6.4 — The vocal organs of an orang-outang, drawn by the Dutch anatomist Petrus Camper for the **Philosophical Transactions** of the Royal Society of London in 1779. Camper was particularly anxious to depict the air-sacks (shown covering the upper chest) which prevent the orang-outang from modulating the voice like a human being. Tyson found no similar structure in the chimpanzee he dissected eighty years earlier; he could see no anatomical reason why his "pygmy" should not speak.

Having dissected the whole organ of voice in the Orang, in apes, and several monkies, I have a right to conclude, that Orangs and apes are not made to modulate the voice like men: for the air passing by the *rima glottidis* is immediately lost in the ventricles or ventricle of the neck, as in apes and monkies, and must consequently return from thence without any force and melody within the throat and mouth of these creatures: and this seems to me the most evident proof of the incapacity of Orangs, apes, and monkies, to utter any modulated voice, as indeed they never have been observed to do (20).

The orang-outang was equally ill-constructed for walking on two feet, for sitting, and for lying on its back, Camper continued. Its thumb was too short to serve well for grasping; its height was but half that of a man. In brief, it was not a man but a four-footed beast. The chances were, he added, that the apes described and figured by Tulp, Tyson, Buffon, and others had been less human in structure, appearance, and behavior than those authors had represented them. Why, for instance, had they portrayed apes with straight knees if not to make them appear human? Such representations served only to mislead other naturalists and to degrade human nature by belittling man's pre-eminence over all other creatures (21).

Another book by Camper, his *Dissertation on the Natural Varieties Which Characterize the Human Physiognomy,* published posthumously in 1792, was destined to exert a profound influence on the development of physical anthropology. His interest in anatomy, joined to his passion for art, led him to compare the heads modeled by the ancient Greek sculptors with those drawn by Dutch and Flemish artists. Observing that the profile of the classical heads was much steeper than that of the Dutch and Flemish heads, he extended the comparison to the animal kingdom and discovered the importance of the jaw structure in determining the shape of the head and face.

Upon placing beside the heads of the Negro and the Calmuck those of the European and the ape, I perceived that a line drawn from the forehead to the upper lip indicates a difference in the physiognomy of these peoples and makes apparent a marked analogy between the head of the Negro and that of the ape. After having traced the outline of several of these heads on a horizontal line, I added the facial lines of the faces, with their different angles; and immediately upon inclining the facial line forward, I obtained a head like that of the ancients; but when I inclined that line backwards, I produced a Negro physiognomy, and definitively the profile of an ape, of a Chinese, of an idiot in proportion as I inclined this same line more or less to the rear (22).

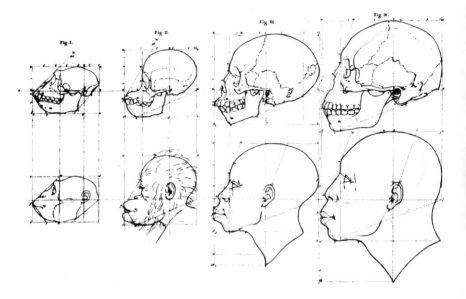

FIG. 6.5 — Skulls and heads of a monkey (Fig. I), an orang-outang (Fig. II), a Negro (Fig. III), and a Kalmuck (Fig. IV) arranged by the Dutch anatomist Petrus Camper to illustrate his conception of the **facial line** and **facial angle**. The **facial line** is drawn from the forehead to the upper lip; it forms a **facial angle** with a line drawn from the base of the nose to the auditory opening. This idea of a gradation of head forms in the animal kingdom was used by some writers, though not by Camper himself, to support the thesis that there were several species of men, the Negro being the lowest in the scale of being. The illustration is from Camper's **Physical Dissertation on the Real Differences Presented by the Facial Traits of Men of Different Countries and Different Ages** (Autrecht: 1791).

In these words Camper announced his discovery of the *facial line,* "the line which describes the physiognomy of man and animals." Among human beings, he observed, the angle formed by the facial line and a line drawn through the base of the nose to the auditory opening varied between eighty and seventy degrees. Everything above eighty degrees belonged to the realm of art, everything below seventy degrees to the animal kingdom.

A very satisfying discovery, the *facial line,* but not without disquieting implications. Camper himself noted uneasily that comparisons of the Negro and the ape had led some philosophers to imagine that the Negro might be a hybrid produced by intercourse between white men and apes. Others supposed that some orang-outangs might possibly have been able in the course of time to develop into full-fledged human beings without such intercourse. For the refutation of these wild and superficial speculations Camper referred his readers to his *Natural History of the Orang Outang* (23).

In 1789, one year after Buffon's death, the *Supplement* to the fourteenth volume of his *Natural History* appeared. In this his last word on the subject of apes, Buffon modified his earlier views. He was more convinced than ever that there were at least two species of orang-outang, the *pongo* and the *jocko.* But examination of the remains of a Bornean orang-outang differing considerably from Tyson's ape and his own had shown him that the two species of orang were distinguished by other characters than size. He was now virtually certain that Tyson's, Tulp's, and his own ape were immature *pongos* which would have grown to a height of about five feet had they lived longer. Hence he transferred the name *jocko* to apes like that described by Vosmaer — apes which were covered with red hair, lacked the nail on the big toe, walked upright infrequently, and attained but half the size of the full-grown *pongo* (24). It did not occur to Buffon that the orang-outangs sent from Borneo might also be immature or that the manlike apes might be divided according to habitat, one species inhabiting the East Indies, the other Africa, both species attaining to human size at maturity.

For those who, like Vosmaer and Camper, doubted the *pongo's* existence, Buffon quoted new accounts of the man-sized ape. The Chevalier d'Obsonville reported having seen a four-and-a-half foot ape alive in captivity. Professor Allamand had talked with a Dutch sea captain who had seen a captured African ape five-and-a-half feet high, similar in many respects to the mandrill, a west African baboon. Allamand not only accepted the story as true but quoted a letter from a Dutch physician in Batavia confirming de Bondt's account of the *Homo sylvestris,* adding that the creature was of human dimensions.

Attacked from both sides by the naturalists of Holland, Buffon yielded to neither party though he borrowed from both. The genus orang-outang, he concluded, comprised several species. Besides the *pongo* and the *jocko* there was, perhaps, the creature described by Allamand. Finally, there was the huge ape reported by the Dutch sea captain, "which might possibly constitute the shade between the pongo and the mandrill" (25).

► A MAN-SIZED APE AT LAST

Although Buffon claimed no special merit for his classification of the apes, baboons, and monkeys it was taken over with relatively few changes by the great systematist, Cuvier. Like his predecessor Linnaeus, Cuvier aimed at nothing less than the classification of the whole animal kingdom according to the natural affinities of the creatures comprised therein. In preparation for this huge undertaking, Cuvier devoted himself to the study of comparative anatomy.

Cuvier, 1769 . 1832

His *Tableau of the Natural History of Animals,* published in 1798, was the first systematic presentation of his findings (26).

In this work Cuvier departed from Linnaeus by placing man in a separate order, *Bimanes,* distinguished from apes and monkeys, or *Quadrumanes,* by the possession of genuine feet, erect posture, biped gait, and other related characteristics. To divide the *Quadrumanes* Cuvier used, in addition to the characters noted by Buffon, head form and the degree of prognathism of the jaw. Camper had called attention to the cranial features and their interrelation without exploring the matter much farther. Cuvier, endeavoring to understand animal structure in relation to the "conditions of existence," hit upon the idea that the proportion of the facial structure to the cranium reflected the relative importance of the external senses as compared with the "interior faculties" in the animal's dealings with its environment. In his

Lessons in Comparative Anatomy (1800–1805), he described three methods of measuring the proportions of facial structure to cranium. Camper's method of the *facial line* he declared to be useful only in the case of human beings and the quadrumanes; in other animals the frontal sinuses were often greatly developed or the nose projected above the line. Cuvier's own method was to divide the head with a vertical and a longitudinal section. "Relatively to their respective proportions, the cranium occupies in this section a space sometimes larger, sometimes almost equal to that which the face occupies" (27). By this method Cuvier obtained the same gradation from European to Calmuck to Negro to orang-outang to monkey that Camper had derived by means of the *facial line*.

In Cuvier's *Tableau* the orang-outang and the chimpanzee were clearly differentiated and assigned to their respective habitats in East Asia and Africa. The orang-outang was declared to resemble man more closely but to differ from him in the prognathism of its jaw, the length of its arms, the awkwardness with which it walked on two feet, and in the possession of air sacks which connected with the larynx and made its voice sound hollow. The gibbon and the "Wouwou" completed the list of "apes properly so-called" (28).

Among the baboons appeared an animal called the *pongo* — "an ape from the island of Borneo which has the height of a man, arms as long as its body, very powerful jaws, enormous canine teeth and which lacks a tail. It resembles the mandrill in its head" (29). An animal of this kind, an adult orang-outang, had been captured about 1780 by a Dutch merchant travelling in Borneo. The creature was killed because of its fierce resistance and sent preserved in arrack (spirits of rum), to the Batavian Society of Arts and Sciences. The Secretary of the Society, one Baron von Wurmb, a German officer in the employ of the Dutch East India Company, prepared a detailed description of its external appearance for the *Transactions* of the Society, after which the cadaver was shipped to Vosmaer for the collection of the Prince of Orange. Whether the carcass ever reached Holland seems doubtful, for the ship was wrecked enroute. In any event, Wurmb's account of the animal was translated into French in 1796 and appeared in English in the *Philosophical Magazine* in 1798 (30). In this account Wurmb identified the animal with the *Homo sylvestris* of de Bondt, supposing that de Bondt had erred only in delineating

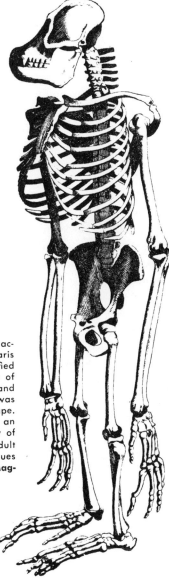

FIG. 6.6 — Skeleton of an adult orang-outang, acquired by the Museum of Natural History in Paris about 1796 (probably from Holland) and classified by Cuvier and Geoffroy Saint-Hilaire as a species of baboon because of its huge, prognathous jaw and other facial and cranial features. Here, at last, was tangible evidence of the existence of a man-sized ape. Not until 1829 did Cuvier list this creature as an orang-outang. See Figure 6.7, p. 197, for a view of the differences between the immature and the adult orang-outang which led Cuvier and his colleagues astray. The illustration is from the **Philosophical Magazine,** September, 1798.

the figure and in repeating tall tales of the orang-outang's human traits. This large ape, Wurmb concluded, was Buffon's *pongo* discovered at last; it was a species distinct from the small orang, or *jocko*, described by Vosmaer in 1778.

But the matter did not rest there. A month later the *Philosophical Magazine* printed a translation of Geoffroy Saint-Hilaire's "Observations" on the subject of Wurmb's ape. In the Museum of Natural History at Paris, the French zoologist had found a skeleton which he supposed to be that of the animal shipped to Holland by Baron von Wurmb. On examining the skeleton, he was struck by the differences,

Etienne Geoffroy Saint-Hilaire, 1772 · 1844

particularly in facial and cranial structure, between this ape and the apes described by Tyson, Buffon, and Vosmaer. Adopting Cuvier's principle that "the size and convexity of the cranium indicate sensibility, as the prolongation and size of the muzzle indicate brutality," Geoffroy argued that Wurmb's ape must rank lower among brutes than the orang-outang, the chimpanzee, and the gibbon. Yet it was of human dimensions, and although its projecting muzzle and the position of the occipital foramen (opening to the spinal canal at the posterior base of the skull) seemed to deprive it of the equilibrium necessary for walking erect, the compensating construction of the spine, pelvis, and arms undoubtedly enabled it to support its head and run upright. All in all, it was a new species, resembling the orang-outang in some respects and the mandrill in others. As for Buffon's *pongo,* it was, said Geoffroy, "an imaginary being to which Buffon has assigned a form and characterising marks, by confounding, under the same name, and in the same description, six different species of apes described by travellers" (31).

Geoffroy's article makes clear the considerations which led his colleague and collaborator Cuvier to place Wurmb's ape among the baboons. Cuvier repeated this classification in the first edition of his *Animal Kingdom* (1817). Not until the second edition, in 1829, did Wurmb's ape appear among the orang-outangs and even then with reservations and qualifications (32). The large chimpanzee, Cuvier reported, was still known only from travellers' accounts. According to these, it equalled or surpassed man in stature, constructed shelters made from leaves, used sticks and stones to drive away men and ele-

FIG. 6.7 — Comparison of the skeletons of an immature and an adult orang-outang, from Richard Owen's article in the **Transactions** of the Zoological Society of London for 1835. Note the more bestial features of the adult cranium, jaw, and teeth, which led able anatomists like Cuvier to classify the first available adult skeleton of an orang-outang as that of a species of baboon, which Cuvier named **pongo**.

FIG. 6.8 — The skeleton of a chimpanzee (left) compared with that of an orang-outang (right), figured by Richard Owen in the **Transactions** of the Zoological Society of London for 1835. This was apparently the first skeleton of an adult chimpanzee available to the anatomists of Europe.

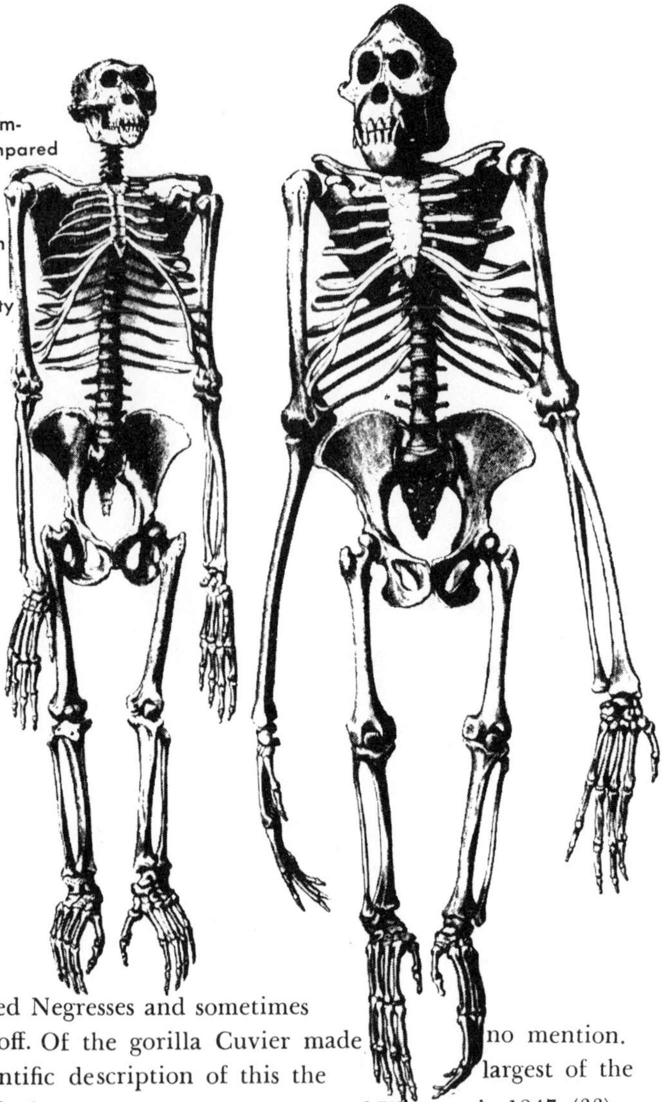

phants, pursued Negresses and sometimes carried them off. Of the gorilla Cuvier made no mention. The first scientific description of this the largest of the apes was made by two Americans, Savage and Wyman, in 1847 (33) .

Meanwhile Lamarck's suggestion that a race of apes "impelled by the need to dominate and to see far and wide" might gradually acquire the conformation and capacities peculiar to man lay neglected in the pages of the *Philosophie zoologique*. Not until Darwin had rendered the "development hypothesis" scientifically respectable was the idea of human evolution taken seriously by naturalists (34) .

198

CHIMPANZEE	ORANG-OUTANG	GORILLA	GIBBON	OTHERS
		1613 (1625) Battell's account of the gorilla (?)		
1641 Tulp's "Indian Satyr"				
	1658 de Bondt's *Homo sylvestris*			
1699 Tyson's "pygmy"				
				1735 Linnaeus on the satyr and tailed men
1738 Scotin's figure.				
1758 Edwards' "Man of the Woods"				1758 Linnaeus *Homo troglodytes* 1760 Hoppius' *Homo lucifer*
1766 Buffon's *jocko* (citing Tulp, Tyson)		1766 Battell's story re- told by Buffon	1766 Buffon's descrip- tion and figure	
	1776 Vosmaer's small orang 1779–82 Camper on the orang 1796–98 Wurmb's account, with Geoffroy's comment on it		1769–84 Good descriptions by several writers	
1798 Cuvier distinguishes the chimpanzee from the orang, classifies "Wurmb's ape" as a baboon				
	1832 Cuvier classifies "Wurmb's ape" as an orang			
1835 Owen gives a full account of both the chimpanzee and the orang-outang, infant and adult				
		1847 Savage and Wy- man describe the gorilla		

And God created man in His own image.

Genesis 1:27

And if we rightly consider the matter, we shall find that our nature is chiefly constituted of acquired habits, and that we are much more creatures of custom and art than of nature. It is a common saying that habit (meaning custom) is a second nature. I add, that it is more powerful than the first, and in a great measure destroys and absorbs the original nature: For it is the capital and distinguishing characteristic of our species, that we can make ourselves, as it were, over again, so that the original nature in us can hardly be seen; and it is with the greatest difficulty that we can distinguish it from the acquired (1).

LORD MONBODDO, 1774

7.

The Perfectible Animal

ALTHOUGH THE PROGRESSIVE DISCOVERY of the anthropoid apes did little to advance the idea of organic evolution among comparative anatomists, it gave a powerful impetus to theories of cultural evolution. In social thought the idea of progress was beginning to take firm hold, aided by evident signs of improvement in Western science, technology, and government and by growing familiarity with the customs of savage peoples. According to traditional concepts, the Indians of America, the Hottentots of Africa, the bushmen of Australia, and all the other savages encountered in the course of European expansion were "degenerate sons of Adam." They were tribes who had sunk into barbarism after the Fall or in the general dispersion of mankind after the destruction of the tower of Babel. But as the eighteenth century advanced, another interpretation began to find favor. Should not the history of man be conceived as an ascent from savagery rather than as a decline from original perfection? Might not the enlightened society of the age of Voltaire and Franklin represent the highest point to which mankind had yet risen? If so, had not man's original nature been akin to that of the brutes, more so even than that of the crudest savage yet encountered? These questions were of immense importance to political philosophers, for they had become accustomed since the time of Thomas Hobbes and John Locke to think of government as based on a social contract into which men living in a state of nature had entered for the protection of their natural rights. In its original form this notion of the "state of nature" had had no particular historical content. By it Hobbes had meant simply a condition in which men lived together without acknowledging any common judge or superior authority. But it was inevitable that the state of nature should also come to be thought of in historical terms. Locke himself declared that "in the beginning all the world was America." It was natural, therefore, that the customs of savage tribes should be studied for the light they might throw on the state of nature and that the reports of travellers and naturalists concerning "wolf-girls," "bear-boys," and anthropoid creatures in the jungles of Africa or East Asia should be received eagerly in the hope of discovering the true natural man. As Buffon put the matter in the first volume of his *Natural History* (1749) :

> An absolute savage, such as the boy educated among the bears, as mentioned by Conor, the young man in the forests of Hanover, or the little girl in the woods of France, would be a spectacle full of curiosity to a philosopher: in observing this savage he might be able precisely to ascertain the force of the appetites of nature; he might see the soul undisguised, and distinguish all its natural movements. And who knows whether he might not discover in it more mildness, more serenity and peace than in his own; whether he might not perceive, that virtue belongs more to the savage than to the civilized man, and that vice owes its birth to society (2).

A Rousseau-like passage this, but it was not typical of Buffon's thinking about man's natural condition. In Buffon's estimation, man was a creature of society, deriving from it all his truly human attributes. Apart from his intellectual powers, man was a defenseless animal, but these powers required society for their development. The longer Buffon considered the matter the more he became convinced that society was natural to man, being imposed on him by his physical and mental endowment and by his struggle to survive in a hostile environment.

In his essay entitled "Nomenclature of the Apes," Buffon considered the problem of man's mental development anew in the light of the similarities and differences between men and apes. The Hottentot, he declared, might seem a different creature from the European, but the difference was much less than that which separated the Hottentot from man in the pure state of nature. And this latter gulf was nothing compared to the abyss which divided the natural man from the ape. For man was endowed with powers of thought and speech, powers deriving from a higher principle than material organization. Yet mind, though independent of matter, was conditioned in its action by the state of the body. The education of the mind and the training of the body went hand in hand. Indeed, said Buffon, it was the prolonged dependence of the human infant on its parents which insured the unfolding of its distinctively human powers. If the child were capable of taking care of itself after a few months, as the infant animal was, there would be no need for family life and hence no basis for the development of language and culture. Man's physiological peculiarity necessitated the family, the family gave birth to language and society, and society created culture, a uniquely human phenomenon. A pure state of nature, in which man neither thought nor spoke, could never have existed (3).

But although early man was not a stupid animal, neither was he the finished work of the Creator's hand. In the glowing language of the *Epochs of Nature* (1778):

> The first men were witnesses of the convulsive motions of the earth, which were then frequent and terrible. For a refuge against inundations, they had nothing but the mountains, which they were often forced to abandon by the fire of volcanoes. They trembled on the ground which trembled under their feet. Naked in mind as well as in body, exposed to the injuries of every element, victims to the rapacity of ferocious animals, which they were unable to combat, penetrated with the common sentiment of terror, and pressed by necessity, they must have quickly associated, at first to protect themselves by their numbers, and then to afford mutual aid to each other in forming habitations and weapons of defence. They began with sharpening into the figure of axes those hard flints, those *thunderstones,* which their descendants imagined to have fallen from the clouds, but which, in reality, are the first monuments of human art. They would soon extract fire from these flints by striking them against each other.... (4).

With these few bold strokes Buffon sketched an essentially modern view of human history, placing it against the awesome backdrop of the history of nature — the planet born from a collision of a comet with the sun, the long series of revolutions on the earth's surface, the change of climates and the transformation and extinction of flora

203

and fauna as the great globe cooled. Upon this scene man made his eventual unprepossessing appearance, managed to survive by slowly sharpening his dull wits, and learned at last how to conquer and control a hostile or indifferent nature. For Buffon, as for Laplace, science was man's crowning achievement, the last creation of society and man's deliverance from the hardship and superstitious terror of his early life on earth (5).

► MAN IN THE STATE OF NATURE

Rousseau, too, set out to disentangle the original from the artificial in man's nature, hoping thereby to lay a solid foundation for the theory of natural right. In his famous *Discourse on the Origin and Foundations of Inequality Among Men,* first published in 1755, he took Hobbes and his successors to task for seeking to discover the true nature of man by studying his behavior in historical times. In so doing, Rousseau complained, they had overlooked the transformations wrought in human nature by society and the progress of civilization. They had mistaken the end product of a long social evolution for the original handiwork of nature and nature's God.

But how to discover man's original condition and endowment? Not by studying savage tribes — Rousseau agreed with Buffon that the Hottentots and Caribs were far removed from the pure state of nature. Not by questioning the comparative anatomist — his science was too immature to shed much light on the alterations which new habits and new foods might have produced in man's physique over a long series of ages. Not by reading travellers' descriptions of manlike creatures dwelling in the forests of Europe, Africa, and Asia — some of these creatures were doubtless real men who, owing to their dispersed condition, had never had occasion to develop their human potentialities, but the accounts of them were too sketchy and contradictory to afford a solid basis for reasoning about man's original nature. In the absence of reliable evidence as to man's primitive condition, the philosopher could only construct a reasonable hypothesis which accorded with all the known facts. This hypothetical method was particularly appropriate, Rousseau declared, because Scripture had pronounced

Jean Jacques Rousseau, 1712 • 1778

authoritatively on the questions at issue. But although there was no appeal from Scripture, the philosopher was free to speculate as to what *might* have happened if the human race had been left to itself immediately after its creation. Moreover, even if the state of nature had never existed, it was a useful concept in political theory (6).

After stripping man of all his acquired characteristics, Rousseau found him a healthy animal, less strong and agile than some but more adaptable and versatile than any other. The specific characteristic which distinguished him from the apes was not the power of thought nor of speech but the ability to perfect himself, "a faculty which, aided by circumstances, develops all the others successively" (7). Other animals were slaves to instinct; man alone could choose his course of action and thereby alter his condition and, eventually, his nature. Thus, natural man differed from the beasts more in potentiality than in actuality. Like them, he wandered in search of food. Like them, he was dominated by the instinct of self-preservation. He thought, but only to discover a way of satisfying his physical needs. His emotions and ideas were few. Natural pity restrained him from harming his fellow men, but he lacked the knowledge and hence the fear of death. Since he lived alone, he had no occasion to speak. Gestures sufficed for his occasional encounters with his own kind. Children were born from random intercourse, but no families resulted. The male had long since vanished when the child was born, and the period of infancy was too brief to produce a lasting attachment between mother and child. Society, far from being natural to man, was an artificial association into which man was driven by force of circumstances.

When Rousseau attempted in imagination to close the gap between the solitary, mute, beastlike man of nature and the civilized man of historical times, he was impressed by the length of time required to bring this transformation about by natural causes.

> The more one meditates on this subject the more the distance from pure sensations to simple knowledge swells in our imagination...How many ages may have elapsed before men were in a position to see any other fire than that from the sky! how many different accidents must have happened to them to teach them the most common uses of this element! how many times must they have let it die out before they acquired the

art of kindling it! and how many times may each of these secrets have perished with its discoverer (8).

If to these considerations were added the difficulties involved in the invention of language, the conclusion was inevitable that "thousands of ages" had been needed to develop in the human mind what were now deemed its most ordinary capacities.

With eloquence and imagination no less glowing than Buffon's, Rousseau depicted the long series of fortuitous events, of trials and errors, by which human reason had been perfected. But unlike Buffon, he could not view this progress of the mind with unmixed satisfaction. In considering the changes civilization had made in human nature, he was nearly as ambivalent as Buffon had been when he contemplated the effects of domestication on animal nature. On the one hand, Rousseau seemed to share the traditional view of first things as best, as norms from which any change must constitute degeneration. Thus, the theory of natural right found its ethical basis in man's primitive nature as a solitary being seeking to preserve himself. On the other hand, the natural man turned out to be more animal than human. His distinguishing quality consisted not in what he was but in what he might become, hence the perfection of his nature was not primordial but potential and prospective. History was not a decline from perfection but an unfolding of potentialities. Indeed, the word *perfection* took on a new meaning. When Rousseau spoke of man's faculty of perfecting himself, he did not mean man's capacity to achieve an ideal of either individual or social life. He meant rather man's ability to develop his intellectual powers and thereby to transform nature, animate and inanimate. The transformation of human nature which followed was a by-product, an unforeseen and unplanned consequence of the efforts of individuals to improve their lot. Wants, intellect, and passions gave rise to society, and society in turn created new wants and thereby new passions and new occasions for sharpening the intellect. But this progress of the mind was attended by more evil than good in terms of human happiness and well-being. The perfection of the intellect was achieved at the expense of health, liberty, fraternity, and peace of mind. Yet, given man's freedom of action and his desire to preserve himself and improve his lot, it was hard to see how the course of history could have been different.

206

Rousseau wavered in his conception of the forces which had brought mankind to this dreadful impasse. In the *Discourse on Inequality* he regarded the whole development as produced by "fortuitous causes," that is, as an unplanned consequence of the interplay of events and purposes. Like Buffon, he thought that geologic upheavals might have brought about the formation of the first societies. In his *Essay on the Origin of Languages* (9) , he called these catastrophes "accidents of nature" but at the same time suggested that they were instruments of Providence to compel human beings to associate. It occurred to him, however, that catastrophes of this kind, if they had continued unabated in frequency and force, might have dispersed the very societies they had once brought together. Perhaps the tranquil state of the earth in recent times was a wise provision of nature designed to forestall the dissolution of society. But Rousseau did not really believe this. His dominant feeling was that nature, apart from human control, tended toward confusion and decay. The rough equilibrium which nature maintained was accomplished at the expense of whole species of plants and animals and without regard to human purposes. It was only when men organized to control and cultivate nature that a degree of order and stability was introduced on the surface of the globe.

Thus Rousseau vacillated between the traditional view that nature is perfect and subservient, only man being vile, and the modern view that nature is indifferent to man's purposes but capable of being turned to his service by the power of organized intellect. On the whole, he adopted the modern view: by conceiving man's early condition as brutish, by focusing attention on the historical process in which man's powers had been unfolded, by imagining a time scale appropriate to that process, by abandoning the assumption of intelligent design, and, above all, by developing a truly evolutionary view of human nature, involving the emergence of real novelty out of the interplay between man and his changing environment. He was not, however, a believer in organic evolution. On the contrary, he regarded the capacity to evolve in response to changing conditions as a distinctively human faculty. Man's ability to perfect himself, he declared, "resides among us as much in the species as in the individual; whereas an animal is at the end of several months that which it will be all its life, and its species at the end of a thousand years that which it was during the first year of those thousand years" (10). The orang-outang might be a man, but if he were, it was because he possessed the uniquely human capacity *de se perfectionner.*

▶ **MAN MAKES HIMSELF**

The inquiry into man's original nature and its subsequent modifications was taken up in 1773 by the learned Scottish jurist James Burnet, known to his contemporaries and to posterity as Lord Monboddo. Monboddo was a stalwart champion of ancient learning against the pretensions of the science and metaphysics of his own day. At his classical supper parties, Edinburgh's scientific elite donned the toga, drank wine from garlanded vessels, and engaged in learned debate. Their host, for all his veneration of the ancients, had an original and inquiring mind. He considered contemporary travel books "the most instructive of all the modern reading" for the light they threw on human nature.

Lord Monboddo (James Burnet), 1714 . 1799

These, combined with his classical studies, led him into speculations which, although they might have won the approval of his master, Aristotle, earned him a name among his contemporaries for eccentricity and lunacy. Dr. Johnson probably expressed the general opinion of Lord Monboddo when he said: "It is a pity to see Lord Monboddo publish such notions as he has done; a man of sense and of so much elegant learning. There would be little in a fool doing it; we should only laugh: but when a wise man does it, we are sorry" (11).

Monboddo's anthropological speculations were set forth in two six-volume treatises: *Of the Origin and Progress of Language* (1773–1792) and *Antient Metaphysics* (1779–1799). In the "Preface" to the former work he indicated the reasons which led him to investigate the history of language:

> The origin of an art so admirable and so useful as language...must be allowed to be a subject, not only of great curiosity, but likewise very important and interesting, if we consider that it is necessarily connected with an inquiry into the original nature of man, and that primitive state in which he was, before language was invented; a subject of so much

greater dignity and importance, by how much the works of God are nobler than those of men. For man in his natural state, is the WORK OF GOD; but, as we now see him, he may be said, properly enough, to be the work of man.... Further, if it be true, as I most firmly believe it is, that the state in which God and nature have placed man is the best, at least so far as concerns his body, and that no art can make any improvement upon the natural habit and constitution of the human frame; then, to know this natural state is of the highest importance, and most useful in the practice of several arts, and in the whole conduct of life (12).

Like Rousseau, Monboddo started from the assumption that first forms were best, being at once natural and divine. But he, too, applied this maxim only "so far as concerns the body." In regard to man's mind, he distinguished two meanings of "state of nature":

It may denote either his most perfect state, to which his nature tends, and towards which he either is or ought to be always advancing, I mean the perfection of his Intellectual Faculties, by which, and which only, he is truly a man... Or it is the state from which this progression begins. It is in this sense that I use the term, denoting by it the original state of Man, before societies were formed, or arts invented. This state, I think, may also be called a state of Nature, in contradistinction to the state in which we live at present, which, compared with it, is certainly an artificial state (13).

Thus, although man's original physical condition was natural in both the chronological and the normative sense, the primitive state of his mind was natural only in the first sense. His body had degenerated step by step with the improvement of his mind.

This paradox of physical degeneration and cultural improvement, Monboddo assured his readers, was quite in accord with nature's way of doing things. She equipped her creatures with everything necessary to their survival but with nothing more. Thus primitive man must have been created strong and agile in order to survive the inclemencies of the weather and the attacks of wild beasts. Of mental faculties, however, he received only the simplest elements, since he required nothing more to live. Was it not much wiser of nature to let man develop his intellectual powers in the struggle to improve his lot than to spoil him with a superfluity of endowments?

Monboddo thus gave a new twist to the old idea that "nature does nothing in vain." He converted the principle of parsimony in nature's economy into a principle of bare survival in the struggle for existence. The subservience of nature to man in providing for his wants was not something original and natural but rather the result of man's conquest of nature in the course of history. It was something to be achieved by man, not something provided for him by the inexhaustible bounty of a benevolent Creator. Monboddo apparently sensed the difference between his own and the traditional view of man's relation to nature, for he added that he was speaking only of man's condition since the Fall (14).

He then took up a second principle of nature's operations, "that no species of thing is formed at once, but by steps and progression from one state to another" (15). The life cycle of plants and animals was Monboddo's favorite illustration of this principle — a poor illustration, since it exhibited the evolution of the individual organism, not the progress of the species. Monboddo seems to have had no conception of a general organic evolution. True, he regarded nature as a chain of being rising from mere matter to sense, memory, intellect and imagination, but only in the case of man did he regard the transition to a higher order of being as a temporal one. Animals, he thought, had been given a definite form and endowment suited to their environment. They were governed by instinct and had no capacity for development; they tended to degenerate when domesticated. Man was the great exception to the general rule of constancy of form, endowment, and environment. He had been constructed to inhabit all of the various climates of the globe. In the development of his intellectual powers he had transformed the environment in which he lived; in the process his own nature had been transformed. "For it is the capital and distinguishing characteristic of our species, that we can *make* ourselves, as it were, over again, so that the original nature in us can hardly be seen, and it is with the greatest difficulty that we can distinguish it from the *acquired*" (16).

It turned out, then, that man's progression from the state of nature to a civilized state, far from exhibiting nature's general mode of operation, was quite unlike anything else in nature. The characteristic of nature, the traditional view asserted, was to remain as it was.

The characteristic of human nature, Monboddo and Rousseau discovered, was to develop and, in developing, to transform itself and its natural environment. Thus the idea of evolution got its first foothold not in the field of natural history, where both organic form and physical environment appeared immutable, but in the study of man in his most mutable aspect, his mind and culture. Natural history hastened the development of the idea, however, by posing the problem of man's specific difference from the apes. It would be interesting to know whether Lamarck, like Charles Darwin, discovered a clue to the problem of changes in organic form from reading the social theory of his day.

For his part, Monboddo had no intention of overthrowing established beliefs. His views, he insisted, were simply those of various classical authors, illustrated and supported by the testimony of ancient and modern travellers and naturalists. Had not Aristotle viewed man as developing from a brute condition both in body and mind? Had not Horace written:

> When living beings first crawled on earth's surface, dumb brute beasts, they fought for their acorns and their lair with nails and fists, then with clubs, and so from stage to stage with the weapons which need thereafter fashioned for them, until they discovered language by which to make sounds express feelings. From that moment they began to give up war, to build cities, and to frame laws that none should thieve or rob or commit adultery (17).

There was nothing subversive, disconcerting, or degrading in this concept of man's nature and history, Monboddo argued. It preserved the separation between mind and body and derived the mind's development from its own powers aroused by "the necessities of human life and the social intercourse required to supply those necessities." It was perfectly reconcilable with Scripture, since it applied only to man's condition after the Fall. It enhanced rather than diminished man's appreciation of God's wisdom and benevolence. Those who placed all their hopes in this life and denied the life to come might find the picture of man's physical degeneration in the course of his intellectual improvement depressing. But the true philosopher, convinced of the spirit's immortality and aware of its eternal warfare with the flesh, would rejoice to see the mind weaned from the body as its powers developed. Indeed, said Monboddo, "I could not reconcile the miserable state in which Men are now to be found in almost all the nations of the world, the more miserable the more the nations are civilized, with the administration of a wise and good God, otherwise than by showing that Man is in this life in a state of progression, from the mere Animal to the Intellectual Creature, of greater or less perfection, and a progression not to end in this life; from which progression I propose to show that Moral Evil is as necessary as Physical, if the Moral World be a System, as well as the Natural, and consequently both governed by general laws: And, if it be true, as I believe it is, that this Scene of Man is to have an end, as well as the present System of Nature, and that Man is to appear again in some other form, as we are told the Heavens and the Earth will do, it is according to the order of Nature that this change of his state should not happen at once, but should come on by degrees, and, consequently, that the species should decline, degenerate, and become old, as we see the Individual does, before its extinction." Thus Monboddo's pessimistic view of the progress of civilization was tempered by a Christian faith in God's eternal purposes for the individual human soul (18).

When Monboddo descended from general reasoning to empirical proofs of man's primitive state, he drew heavily on the travel literature of his own day, finding there a complete confirmation of the doctrines of the ancients concerning man's nature and early history. Of the various manlike creatures described by travellers he cited the wild children of Europe and the orang-outang as unmistakable specimens of primeval man. The orang-outang, said Monboddo, was described as having a human form and temperament, walking on two feet, living in society, constructing shelters of leaves and branches, defending itself with sticks and stones, burying its dead, and manifesting a desire for intercourse with human beings. Why, then, was it not a man? Because it could not speak, said Buffon. But this was not conclusive. Persons born deaf did not ordinarily speak, but they could with sufficient pains be taught to do so. The wild creatures caught in the forests of Europe could not speak when captured, but some of them had later been taught the art. Why, then, should not the orang-outang learn to speak? His vocal organs were like man's, and he manifested a

212

human intelligence. It might take him a long time to learn, but so likewise it took the first men a long, long time to form abstract ideas and connect them with articulate sounds (19).

Monboddo then turned to Linnaeus' views. Linnaeus, he noted, classified the feral children of Europe as members of the human species despite the fact that they were "four-footed, mute, rough, and hairy," but he placed de Bondt's orang-outang and Pliny's· troglodyte in a separate species of the human genus because they had extremely long arms and a nictating membrane in their eyes. This was not a sound classificatory procedure. In defining any species, one must look for its dominant characteristic, a character from which most of its other properties flowed. In every case, the distinguishing character should be drawn chiefly from the creature's mind, or "internal principle." This it was which governed the animal's motion and action, and, "as every animal is by nature destined for a certain course of action, and a certain oeconomy and manner of living, whatever prompts him to that, must be accounted principal in his frame and constitution" (20). The dog, for example, was distinguished from the fox and the wolf not so much by its external appearance or internal conformation as by its docile and friendly disposition, wherein it seemed formed by nature for a companion to man.

Applying these principles of classification to the case of man, Lord Monboddo found that Aristotle had defined him correctly as "a mortal, rational animal capable of intellect and science." In terms of this definition there was no reason to exclude the orang-outang from the human species. By his use of sticks to defend himself the orang showed that he possessed the rational faculty, that is, the power of comparing sense perceptions and thereby of perceiving relations between things. The presumption was, then, that this creature was capable, given sufficient time and favorable circumstances, of arriving at intellect and science. Indeed, he represented the second, not the first, stage in man's progress from the state of nature, surpassing in several respects such purely natural men as Peter, the wild boy of Hanover. A third stage was represented by the wild girl brought to France from the coast of Labrador. Monboddo declared that he had seen with his own eyes representatives of all three stages. In London he had seen Peter and two orang-outangs. In Paris he had conversed with the girl from Labrador. She told him that the people of her country did not use fire but that they had, besides language, a kind of music which they formed in imitation of the birds. She said that she could formerly climb like a squirrel and leap from tree to tree but had long since lost the faculty.

It seems unfortunate that Monboddo died just before Jean Itard, physician at the National Institute of the Deaf and Dumb in Paris, began his famous attempt to civilize the wild boy caught in the woods of Aveyron in 1799, the year of Monboddo's death. With what interest would Monboddo have viewed the boy as he first appeared in Paris: "A disgustingly dirty child affected with spasmodic movements and often convulsions who swayed back and forth ceaselessly like certain animals in the menagerie, who bit and scratched those who opposed him, who showed no sort of affection for those who attended him; and who was in short, indifferent to everything and attentive to nothing." With what joy would he have learned of the steady progress of the boy's education, with what heartbreak of Itard's failure to teach him to speak! And with what satisfaction would Monboddo have read Itard's conclusions:

> First, that man is inferior to a large number of animals in the pure state of nature, a state of nullity and barbarism that has been falsely painted in the most seductive colors; a state in which the individual, deprived of the characteristic faculties of his kind, drags on without intelligence or without feelings, a precarious life reduced to bare animal functions.
>
> Second, that the moral superiority said to be *natural* to man is only the result of civilization, which raises him above other animals by a great and powerful force. This force is the pre-eminent sensibility of his kind, an essential peculiarity from which proceed the imitative faculties and that continual urge which drives him to seek new sensations in new needs (21).

Monboddo's remarks on classification reveal his essential conservatism in matters of natural history. A disciple of Aristotle, he was alarmed at the tendency in modern science to derive everything from matter and motion. In his *Antient Metaphysics* he attacked Newton's natural philosophy and attempted to show that "Motion can neither be begun nor continued by any power in Matter, but Mind only." In the field of natural history, Monboddo rejected the Linnaean system

2 1 4

FIG. 7.1 — The Wild Boy of Aveyron, caught in the woods of Aveyron in southern France in 1799. A French physician, Jean Itard, acting on the supposition that the boy was potentially a normal human being, attempted to civilize him but was only partly successful. The portrait is from Itard's fascinating account of the experiment: **De l'Éducation d'un homme sauvage** . . . (Paris: 1801).

of species, genera, orders, and classes based on physical characters only and reasserted the Aristotelian system, in which, according to Monboddo, man was a species of the genus animal, animal a species of the genus body, and body a species of the genus substance. There was nothing of evolution in this system. Every creature was regarded as having been "formed by nature" for its particular abode and manner of life. For Monboddo, as for Rousseau and Buffon, the question with respect to the orang-outang was not whether subhuman creatures had been transformed into human beings but whether some creatures called apes might not deserve to be called men. Was the orang-outang "a variety of the human species?" According to Monboddo, this question was to be settled by reference to mental rather than physical qualities. Since the orang-outang displayed the intelligence and the disposition of a man, he *was* a man. Baboons, gibbons, and monkeys were creatures of different natures, though some of them might be hybrids produced by intercourse between men and apes (22).

Buffon's environmentalism was as distasteful to Monboddo as

Linnaeus' preoccupation with physical characters, and for much the same reason. Instead of taking the physical and mental differences among men and animals for granted, Buffon attempted to explain how they had been produced by the action of climate, diet, and manner of living. Monboddo rejected this conception as wrong both in principle and in fact. In principle it made form and mind derivative from matter, whereas, in truth, mind "constitutes the essence of everything, making it that which it is, and distinguishing it from everything else." As to the facts, since men and animals differing widely in temperament and physique were found to coexist in the same climate, climate could not be the factor determining their natures. Moreover, since the characteristics of the parents were transmitted to the offspring quite independently of the environment, the environment could not be the cause of those varieties. Rather they must be attributed to Mind, "directly and immediately to those inferior Minds which animate everything in the universe, but ultimately to that Supreme Mind, who has willed that there should be such a variety in his creation" (23).

Monboddo thus refused to accept the sharp distinction between species and varieties which characterized the natural history of his day. According to Linnaeus, the form of the species was to be ascribed to God; varieties were to be explained as products of time and circumstance. Perhaps Monboddo sensed that if material causes were admitted to explain varieties they would eventually be invoked to account for specific differences. Already in Buffon's thought the distinction between species and variety had begun to blur. In Lamarck's system it would become unimportant, all organic forms being regarded as produced in response to changes in the material environment.

Although Monboddo disagreed with Buffon as to the causes of variety in nature, he was even more insistent than Buffon on the luxuriance of that variety. "Everything which can exist either does exist or has existed." Hence no one should be incredulous when travellers reported the existence of mermaids and mermen, men with one leg, giants and pygmies, men with tails, sphinxes, and the like. In all these matters one should be guided by Linnaeus' reply when he was asked his opinion of Koeping's story of the tailed men of the Bay of Bengal: "The testimony of one eyewitness is more to me than the denial of a hundred who have not seen" (24).

Monboddo insisted, however, that his case for man's progression from rude beginnings did not stand or fall with the veracity or mendacity of the travellers he had cited. Besides general reasoning from first principles, there was the testimony of history and the evidence afforded by nations still living in barbarism. The most ancient histories, including the Bible itself, told of a time when man lived on the natural fruits of the earth without the use of fire, without domesticated plants or animals, without any knowledge of the arts and sciences. The fable of Prometheus and the story of the Tree of Knowledge depicted allegorically the fact and the unhappy consequences of man's evolution from his natural state by the gradual development of this intellect. If these allegories were well founded, the human race must have originated in a warm and fruitful climate, perhaps in Africa. Indeed, "the short history of man is, that the race having begun in those fine climates, and having, as is natural, multiplied there so much that the spontaneous productions of the earth could not support them, they migrated into other countries, where they were obliged to invent arts for their subsistence, and, with such arts, language, in process of time, would necessarily come" (25). With the invention of language, only time and circumstances had been required to bring mankind to its present condition.

Like Rousseau, Monboddo seemed now to approve, now to disapprove, the progress of civilization. He agreed with Rousseau that the state of nature was not a state of war. Natural man was solitary, feeding on fruit. Only when population increased, precipitating a struggle for subsistence, did man resort to hunting and fishing, thereby developing a cruel and predatory disposition. Only then were tribes formed to procure food and ward off attack. But Monboddo was not consistent in his account of the causes of overpopulation. On the one hand he argued that, since the number of every species of animals was kept at its natural proportion by the perfect equilibrium of natural forces, man should never have multiplied beyond his natural proportion in relation to other animals, and, indeed, never would have done so if he had not invented means of procuring subsistence beyond those provided by nature. On the other hand, Monboddo recognized that overpopulation had been the *cause* of the invention of the arts, since it had rendered the natural food supply inadequate. The difficulty was, of course, that whereas animals simply died when their usual means of supporting life failed, men used their minds to devise new ways of surviving. Thus natural necessity had evoked man's distinctively human capacities, but the exercise of those capacities had carried him farther and farther away from nature and disrupted the whole economy of nature. Since Monboddo believed in the perfect contrivance of that economy, he could not but view its disruption with horror. On the other hand, he could not help admiring his own species for its prodigious accomplishment. The vulgar, he said, would ever consider it derogatory to man to derive him from brutish beginnings, but the philosopher would account it man's greatest glory "that, from the savage state, in which the Orang Outang lives, he should, by his own sagacity and industry, have arrived at the state in which we now see him" (26).

► RACE AND ENVIRONMENT

Cuvier sided with Buffon in the controversy concerning the orang-outang. To Cuvier the relative brain-size of the orang proved more than a thousand speculations based on reports of travellers. For the rest, he repeated Buffon's arguments concerning man's long infancy, his native sociability, his relative defenselessness even when full-grown. Society, thought, and language, said Cuvier, constituted the means of man's survival and the basis of his perfectibility. In inhospitable climates men were reduced to hunting and fishing for subsistence. Where agriculture was possible, population increased and arts and sciences flourished, but agriculture gave rise to property, and property to money, riches, inequality, vice, and war. That much must be conceded to Rousseau and Lord Monboddo.

Cuvier,
1769 · 1832

An interesting shift of emphasis occurred between the publication of Cuvier's *Tableau of Natural History* in 1798 and the appearance of his *Animal Kingdom* in 1817. In the *Tableau,* Cuvier had been concerned primarily with the "habitudes peculiar to the human species" (27). Various types of human culture had been mentioned, but there was little suggestion that they formed a developmental sequence. In 1817, however, the section dealing with the same matters bore the title "Physical and Moral Development of Man." The various types of culture were now viewed as representing different degrees of human progress. The first hordes, said Cuvier, were too busy gathering food

2 1 8

to make much progress. After man had learned to domesticate animals he had progressed a little faster, but the full development of the arts and sciences had had to await the advent of agriculture, property, money, and exchange, "sources of a noble emulation and of vile passions." But what of those people who had remained in the earlier stages of civilization? In some of the cases, said Cuvier, the natural environment had retarded or prevented improvement. In other instances, however, "intrinsic causes" had arrested the development of certain races, "even in the midst of the most favorable circumstances" (28).

Environment and race — these were the two most potent concepts which social theorists of the nineteenth century would use to attack their favorite problem: the discovery of the laws of historical development. During the first half of the century they would be, like their eighteenth century predecessors, social but not biological evolutionists. Thus Auguste Comte, the founder of modern sociology, undertaking to show "by what necessary chain of successive transformations the human race, starting from a condition barely superior to that of a society of great apes, has been gradually led up to the present state of European civilization," would admit Lamarckian "development" only in the evolution of human nature, invoking Cuvier's arguments against it in the realm of biology. Not until the time of the great English philosopher Herbert Spencer would the idea of social evolution be linked clearly to the idea of organic evolution. Not until then, the middle of the nineteenth century, would man's progress in history be viewed as a continuation of the progress of nature (29).

FIG. 7.2 — Peter, the Wild Boy of Hanover, in middle life. The German naturalist J. F. Blumenbach proved beyond reasonable doubt that he was born dumb and had run away or been driven away from home a year or two before he was found naked in the woods near Hameln in 1724. The portrait is reproduced from Blumenbach's **Contributions to Natural History** (Göttingen: 1811).

Providence has distributed the animated world into a number of distinct species, and has ordained that each shall multiply according to its kind, and propagate the stock to perpetuity, none of them ever transgressing their own limits, or approximating in any great degree to others, or ever in any case passing into each other. Such a confusion is contrary to the established order of Nature.

JAMES PRICHARD, 1813

...those who attend to the improvement of domestic animals, when they find individuals possessing, in a greater degree than common, the qualities they desire, couple a male and female of these together, then take the best of their offspring as a new stock, and in this way proceed, till they approach as near the point in view as the nature of things will permit. But, what is here done by art, seems to be done, with equal efficacy, though more slowly, by nature, in the formation of varieties of mankind, fitted for the country which they inhabit.

WILLIAM WELLS, 1818

The Origin of Human Races

To the eighteenth century mind the basic issue concerning human races was whether they were to be regarded as separate species or as varieties of a single species. The issue was a vital one. Theologically it bore upon the Christian doctrine of the spiritual unity of men in their common descent from Adam. Politically it influenced the white man's conception of his rights and duties with respect to colonial peoples. Scientifically it involved the distinction, enormous in the eyes of eighteenth century naturalists, between species and varieties. If the various types of human beings were separate species, as the polygenists claimed, the task of the natural historian was to classify them according to their specific characters, accepting these as permanent and divinely ordained. But if they were varieties of a single species, as the monogenists claimed, their peculiarities must be accounted for by natural causes. The monogenists, therefore, faced the formidable problem of showing how the observable variety of human types had developed in the traditional six thousand years of human history. In order to do this, they had to assume the mutability of the human species and to speculate concerning the mechanisms governing change of organic form. So while in biology the pressure of tradition tended to place rigid limits on the variability of species, in anthropology it favored the assumption of mutability and stimulated the study of the conditions and causes of natural variability.

The leading naturalists of the period were monogenists (1). The polygenists were, in general, either lesser lights in science or else, as in the case of Lord Kames and Voltaire, men whose reputations rested on other than scientific achievements. Although these writers broke with tradition and Scripture in assuming a plurality of human species and in regarding human history as a progress from a rude and savage condition, their view of nature was highly traditional in other respects, emphasizing the wise adaptation of every creature to its environment, the limits of variability, and the hierarchical arrangement of nature in a great chain of being. They tended to a racial interpretation of history, explaining differences in cultural achievement in terms of differences in racial endowment. Since they accepted the diversity of human types as a primordial fact, their speculations contributed little or nothing to the development of evolutionary conceptions in natural history (2).

Among the monogenists, however, the decision to regard human races as varieties rather than as species brought the problem of race formation to the center of attention. Except for Linnaeus and Cuvier, these writers were more interested in explaining the origin of races than in classifying them. Buffon described a great variety of racial types but made little attempt to group them systematically. Kant and Johann Friedrich Blumenbach, professor of medicine at the University of Göttingen, propounded classifications but went on to grapple with the problem of origins. The Reverend Samuel Stanhope Smith, best known of the early American writers on anthropology, dismissed the question of classification with the observation that it was probably impossible to draw the line precisely between the various races of man and that it was "a useless labor" to attempt it (3).

► **FIRST FORMS ARE BEST**

When the monogenists attempted to explain the formation of human races, they were naturally inclined (given the assumptions of eighteenth century biology) to view the process as one of degeneration from the primordial type of the species. Both Buffon and Blumenbach conceived the origin of racial types in this manner. Blumenbach was an anthropologist, physiologist, and comparative anatomist. Called "the Buffon of Germany," he did much to popularize the study of

Johann Friedrich Blumenbach, 1752 • 1840

natural history in Germany through his teaching and writing. His collection of human skulls was world-famous. His *Handbook of Natural History* went through twelve editions by 1830 and was translated into many languages. Alexander von Humboldt was the most famous of many students whom he inspired to investigate the natural history of the unexplored regions of the earth. In natural history Blumenbach divided his loyalties between Linnaeus and Buffon. A Linnaean in most zoological matters, he could not stomach Linnaeus' anthropology. Dismissing the troglodytes and tailed men as imaginary, he placed man in a separate order, *Bimanes,* thereby anticipating Cuvier's classification. To Linnaeus' four geographical varieties of man — American, European, African, and Asiatic — he added a fifth, the Malayan. Linnaeus' two other varieties — *homo sapiens ferus* and *homo monstrosus* — he rejected as inadmissible. Careful inquiry into the case of Peter, "the wild boy of Hanover," convinced Blumenbach that the boy was a dumb, feeble-minded child who had been abandoned by his parents a year or two before he was found running naked in the fields near Hameln. Investigation of the other accounts of feral children cited by Linnaeus showed them to be either fictitious or exaggerated. On the question of the origin of human races Blumenbach's ideas were closely similar to Buffon's. According to Buffon, the peoples in the vicinity of the Caspian Sea were most perfect in form and feature, hence the progenitors of mankind must have lived there. To that region the natural historian must look to ascertain the "real and natural color" of man, the original white hue from which the shades of yellow, brown, and black had been produced by time and circumstance. In the last stages of degeneration the color white appeared again, as in the white Negroes and white Indians reported by travellers. "But the white of the species, or the natural white, is widely different from the white of the individual, or the accidental white" (4). To Blumenbach, too, the Caucasian type seemed primitive, natural, most beautiful. By the time of the third edition of his *On the Natural Variety of Mankind,* published in 1795, he had assembled the largest anthropological collection in Europe, including eighty-two skulls, several foetuses, some hair samples and anatomical preparations, and a collection of portraits of people of different races "carefully taken from the life by the first artists" (5). In comparing the skulls in his collection he substituted for Camper's method of the facial line a method of his own, "in which skulls are seen from above and from behind, placed in a row on the same plane, with the malar bones directed towards the same horizontal line jointly with the inferior maxillaries." When the skulls of the

223

Mongolian, American, Caucasian, Malay, and Ethiopian races were viewed together in this fashion, said Blumenbach, the Caucasian was seen to have the most beautiful and symmetrical form, "from which, as from a mean and primeval type, the others diverge by most easy gradations on both sides to the two ultimate extremes (that is, on the one side the Mongolian, on the other the Ethiopian)" (6). In like manner, the white color of the Caucasian skin was the norm from which degeneration toward darker shades had taken place.

Assuming the hypothesis of degeneration from a primitive stock, there was still the problem of explaining the mechanisms by which degeneration had taken place. A convenient explanation lay at hand in the timeworn environmentalist theory, which attributed racial peculiarities to the influence of climate, diet, and mode of life and assumed the transmissibility of acquired characters. To this theory both Buffon and Blumenbach turned, Buffon leading the way in the third volume of his *Natural History,* published in 1749. According to Buffon,

> ...there was originally but one individual species of men, which after being multiplied and diffused over the whole surface of the earth, underwent divers changes, from the influence of the climate, from the difference of food, and of the mode of living, from epidemical distempers, as also from the intermixture, varied *ad infinitum,* of individuals more or less resembling each other;...at first, these alterations were less considerable, and confined to individuals;...afterwards, from continued action of the above causes becoming more general, more sensible, and more fixed, they formed varieties in the species;...these varieties have been, and are still perpetuated from generation to generation, in the same manner as certain deformities, and certain maladies, pass from parents to their children; and...in fine, as they would never have been produced but by a concurrence of external and accidental causes, as they would never have been confirmed and rendered permanent but by time, and by the continued action of these causes, so it is highly probable, that in time they would in like manner gradually disappear, or even become different from what they at present are, if such causes were no longer to subsist, or if they were in any material point to vary (7).

FIG. 8.1 — Portraits of representative specimens of the five races of man, from J. F.
Blumenbach's **Abbildungen Naturhistorischer Gegenstände** (Göttingen: 1810).
    Mongolian — Feodor Imanowitsch, a talented Kalmuck servant presented to the Crown
    Princess of Baden by the Czarina of Russia. Self portrait.
    American Indian — Tayadaneega, a Mohawk chief more commonly known as Joseph
    Brant. Portrait done by Romney.
    Caucasian — Mahommed Jumla, vizier to the Mogul emperor Aurungzeb, from a
    Hindustani painting.
    Malay — Omai, a Tahitian page boy brought to London from the South Seas by
    Captain Furneaux and subsequently returned to his native land by Captain Cook.
    From a portrait by Sir Joshua Reynolds.
    Ethiopian — Jac. Jo. Eliza Capitein, described by Blumenbach as a Negro widely
    known for his sermons and other writings, published in both Latin and Dutch.
    The likeness was done after one by Philip Van Dyck.

As to exactly how these environmental influences operated to modify the human constitution, Buffon was rather vague. He confined himself to asserting a general correlation of climate and racial type and to explaining away apparent exceptions to the rule. The Hottentots, he said, were lighter than their African neighbors in the tropical zone because their peculiar diet modified the influence of the climate. The American Indians displayed relatively little variety from Canada to Tierra del Fuego partly because they had remained uncivilized, partly because their modes of life were similar, partly because the climatic variation in the western hemisphere was not so great as that in the eastern, and partly because they had arrived in America from Asia too recently to have undergone marked diversification. Likewise, Negroes imported into North America from Africa had not lived there long enough to experience a transformation of skin color except through interbreeding with the whites.

Buffon,
1707 • 1788

Blumenbach's theory of race formation was similar to Buffon's, although it differed in some particulars. Whereas Buffon found the proof of the unity of the human species in the capacity of all human races to interbreed successfully, Blumenbach found it in the possibility of explaining all racial differences by "known causes of degeneration." Whereas Buffon supposed a *moule intérieur,* or interior mold, capable of being modified by climate, diet, and mode of life, Blumenbach posited a *nisus formativus,* or purposive force, which guided the growth of organisms toward determinate forms but which could be deflected from its goal by internal and external influences. Both writers assumed the transmissibility of acquired characters, but Blumenbach was more cautious than Buffon on this point. He was puzzled "why peculiarities of the same sort of conformation, which are first made intentionally or accidentally, cannot in any way be handed down to descendants, when we see that other marks of race which have come into existence from other causes which up to the present time are unknown, especially in the face, as noses, lips, and eyebrows are universally propagated in families for few or many generations with less or greater constancy, just in the same way as *organic* disorders, as deficiencies of speech and pronunciation, and such like; unless perhaps...all these occur also by chance" (8). There must be,

226

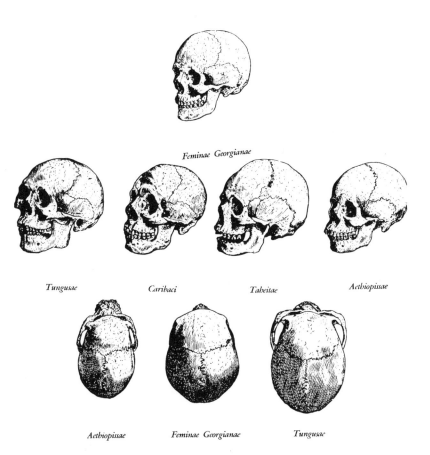

Feminae Georgianae

Tungusae   *Caribaci*   *Tabeitae*   *Aethiopissae*

Aethiopissae   *Feminae Georgianae*   *Tungusae*

FIG. 8.2 — Crania of the five races of man as portrayed by J. F. Blumenbach of Göttingen University in his account of the crania in his collection. The Caucasian skull, top center, is placed in the middle as the original, most perfect type, from which other races are supposed to have degenerated in the course of time. Degeneration in one direction produced the race of American Indians, represented by a Carib skull, and the Mongolian race, represented by a Tungus skull. Degeneration in another direction produced the Malay race, represented by a Tahitian skull, and the Ethiopian race, represented by a Negro skull. Below, three skulls — Ethiopian, Caucasian, and Tungus — are arranged to illustrate Blumenbach's method of comparing crania, "in which skulls are seen from above and from behind, placed in a row on the same plane, with the malar bones directed towards the same horizontal line jointly with the inferior maxillaries."

he concluded, hidden agencies which mediate the influence of climate and mode of life and account for the constancy of skull form, eye color, and the like in different racial types.

Despite his emphasis on the susceptibility of the human constitution to modification by environmental influence, Blumenbach never doubted the fixity of species. His major purpose as an anthropologist was to demonstrate the biological unity of mankind by showing that the range of variation among human beings was no greater than that in animal species generally. That variation might proceed so far as to produce new species of either men or animals he would not concede. Despite his talk of geologic revolutions, despite his censure of the advocates of design for carrying their illustrations of divine contrivance to the point of absurdity, despite his frank acceptance of the extinction of species and even of the occasional appearance of new species, he never seriously questioned the traditional concept of nature as a stable framework of structures fitted as a stage for the activities of intelligent beings. After describing the remarkable transformations produced in domestic animals by the combined effects of climate, food, mode of life, and artificial selection, he nevertheless concluded that there must have been more than one prototype of the existing kinds of dogs, "because many, as the badger dog, have a build so marked, and so appropriate for particular purposes, that I should find it very difficult to persuade myself that this astonishing figure was an accidental consequence of degeneration, and must not rather be considered as an original purposed construction to meet a deliberate object of design" (9). In his time scale, too, Blumenbach remained a traditionalist. Although he recognized that the evidence afforded by mummies seemed to indicate that the Egyptian people had changed very little in physical make-up in the course of several thousand years, he seems never to have wondered how much time had been required to produce the diversity of type exhibited by the various races of the human species. In this, as in most matters, Blumenbach had the defects of his virtues. The same sturdy common sense which made him distrust exaggerated accounts of "wild boys" kept him from exploring the implications of the anomalies which he so carefully recorded.

Buffon was more imaginative than Blumenbach, less prone to prescribe limits to the variability of nature, more sensitive to the theoretical implications of unusual facts. Although he suggested in 1749 that five feet was the normal height of mankind and that extremes of tallness or shortness were to be regarded as examples of individual rather than of racial variation, enough evidence had accumulated by 1777 to convince him of the existence of tribes of giants and pygmies. He estimated seven feet as the norm for giant races, such as the Patagonians, and three feet for dwarfish peoples like the Quimos of Madagascar, but he concluded that this range of variation was no greater than that commonly observed in animal species. In his *Epochs of Nature,* published in 1778, he examined at considerable length the historical evidence attesting the existence of races of giants in primitive times. Influenced, perhaps, by his own theory of the gradual cooling of the globe and the consequent decrease in the violence of terrestrial upheavals, he conjectured that, just as there had once been species of huge animals, so there had been "permanent and successive" races of giants, of which the Patagonians were the only surviving example. In that case man's gradual decrease in stature was but a consequence of the general decline in the energy of the natural forces operating on the earth's surface (10).

In general, however, Buffon was less daring in his anthropological than in his zoological speculations. The notion of natural selection, by which he sometimes accounted for the extinction of animal species, played little part in his discussion of the origin of human races. Nor did he press the implications of his environmentalism for the stability of the human species. His terrestrial time scale of 70,000 to 100,000 years, breath-taking as it seemed to his contemporaries, was too limited to allow for the evolution of new species. Man, he declared, must be older than the separation of the eastern and western continents, since the human type was essentially the same in both hemispheres. But this substantial identity, evidenced by the capacity of all races to interbreed, made it equally clear that man could not be very far removed from the original form of his species. If, by any circumstance, mankind should return to its original habitat, it would eventually resume the features and color of its primitive ancestors. Environment alone would produce this result in due time, but crossbreeding would greatly accelerate the process (11). By the same reasoning, of course, a prolonged difference of environmental conditions could be expected to produce a progressive divergence of human types. Whether new species of men might originate in this way Buffon did not say. He seemed reluctant to explore the full implication with respect to man of his general conception of nature as a system of laws, elements, and forces.

► IMPROVEMENT THROUGH SELECTION

The environmentalism of Buffon and Blumenbach assumed that the effects of climate, diet, and mode of life on one generation of men were inherited by succeeding generations. Naturalists who rejected this assumption — and there were many who did — were forced to conceive some other explanation of the origin of human races. One of the earliest attempts of this kind was made by Maupertuis, the brilliant French scientist who did so much to introduce Newtonian principles on the Continent. In 1745, one year before his appointment as President of the Berlin Academy of Sciences by Frederick the Great, and four years before Buffon's volume on the natural history of man appeared, Maupertuis propounded a particulate theory of inheritance designed to explain the occurrence and recurrence of such anomalies as six-fingered individuals and white Negroes.

Pierre de Maupertuis, 1698 · 1759

In order to explain all these phenomena — the production of accidental varieties, the continuation of these varieties from one generation to another, and finally, the establishment or the destruction of types [espèces] — it seems necessary to suppose the following:

That the seminal liquor of each kind of animal contains an innumerable multitude of parts appropriate to form by their assemblage animals of the same kind.

That within the seminal liquor of each individual, the parts appropriate to form traits similar to those of that individual are those which are ordinarily greatest in number and which have the greatest affinity; although there are many others for different traits...

The parts analogous to those of the father and the mother being most numerous and having the greatest affinity for each other are those which will unite most frequently: and they will form ordinarily animals similar to those from which they are born.

230

Chance, or the scarcity of family traits, will, however, form other assemblages; and one will see a white child born of black parents, or perhaps even a black of white parents. . .

These productions are at first only accidental: the original parts of the ancestors become again more abundant in the seminal fluid; after some generations, or in the following generation, the ancestral type comes to the top, and the child, instead of resembling his father and mother, resembles more distant progenitors. In order to create races which perpetuate themselves, it is probably necessary that these generations be repeated several times; the particles appropriate to the original [parental] traits must become less numerous at each generation, dissipate, or remain so few in number that a new chance operation would be necessary to reproduce the original [parental] type.

One thing is certain, that all the varieties which might characterize new kinds of animals and plants tend to extinction; they are departures from nature and can be maintained only by art or regimen. These works of nature tend always to revert to the original type (12).

The bearing of this hypothesis on the origin of human races was indicated only in a general way. From the sudden appearance of white children in Negro families and the apparent absence of the reverse phenomenon in white families Maupertuis concluded that white was the original color of mankind and that black was "only a variety which has become hereditary during several ages but which has not entirely effaced the white color." As to exactly how the black color had become established in certain regions of the world Maupertuis was not clear. The combinations and permutations of the hereditary mechanism would provide the necessary dark variations from time to time, but some selective agency seemed required to establish the new type and prevent reversion to the original model. In breeding animals selection was accomplished by human art. Among human beings artificial selection, though possible, was unusual, the more usual selective factor being esthetic aversion to ugly and malformed types. Perhaps this factor could explain the repulsion of certain human types to the remote quarters of the globe.

Supposing the Giants, the Dwarfs, and the Blacks to have been born among other men, pride or fear will have armed the greatest part of mankind against them, and the most numerous type will have relegated these dissimilar races to the least habitable climates of the Earth. The Dwarfs will have retired toward the arctic pole, the Giants will have inhabited the lands about the Straits of Magellan, the Blacks will have peopled the torrid zone (13).

In a later work Maupertuis showed that his theory of generation could throw light on the origin of species as well as on the derivation of human races (14). These were bold ideas, far in advance of their time.

Maupertuis had a worthy successor in the philosopher Immanuel Kant, whose anthropological speculations were no less daring than his nebular hypothesis in astronomy. In anthropology, Kant seems to have derived his basic approach more from Buffon than from Maupertuis, however. Like Buffon, Kant regarded genetic relationship as the key to natural history. All men were of one species precisely because they could all interbreed and produce fertile offspring. But Kant did not conclude, as Buffon had, that racial classification was unimportant. Instead, he inferred that a sound racial classification must be based on characters which were invariably hereditary. Upon surveying the array of human physical traits, he found only one — skin color — that seemed to be transmitted invariably from generation to generation, reproducing itself in all matings between persons of the same color and blending to form an intermediate shade in unions between persons of different hue. Moreover, there seemed to be four basic skin colors from which all others could be derived by mixture — the white of the northern European, the copper-red of the American Indian, the black of the Senegambian, and the olive-yellow of the oriental Indian — and each of these four basic colors was found to predominate in one of four regions of the earth. What could be more natural, then, than to assume that these were the four basic races from which the great diversity of human types had sprung by crossing? (15).

This conception would have been indistinguishable from the polygenist idea of distinct human species created in the various regions of the world but capable of crossing to form intermediate types, had Kant not insisted that the four primary races were sprung from a common stock. The polygenist theory was objectionable to him for

Immanuel Kant, 1724 • 1804

several reasons. For one thing, it involved postulating more causes, or creations, than were necessary to explain the variety of mankind. Again, if the four basic types were distinct species, it was hard to understand how they were able to interbreed successfully and why the skin color of each type should be transmitted so invariably in each case of crossing.

For animals whose variety is so great that an equal number of separate creations would have been necessary for their existence could indeed belong to a nominal family grouping (to classify them by certain similarities) but never to a real one, other than one as to which at least the possibility of descent from a single common pair is to be assumed... [Otherwise] the singular compatibility of the generative forces of two species (which, although quite foreign as to origins, yet can be fruitfully mated with each other) would have to be assumed with no other explanation than that nature so pleases. If, in order to demonstrate this latter supposition, one points to animals in which crossing can happen despite the [supposed] difference of their original stems, he will in every case reject the hypothesis and, so much the more because such a fruitful union occurs, infer the unity of the group, as from the crossing of dogs and foxes, etc. The unfailing inheritance of peculiarities of both parents is thus the only true and at the same time adequate touchstone of the unity of the group from which they have sprung: namely, the original seeds [*Keime*] inherent in this group developing in a succession of generations without which those hereditary variations would not have originated and would presumably not necessarily have become hereditary (16).

When Kant came to explain how the four basic races had developed from one ancestral stock, he found himself unable to accept the environmentalist hypothesis. Not only was there no evidence to suggest that acquired characters were transmitted to succeeding generations, but the very idea that human art or external circumstance could alter the deep-laid design of nature incorporated in the generative force seemed to Kant a contradiction of one of the soundest maxims of reason, "that in all organized Nature, despite all changes in single individuals, their species maintains itself unchanged (according to the school formula: *quaelibet natura est conservatrix sui*)." Environment, said Kant, might provide the occasion for new hereditary developments, but it could not directly cause them.

233

For external things can be causes of an occasion, but not evocative causes, of that which is necessarily inherited and makes for resemblance. Just as chance or physical mechanical causes cannot produce an organic body, no more can they add something to its generative force, i.e., effect something that can reproduce itself, if it be a special configuration or a relationship between the parts (17).

How, then, had the races become differentiated? According to Kant, the ancestral human stock had been endowed with a variety of latent powers which could be evoked or suppressed as new conditions of life required. The process by which the Negro race originated he described as follows:

> We know now, for example, that human blood turns black (as is to be seen in blood coagulum) when it is overloaded with phlogiston. Now the strong body odor of the Negroes, not to be avoided by any degree of cleanliness, gives reason to suppose that their skin absorbs a very large amount of phlogiston from the blood, and that nature must so have designed this skin that in them the blood can dephlogisticate itself through the skin to a far greater degree than is the case with us in whom the latter function is mostly performed by the lungs. But the true Negroes live in regions in which, by thick forest and areas that have become swamps, the air is so phlogisticated that according to Lind's report there is really danger there for English sailors if they travel even for a day up the Gambia river to buy meat. Hence it was a very wisely designed device of Nature so to constitute their skin that the blood, as it cannot dispose of enough phlogiston through the lungs, can dephlogisticate itself through the skin much better than it can in our case. It must then carry very much phlogiston into the ends of the arteries, and thus here — that is, just under the skin itself, be overloaded with it, and so show through black, although in the internal parts of the body it is red enough. The difference of constitution of the Negro's skin from ours is also noticeable to the sense of touch. As to the purposefulness of the physical constitution of the other races, as it may be determined by their color, one cannot, to be sure, establish it with the same probability; but grounds of explanation of skin color which could support that theory of purposefulness are not entirely lacking (18).

In this passage Kant anticipated William Wells' suggestion, described later in this chapter, that skin color was correlated with physiological processes essential for survival. Kant, however, viewed the adaptive process as a positive and preordained response of the organism to environmental demands, whereas Wells was to regard it as consisting in the elimination of those individuals who did not happen to vary in an adaptive direction. Kant's theory of intelligent design broke down, however, when he faced the problem of explaining why the American tropics had not produced a black-skinned type. The reason, he conjectured, was that the inhabitants of those regions came there from the Old World by an arctic route and that, having undergone adaptive exspeciation in northern latitudes, they were thereby precluded from exspeciating again in response to a warmer environment. This explanation may have served to account for the absence of Negroes in tropical America, but it placed a severe strain on the idea of nature's wise foresight.

► MAN IS OF RECENT ORIGIN

The writings of Buffon, Blumenbach, Maupertuis, and Kant provided Cuvier with three different approaches to the problem of explaining how human races could be derived from a single stock: first, Buffon's and Blumenbach's theory of degeneration through direct environmental influence; second, Maupertuis' idea of random variation and the establishment of new types through the operation of some selective agency; and third, Kant's conception of preformation and subsequent adaptive exspeciation. In the interim, moreover, Erasmus Darwin and Lamarck had propounded a general theory of evolution, postulating the development of new biological types through the adaptive response of organisms to the changing demands of the environment.

But Cuvier was not much interested in the problem of racial origins. He recognized three main races — Caucasian, Mongolian, and Negroid — of which he considered the Caucasian the most beautiful, the most enterprising, and the most cultured. The Negroes, he declared, Cuvier, 1769 • 1832 constituted "the most degraded race among men, whose forms approach nearest to those of the inferior animals, and whose intellect has not yet arrived at the establishment of any regular form of government, nor at anything which has the least appearance of systematic knowledge" (19). Yet Caucasian, Mongol, and Negro were all of one species,

235

since they could all interbreed. Geographical isolation had probably brought about the differentiation of mankind into three strikingly different races. Perhaps the ancestors of the yellow and white races had escaped from the last grand catastrophe of nature in opposite directions. As for the Negroes, their ancestors had probably been isolated from the rest of mankind even before the upheaval known as the Deluge, for their race had continued submerged in barbarism while the white race had progressed steadily and the yellow race had built imposing (if entirely stationary) civilizations.

In Cuvier's opinion, this process of racial differentiation had required no great length of time; man was a recent product of creation, barely antedating the last great geologic revolution. Cuvier denied that any human bones had been found in a fossil state. The supposed discoveries of human fossils were all fakes, he declared. Of the scores of bones found in the gypsum quarries near Paris, none were those of prehistoric men. As for Lazzaro Spallanzani's collection of fossils from the island of Cerigo in the Mediterranean Sea, Cuvier was convinced from personal inspection that they lent no support to the idea of man's antiquity. Nor did the fragment of a jaw dug up at Cronstadt together with articles of human manufacture prove anything, since the digging was done without recording the depth at which each item was found.

Cuvier seems not to have known John Frere's "Account of Flint Weapons Discovered at Hoxne in Suffolk," read before the London Society of Antiquaries in 1797 and published in their *Archaeologia* in 1800. These flint instruments had been dug up by workers excavating clay for bricks. Frere took careful notes on the position of the artifacts in the strata of earth, sand, and gravel. Upon learning that several large bones, including a huge jawbone with some teeth remaining, had been found alongside the flints, he tried to recover them but was unable to. From the presence of these bones and from the position of the stratum in which they and the flints had been found, Frere concluded that the artifacts must be relics of "a very remote period indeed; even beyond that of the present world" (20).

What Cuvier might have thought of this account is impossible to say. He acknowledged that men had been on earth before the last great catastrophe and that some had probably survived it (solicitude for the credibility of Scripture required this much), but he insisted that the human race was no more than a few thousand years old.

FIG. 8.3 — Two views of a human artifact found associated with the bones of pre-historic animals in Suffolk, England, in 1797 by John Frere, Fellow of the Royal Society of London. Frere considered such relics to be of "a very remote period indeed; even beyond that of the present world," but more than half a century was to elapse before the idea of the antiquity of man was taken seriously. From **Archaeologia,** the journal of the Society of Antiquaries, 2nd ed., XIII (1807).

From the Bible and other ancient books and traditions Cuvier drew additional evidence in favor of man's recent origin. The Pentateuch, he declared, had existed in its present form for two thousand eight hundred years. If the book of Genesis was written by Moses, as it seemed to have been, the record was carried back another five hundred years to a time not far removed from that of the last geologic upheaval. Moreover, the substance of the Biblical account was confirmed by the traditions and sacred books of the Chinese, Hindus, and Assyrians if the fabulous elements in their legends were discounted. Traces of the universal tradition of a general deluge and upheaval had also been found in the remains of Aztec and Incan civilizations. Such a remarkable correspondence of traditions must have some basis in fact (21). In anthropology, as in geology and biology, Cuvier was a thoroughgoing conservative.

► **FROM NEGRO TO EUROPEAN**

In Cuvier's hands anthropology had little to contribute to the development of evolutionary ideas. In other hands, however, the search for a key to the origin of human races produced hypotheses which foreshadowed Charles Darwin's theory of the origin of species. In 1813, the year following the publication of Cuvier's "Discourse on the Revolutions of the Surface of the Globe," James Prichard published the first edition of his *Researches into the Physical History of Man* (22). Prichard had become interested in anthropology at an early age from hearing its incompatibility with Scripture asserted and

James Cowles Prichau.
1786 • 1848

denied. He wrote his medical dissertation at Edinburgh on the varieties of the human species and continued thereafter to combine anthropological research with the study and practice of medicine. To the end of his life he was a vigorous defender of the unity of the human species and of the inspiration of Scripture.

In the opening pages of his *Researches* Prichard foreswore all a priori argument, whether from Scripture or from the nature of things. Scripture would best be served by an impartial investigation of the facts of natural history. For the rest, "All speculations concerning

238

the system of the world, which are founded on arguments from probabilities and the supposed fitness of things, demand a greater share of intelligence than has been given to the human mind, and become not the humble interpreter of nature" (23). Sir William Jones' argument that God would not create more than one human species if one would suffice to populate the world was no more to be depended on than Lord Kames' contrary contention that God would not leave a vacant world to be peopled by the accidental wanderings of one primitive stock. Only a careful investigation into the facts of natural history could decide the question of the specific unity or diversity of mankind.

Having denounced the intrusion of a priori conceptions into scientific argument, Prichard proceeded to use them liberally in his definition of a species. Linnaeus' classes, he declared, were arbitrary and artificial.

> But it is not so in the case of species. Here the distinction is formed by nature, and the definition must be constant and uniform, or it is of no sort of value. It must coincide with Nature.
> Providence has distributed the animated world into a number of distinct species, and has ordained that each shall multiply according to its kind, and propagate the stock to perpetuity, none of them ever transgressing their own limits, or approximating in any great degree to others, or ever in any case passing into each other. Such a confusion is contrary to the established order of Nature.
> The principle therefore of the distinction of species is constant and perpetual difference (24).

It was easier to state this criterion than to apply it, Prichard admitted. Since varieties, once formed, tended to perpetuate themselves indefinitely, it was difficult to distinguish them from true species, that is, "species distinct from their first creation." Buffon's suggestion that "permanent varieties" might be distinguished from true species by their capacity to interbreed freely had considerable merit, for if Providence had not taken care to render mixtures between species sterile, "an universal confusion must have ensued, and there would not be at this day one pure and unmixed species left in existence." But Buffon's criterion, though plausible, could not be considered sufficiently well established to decide in itself the question of the specific unity or diversity of human types. For this purpose, said Prichard, the natural historian must resort to Blumenbach's method of comparing the varieties observed among men with those found among animals, "to determine whether they are of a nature analogous to the diversities which other species have a tendency to assume, and therefore to be referred according to our rule to the principle of natural deviation; or on the contrary peculiar, and such as must be held to constitute specific differences" (25).

Applying this method to variations in color, stature, and the like, Prichard concluded that the range of variation found among human beings in these respects was no greater than that which obtained within the species of animals. He was particularly concerned with variations in cranial and facial structure, since both Camper and Cuvier had reported a regular gradation from Negro to Calmuck to European in degree of prognathism, and some writers had cited this gradation as proof of the specific diversity of human races, differences in head form being presumed to reflect differences in intellectual endowment. In some passages Prichard denied the existence of the supposed gradation; in others he admitted its reality but denied that it implied diversity of species. Differences in head form need not imply differences in mental capacity, he argued, unless it could be shown on other grounds that the types exhibiting this range of variation were specifically different; in that case the presumption would be that the Creator had designed the head to match the degree of intelligence. But such was not the case. Instead, as Camper and Blumenbach had already made clear, the variety of cranial structure among human beings was a product of natural causes operating to modify the original form of the species. Since the skulls of animals belonging to one and the same species frequently exhibited an even greater variation in size and form than was to be found among the various human races, the obvious conclusion was that the differences in both cases were sufficiently accounted for by the "principle of natural deviation" (26).

When Prichard turned to the problem of explaining how these deviations had taken place, however, he parted company with Buffon, Blumenbach, Camper, and the whole school of environmentalists. He

240

denied the supposed correlation of climate and skin color and scoffed at the idea that the Negro's features were the product of climate or of the savage practice of shaping the heads of infants. All these explanations assumed that characteristics acquired during the life of the individual were transmitted to the offspring. Nothing could be farther from the truth, said Prichard. Observation and experience showed that the production of varieties in the race was governed by laws quite different from those which controlled changes in appearance during the individual's lifetime.

> In the former instance certain external powers acting on the parents influence them to produce an offspring possessing some peculiarities of form, colour, or organization; and it seems to be the law of nature that whatever characters thus originate, become hereditary, and are transmitted to the race, perhaps in perpetuity. On the contrary, the changes produced by external causes in the appearance or constitution of the individual, are temporary, and in general acquired characters are transient, and have no influence on the progeny (27).

How, then, could the production of varieties in the human species be explained, if not by the agency of environmental influences or of human art? How could the mysterious processes giving rise to peculiarities in the offspring be studied? By observing and generalizing the effects of those processes, said Prichard. If uniformities could be discovered in the variation of plants and animals, these patterns could be extended by analogy to human variation. It was observable, for example, that some species of plants and animals exhibited a greater disposition to vary than others. Where the environment remained constant, the species showed little tendency to vary. When the environment changed, variation increased, reaching its maximum in those species which had undergone domestication. Now the human species inhabited a wider variety of climates than any other in the animal kingdom. Moreover, man was the most domesticated of animals, since civilization was but a kind of self-domestication. It seemed probable, then, that the progress of civilization was the chief cause of the production of varieties in the human species in whatever case.

241

From natural and human history Prichard drew a variety of proofs of this hypothesis. Savage people, he declared, were usually dark in complexion. In the South Sea islands, where the population was divided into different ranks, the lower classes were Negroid in appearance; the upper, more civilized classes were lighter in complexion, some individuals displaying an almost Teutonic fairness. Again, research into the history of the Hindus and Egyptians indicated that these peoples had originally been Negroid in color and feature. From these and other facts, said Prichard, "it must be concluded that the process of Nature in the human species is the transmutation of the characters of the Negro into those of the European, or the evolution of white varieties in black races of men" (28). This conclusion was quite opposite to that reached by Buffon and Blumenbach. Instead of conceiving human variation as a degeneration from an original model of the species, Prichard conceived it as a progress toward the perfection embodied in the European variety.

This ingenious theory was not without its difficulties. Prichard seems not to have reflected that his argument might be reversed (as, indeed, it frequently was) so as to derive the superior civilization of the white peoples from their superior racial endowment instead of explaining their light color by their more civilized condition. But Prichard was preoccupied with other difficulties, in particular the problem of explaining why progress in civilization should produce the light skin and other characteristic features of the European. Perhaps, he suggested, this set of characters was better adapted to the civilized mode of life than were Negroid traits, the latter being adapted to a primitive state of life. "All the laws of nature have a beneficial tendency, and among others this law of deviation in the species of animals. It is a principle of amelioration and adaptation...that the conformation and disposition or instinct of animals varies in domestication in such a way, as to render them more fitted for their new conditions" (29). This argument assumed a directing purpose in human evolution: the variations which civilization produced in human beings were those which were needed in civilized society. But why should this be so? Why, indeed, should the variability of nature

242

exhibit any trend? Unless Prichard was ready to abandon science for natural theology, he had to find some selective factor by which the random variability of nature could be given a definite direction of development.

Prichard was well aware of the importance of artificial selection in producing varieties of domesticated animals, and it occurred to him that a kind of unconscious selection went on in human society, namely, the selection of mates according to esthetic preference.

> The perception of beauty is the chief principle in every country which directs men in their marriages...It is very obvious that this peculiarity in the constitution of man, must have considerable effects on the physical character of the race, and that it must act as a constant principle of improvement, supplying the place in our kind of the beneficial control which we exercise over the brute creation. This is probably the final cause for which the instinctive perception of human beauty was implanted by Providence in our nature. For the idea of beauty of person, is synonymous with that of health and perfect organization (30).

In this argument Prichard assumed that there was but one standard of human beauty operative in the marriage choices of mankind in every part of the earth. He conceded, however, that "the natural idea of the beautiful in the human person has been more or less distorted in every nation" and concluded that the resulting variety of esthetic standards might have operated to produce the existing variety of physical types. Thus he wavered between trying, on the one hand, to explain the evolution of white varieties from black by the continuous influence of a universal standard of beauty and, on the other, to account for the diversity of human types by the influence of varying esthetic standards. To make matters worse, he placed the whole concept of a natural standard of beauty in doubt by conceding that variety in esthetic preference might be as much the effect as the cause of variety in feature and physique, since every people regarded its own physical type as natural and perfect. All in all, Prichard was more successful in explaining the diversity of mankind than in showing why the civilizing process should produce an evolution from black to white types. Selection according to varying esthetic standards, geographic isolation of populations, and the like helped to account for the racial differentiation of the human species, but they gave no reason to expect "improvement" or, indeed, any universal trend of variation. For proof of his grand hypothesis Prichard was thrown back on the asserted correlation between the degree of savagery and the proportion of Negroid traits among the peoples of the world.

Despite their obvious shortcomings, Prichard's speculations are of great historical interest. For Prichard applied the concept of evolution and progress to man's physical as well as to his mental development. Like Lamarck, he faced the problem of accounting for a postulated evolutionary development and for the apparent exceptions to it, but, whereas Lamarck assumed the transmissibility of acquired characters, Prichard rejected this assumption and looked instead for some selective agency which could establish a trend of variation. He recognized the importance of artificial selection in animal breeding and explored some of the implications of sexual selection among human beings, but he stopped short of the idea of natural selection through elimination of varieties poorly adapted to survive. He grasped clearly the notion of random variation in the hereditary constitution, but he was too deeply imbued with the traditional concept of nature to conceive varieties as products of a naked struggle for existence.

► RANDOM VARIATION AND NATURAL SELECTION

The application of the idea of natural selection to the problem of the origin of human races was the work of an Anglo-American, William Wells. Wells was born in Charleston, South Carolina, in 1757, the son of a Scottish printer. His father dressed him in a tartan coat and

William Charles Wells, 1757 · 1817

a blue bonnet and sent him to Scotland for his schooling. After studying medicine at Edinburgh, he returned to Charleston and apprenticed himself to Alexander Garden, one of the foremost of Linnaeus' botanical correspondents in America. Alienated from America by the Revolution, Wells settled in London in 1784, there to practice medicine and pursue his scientific researches. The best known of these were his "Essay on Single Vision with Two Eyes," which won him election to the Royal Society, and his "Essay on Dew." In 1813, the same year that Prichard's *Researches* appeared, the Royal Society heard Wells' "Account of a Female of the White Race of Mankind, Part of Whose Skin Resembles That of a Negro." In this account, published in 1818, Wells described in detail certain patches of black skin on the otherwise white body of one Hannah West. From their close resemblance to Negro skin he drew two inferences: "that the blackness of the skin in negroes is no proof of their forming a different species from the white race," and, "that great heat is not indispensably necessary to render the human colour black" (31). There was no evidence, said Wells, that strong sunlight rendered human beings permanently darker. The deepest tanning was not transmitted to the next generation. On the other hand, there was some evidence that the Negro skin was rendered temporarily lighter by exposure to strong sunlight.

To what causes, then, was the color of the human skin to be ascribed? Wells was ready with a cause which, so far as he knew, had not yet been suggested. Suppose, as observation seemed to show, that resistance to certain diseases was correlated with, though not caused by, darkness of skin color. What would happen to a population gradually dispersing itself over the African continent?

> Of the accidental varieties of man, which would occur among the first few scattered inhabitants of the middle region of Africa, some one would be better fitted than others to bear the diseases of the country. This race would consequently multiply, while the others would decrease, not only from their inability to sustain the attacks of disease, but from their incapacity of contending with their more vigorous neighbors. The colour of this race I take for granted, from what has been already said, would be dark. But the same disposition to form varieties still existing, a darker and a darker race would in the course of time occur, and as the darkest would be the best fitted for the climate, this would at length become the most prevalent, if not the only race, in the particular country in which it had originated (32).

Wells gave no explanation of the "disposition to form varieties." It was an observable fact, he said, that varieties of greater or less magnitude occurred constantly through the animal kingdom. In a freely interbreeding population the varieties produced tended to disappear through intermixture. But in regions cut off by geographic or other barriers, accidental peculiarities in appearance might become established and persist over generations. Among domestic animals, breeds were established by artificial selection. But the selection which man had practiced on domestic animals might have been practiced by nature on the human race, "chiefly during its infancy, when a few wandering savages, from ignorance and improvidence, must have found it difficult to subsist throughout the various seasons of the year, even in countries the most favourable to their health" (33).

If Wells had been a zoologist and geologist as well as a physician, Charles Darwin's theory of the origin of species might have been anticipated by almost fifty years. All the elements of the theory were present in the scientific world by 1818. Buffon, Kant, and Laplace had derived the origin of the solar system from the operation of a universal system of laws, elements, and forces. Hutton had conceived the surface of the earth as a system of matter in motion millions of years old. Cuvier had applied the resources of comparative anatomy to the reconstruction of extinct species and, with William Smith, had discovered how to read the fossil record embedded in the globe's crust. Buffon had suggested the variability of organic forms, and Lamarck had postulated their gradual evolution from monad to man. Buffon had seen that the extinction of species was related to the struggle for survival among the various creatures produced by nature's endless combinations. Maupertuis, Prichard, and Wells had sensed the possibility that new types might be formed from chance variations thrown up in the course of procreation, and Wells had used the notion of natural selection to explain the origin of the Negro race. Even Malthus' *Essay on the Principle of Population,* the book which Darwin said gave him the clue to the origin of species, was available.

But although the elements of Darwin's theory lay at hand, they were not embraced in one powerful and well-informed mind. Moreover, the traditional view of nature, though greatly weakened by these developments, still exerted a powerful influence on scientific thought. Lamarck viewed the organic process as an evolution toward higher forms of life, but even he stopped short of making chance and struggle

246

the engines of that progress. He believed that nature had endowed organisms with a capacity to adapt to their changing environment through their own efforts to survive. The Western world was not yet ready to surrender its belief in the stability, perfect harmony, and wise contrivance of nature.

What a book a devil's chaplain might write on the clumsy, wasteful, blundering, low, and horribly cruel works of nature!

DARWIN (to Hooker, 1856)

How fleeting are the wishes and efforts of man! how short his time! and consequently how poor will be his results, compared with those accumulated by Nature during whole geological periods! Can we wonder, then, that Nature's productions should be far 'truer' in character than man's productions; that they should be infinitely better adapted to the most complex conditions of life, and should plainly bear the stamp of far higher workmanship?

On the Origin of Species, 1859

The Triumph of Chance and Change

FRIDAY, OCTOBER 28TH, 1832, was a gloomy, overcast day in the port of Montevideo, on the Rio de la Plata, but there was rejoicing on board *H. M. S. Beagle,* just arrived from Brazil on a voyage of exploration. Mail from England had been distributed to the crew, and none was happier with his share than young Charles Darwin, naturalist to the expedition. Letters from Shrewsbury brought news of his family, and the London newspapers were full of the controversy over the great Reform Bill, England's first step toward democracy. Also in the mail was a precious book, the second volume of Charles Lyell's *Principles of Geology.* Darwin had taken the newly published first volume with him when he boarded the *Beagle.* His beloved botany professor at Cambridge, the Reverend John Henslow, had recommended it and urged him to read it but by no means to believe it. Darwin had heeded the first of Henslow's injunctions but not the second. For Lyell had opened to him the vast and exhilarating prospect of nature — "no vestige of a beginning, no prospect of an end" — first unfolded in James Hutton's *Theory of the Earth* in 1788.

The son of a botanist, Charles Lyell belonged, like Hutton and Darwin, to the class which Darwin once described as "gentlemen by profession." Like Hutton, his interest was eventually drawn to geology, in the study of which he soon became a thoroughgoing uniformitarian. But whereas Hutton had confined himself almost entirely to processes of inorganic change, Lyell defined geology to include the study of organic change as well. This was in keeping with the progress made in the study of the fossil record since Hutton wrote. Catastrophism was still the creed of most geologists, but it was now linked to Cuvier's doctrine of successive creations. Each new flora and fauna was regarded as a step upward in the progressive series leading to man and the world he inhabited.

Lyell could not accept this view. He rejected it for two reasons: first, because he believed that the apparent discontinuity in the fossil record was an illusion arising from the imperfection of the record, and, second, because the supposed progression from epoch to epoch had no firm basis in that record. Dicotyledonous plants had been found in the carboniferous strata supposed to contain only the remains of the first, or monocotyledonous, stage of plant development. The secondary strata had been thought to contain nothing higher than fish and reptiles, but the remains of cetaceans and of opossum-like creatures had recently been found in them. In case after case the supposed rule of progression from simple to complex forms had broken down as more

FIG. 9.1 — The Port of Montevideo as it appeared when the **Beagle** visited it in 1832.

and more evidence came to light. This was to be expected, said Lyell, if the uniformitarian principle applied in the organic realm as well as in the inorganic, if the extinction of species and the creation of others to take their places was a part of the ordinary operations of nature.

Before Lyell could apply the uniformitarian principle to the extinction of species, however, he had to deal with Lamarck's suggestion that so-called "extinct" species were in reality the ancestors of living species, their descendants having become slowly modified in the struggle to survive amid changing circumstances. In his second volume, the volume which reached Darwin at Montevideo in October, 1832, Lyell took up the species question. With characteristic candor and clarity he summarized Lamarck's reasons for believing that species were mutable. In Lyell's opinion, Lamarck had been led to evolutionary conclusions chiefly by observing that systematic distinctions between species, varieties, genera, and the like became progressively blurred as the comparisons on which they were based were extended spatially to include the whole earth and temporally to include fossil forms. At the same time, he had observed the influence of environmental circumstances on organic form, especially the effects of use and disuse on the development of organs; and he had learned from his geological studies that the environment underwent perpetual slow change. From these observations, Lamarck had leaped to the bold conclusion that nature displayed a tendency toward progressive improvement. "Henceforth," wrote Lyell, "his speculations know no definite bounds; he gives the rein to conjecture, and fancies that the outward form, internal structure, instinctive faculties, nay, that reason itself, may have been gradually developed from some of the simplest states of existence, — that all animals, that man himself, and the irrational beings, may have had one common origin; that all may be parts of one continuous and progressive scheme of development from the most imperfect to the more complex; in fine, he renounces his belief in the high genealogy of his species, and looks forward, as if in compensation, to the future perfectibility of man in his physical, intellectual, and moral attributes" (1).

Against this subversive hypothesis Lyell brought forward a powerful battery of arguments, some old and some new. Like Cuvier, he stressed the limits of organic variability, the absence of the intermediate forms presupposed by Lamarck's theory, and the purely hypothetical character of the causes invoked to explain the supposed development from monad to man. Lyell did not doubt but that the effects of use and disuse might be inherited, but to pretend that entirely new organs could arise in response to the "felt needs" of organisms was to set science back a thousand years. That new varieties of plants and animals had been produced by artificial selection could not be denied, but there were limits to the changes which could be effected in this way. Moreover, the fact that all such modified offspring of an original stock could cross with each other successfully showed that the species barrier had not been broken. Crosses between species seldom produced fertile offspring; when they did, sterility ensued in the next generation or two. Geology, likewise, gave no comfort to the transmutationists. Far from revealing a steady trend toward progressive development from epoch to epoch, it showed in many cases the same species continuing through millions of years. In others, the forms embedded in the higher strata were simpler and less developed than those of the lower. The findings of craniology and embryology also had been advanced by the transmutationists in support of Lamarck's theory, but Lyell could see nothing in these sciences to support the idea of progressive development.

Good sense and scientific caution underlay much of Lyell's argument, but there was also a strong admixture of the traditional view of nature. "We must suppose," he wrote, "that when the Author of Nature creates an animal or plant, all the possible circumstances in which its descendants are destined to live are foreseen, and that an organization is conferred upon it which will enable the species to perpetuate itself and survive under all the varying circumstances to which it must be inevitably exposed" (2). In that case, Lyell explained, it was to be expected that some species would be much more variable than others, since some would be exposed to a greater variety of external circumstances than others. It might even happen that the species of a genus destined to inhabit a very limited range of environments would differ from each other less widely than the varieties or races of a single species created to inhabit widely differing environments. This, in turn, would explain the difficulties encountered by systematists in drawing their classificatory lines.

A similar argument served Lyell in accounting for the wide divergence of instinct and form observable in successive generations of domesticated plants and animals. It was inconceivable, he thought, that the various types of dogs were all descended from an animal such as the wolf. The instincts required by the wolf for survival in the state of nature were totally different from those required in a creature which was to be man's companion and servant. The latter could never have been derived from the former by slow modification. "When such remarkable habits appear in races of this species," declared Lyell, "we may reasonably conjecture that they were given with no other view than for the use of man and the preservation of the dog which thus obtains protection. . . . It seems also reasonable to conclude, that the power bestowed on the horse, the dog, the ox, the sheep, the cat, and many species of domestic fowls, of supporting almost every climate, was given expressly to enable them to follow man throughout all parts of the globe — in order that we might obtain their services, and they our protection" (3).

Having demonstrated to his own satisfaction "that species have a real existence in nature, and that each was endowed, at the time of its creation, with the attributes and organization by which it is now distinguished," Lyell was ready to develop his own ideas concerning their extinction and creation. The basic problem was to explain the geographic distribution of plant and animal species throughout the world. Botanists had been gathering information on this subject since the sixteenth century, and Buffon had made an interesting comparison of the quadrupeds of the New World and the Old. As the data accumulated, a remarkable fact had come to light: regions of the world which were similar in climate and general topography were nevertheless inhabited by quite dissimilar plants and animals. The same was true of oceanic life. There did not seem to be one creative plan for animals inhabiting temperate zones, another for those of the tropics, and so on. Instead, each geographical region seemed to have its own flora and fauna along with a few species similar to those of other regions, these last probably immigrants from other parts of the world.

The problem, then, was to form a theory concerning "the first introduction of species" which would explain this state of affairs.

The means by which plants and animals created in one region of
the globe might stray to other regions in the course of time were
many and varied, and Lyell devoted several chapters to this subject,
little suspecting that the facts he reported would play an important
part in a theory of organic evolution soon to take form in the mind of
Charles Darwin. Lyell's immediate problem, however, was not the
origin of species but their dispersion from various centers of creation.
Suppose, he suggested, that every living thing had been destroyed in
the western hemisphere and that man had been given permission to
stock the New World with plants and animals from the Old, being
forbidden, however, to introduce more than one pair of each species.
The planning of such an operation would require great care and foresight in adjusting the various species to the physical environment and
to other living forms. If carried out successfully, it should eventually
result in a pattern of geographic distribution of organic beings similar
to that actually existing in the New World.

> Thus, for example, almost all the animals and plants natur-
> alized by us towards the extremity of South America, would
> be unable to spread beyond a certain limit, towards the east,
> west, and south, because they would be stopped by the ocean,
> and a few of them only would succeed in reaching the cooler
> latitudes of the northern hemisphere, because they would be
> incapable of bearing the heat of the tropics, through which
> they must pass. In the course of ages, undoubtedly, excep-
> tions would arise, and some species might become common
> to the temperate and polar regions, or both sides of the equa-
> tor; for we have before shown that the powers of diffusion
> conferred on some classes are very great. But we should con-
> fidently predict that these exceptions would never become so
> numerous as to invalidate the general rule (4).

It should not be supposed, however, Lyell continued, that the
"original centers of creation" could be inferred from the present distribution of plants and animals. If geology taught anything, it taught
that the surface of the globe was subject to perpetual change, and these
changes would, in turn, alter the distribution of organic life. They
would do so partly by erasing and erecting barriers to migration and

partly by hastening the extinction of some species and promoting the spread of others. Modifications in the physical environment would cause some extinction directly, as when the climate became too dry for some species of plants, but their greatest influence would be indirect, affecting the mutual relations of organisms in the struggle for existence. The severity and importance of this struggle for-life Lyell illustrated at length in a series of examples that were to work powerfully on young Darwin's mind. From these examples Lyell concluded that extinction must be a common occurrence, "that the destruction of species must now be part of the regular and constant order of nature." Lamarck had imagined that organisms responded to changes in the environment by developing new habits and propensities, thus, eventually, new organs. But the struggle for existence was too harsh and immediate to permit slow changes of this kind, said Lyell.

> Suppose the climate of the highest part of the woody zone of Etna to be transferred to the sea-shore at the base of the mountain, no botanist would anticipate that the olive, lemon-tree, and prickly pear...would be able to contend with the oak and chestnut, which would begin forthwith to descend to a lower level, or that these last would be able to stand their ground against the pine, which would also, in the space of a few years, begin to occupy a lower position. We might form some kind of estimate of the time which might be required for the migrations of these plants; whereas we have no data for concluding that any number of thousands of years would be sufficient for one step in the pretended metamorphosis of one species into another, possessing distinct attributes and qualities....
> It is idle to dispute about the abstract possibility of the conversion of one species into another, where there are known causes so much more active in their nature, which must always intervene and prevent the actual accomplishment of such conversions. A faint image of the certain doom of a species less fitted to struggle with some new condition in a region which it previously inhabited, and where it has to contend with a more vigorous species, is presented by the extirpation of savage tribes of men by the advancing colony of some civilized nation. In this case the contest is merely between two different *races*....Yet few future events are more certain than the speedy extermination of the Indians of North America and the savages of New Holland in the

course of a few centuries, when these tribes will be remembered only in poetry and tradition (5).

But if the destruction of species was an incidental outcome of everyday processes rather than a catastrophic event heralding the arrival of a new epoch of nature, must there not be some equally regular means of introducing new species into nature to take the place of the old? Did not the uniformitarian doctrine, based on the assumption that nature was a divinely ordained, self-balancing system of laws, elements and forces, require that natural causes should produce new species as rapidly as the old were eliminated? Lyell felt the force of the argument, but he could not accept the only theory of species formation by natural causes then known to him — the theory of Lamarck. He was not opposed to the idea that species might originate "through the intervention of intermediate causes," he wrote the astronomer Sir John Herschel, but he thought it unwise to offend public sentiment by suggesting such a possibility without being able to indicate the nature of those "intermediate causes" (6). Faced with this situation, Lyell contented himself with suggesting that new species were "called into being" from time to time as needed, leaving it to future research to determine the time, place, and manner of their appearance. From the researches of geologists, he assured his readers, it would eventually become clear "which of the species, now our contemporaries, have survived the greatest revolutions of the earth's surface; which of them have co-existed with the greatest number of animals and plants now extinct, and which have made their appearance only when the animate world had nearly attained its present condition."

What young Darwin thought of all this as he pondered it in the seclusion of his cabin or on his long treks across the pampas we cannot know for sure. But we do know that Lyell's book stirred his imagination. Only a few days before receiving it he had dug the bones of an extinct megatherium from the soil at Punta Alta with his own hands. As the *Beagle* proceeded southward, he was struck by the general similarity of the plants and animals inhabiting South America from one end to the other and by the resemblances between the extinct creatures entombed in its strata and the living forms which presently inhabited it. Surveying the havoc wrought by the great earthquake at Concepcion, he felt more deeply than words could express the instability of the earth's crust and the vastness of the powers at nature's disposal. Standing astride the continent on the Andean ridge above Valparaiso, he viewed with astonishment the signs of the recent origin of this great barrier and mused on the differences between the flora and fauna inhabiting its eastern slope and those on its western flank. "I am become a zealous disciple of Mr. Lyell's views, as known in his admirable book," he wrote to a friend in England. "Geologising in South America, I am tempted to carry parts to a greater extent even than he does." "I look forward to the Galapagos with more interest than any other part of the voyage," he added. "They abound with active volcanoes, and, I should hope, contain Tertiary strata" (7).

The Galapagos Islands were to prove even more interesting than Darwin thought, but not on account of their geology. Instead, it was the flora and fauna of the islands which fascinated him. "I industriously collected all the animals, plants, insects, & reptiles from this Island," he noted in his diary September 26, 1835. "It will be very interesting to find from future comparison to what district or 'centre of creation' the organized beings of this archipelago must be attached." How like, yet how unlike, the flora and fauna of South America were the plants and animals of the Galapagos! Even more surprising, the inhabitants of the different islands themselves differed slightly, as if the plan of creation had been varied somewhat from island to island. Undoubtedly, secrets were hidden here which "future comparison" might someday disclose.

The comparison was made when Darwin returned to England in 1836 and began to organize his materials for publication. No sooner did he undertake this task than "new views" began to crowd into his mind "thickly and steadily," as he wrote Lyell. Geology, which had been his primary interest during most of the voyage, now took a second seat. In a series of notebooks he began to record whatever facts presented themselves with respect to variation in species. Some of the facts were drawn from his *Beagle* notes, but there were important new sources as he began to comb agricultural and horticultural journals for information about variation in domestic stocks. Suddenly, somewhere in the midst of these researches, he found the clue he had been looking for. As he remembered it later, "I came to the conclusion that selection was the principle of change from the study of domesticated productions; and then, reading Malthus, I saw at once how to apply this principle" (8). According to his own statement, Darwin first read Malthus' *Essay on Population* in October, 1838. Either he read it earlier or it had less influence on his thinking than he remembered, for in his notebook for 1837 he had written:

> With belief of transmutation and geographical grouping, we are led to endeavour to discover *causes* of change; the manner of adaptation...instinct and structure becomes full of speculation and lines of observation....My theory would give zest to recent and fossil comparative anatomy; it would lead to the study of instinct, heredity, and mind-heredity, whole [of] metaphysics.
>
> It would lead to closest examination of hybridity and generation, causes of change in order to know what we have come from and to what we tend — to what circumstances favour crossing and what prevents it — this, and direct examination of direct passages of structure in species, might lead to laws of change, which would then be [the] main object of study, to guide our speculations (9).

"My theory," Darwin called it. But what was the theory? It could scarcely have been Lamarck's development hypothesis, for Lyell had

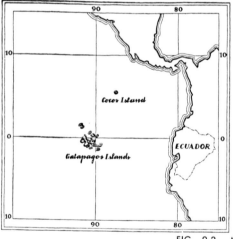

FIG. 9.2 — Map of the Galapagos Islands, located 600 miles west of Ecuador. Cocos Island is not one of the Galapagos group, but it has developed one species of finch, presumed to have come originally from the mainland. Reproduced, with permission, from the **Scientific American**, April, 1953.

FIG. 9.3 — Natural selection illustrated by the finches of the Galapagos Islands, many species of which were collected by Darwin. The six species shown at the foot of the "tree" are ground species, thought to have evolved from a common immigrant ancestor. The eight tree species are believed to.have evolved later in the branching fashion indicated. No. 1 is a woodpecker-like species; 2 inhabits mangrove swamps; 3, 4, and 5 are insect-eaters; 6 is a vegetarian; 7 is a warbler-finch; 8 is an isolated species of Cocos Island finch. The ground finches are mainly seed-eaters, but 13 and 14 feed on cactus. Reproduced, with permission, from an article by David Lack in the **Scientific American**, April, 1953.

condemned that hypothesis, and Darwin had found Lyell's verdict just. It must, then, have been his own hypothesis of organic modification by natural selection in the struggle for existence. Whether Malthus' *Essay* or some other work was the catalyst that precipitated the idea of natural selection in Darwin's mind is not of major importance. Maupertuis, Buffon, William Wells, Charles Lyell, or Darwin's friend Edward Blyth might have done it just as well (10). If it were indeed Malthus, it may have been because Malthus drew a striking picture of the pressure of population on the food supply. The idea of a selective agency was already in Darwin's mind from his studies of artificial selection. What was needed in addition was the conviction that selection was not a sporadic and occasional process, as Buffon had conceived it, but an omnipresent power acting at every moment and in every situation to check the irrepressible fecundity of organic life. It may well have been that Malthus' long, doleful account of the effects of vice and misery in eliminating the poor and vicious members of society did for Darwin what Lyell's portrayal of the struggle for existence had failed to do. There is a difference between being exposed to an idea and having it deeply impressed upon one's mind by constant repetition and a detailed elaboration of its consequences.

By 1842 Darwin was ready to commit a rough draft of his entire theory to paper. Two years later this rough draft was converted into a fairly finished essay containing most of the ideas and arguments embodied later in the *Origin of Species*. The essay of 1844 remained unpublished during Darwin's lifetime. But it is of extraordinary importance as indicating the form in which the theory of evolution by natural selection first took shape in his mind. One is struck immediately by Darwin's extreme caution in advancing his bold hypothesis. Where Buffon or Lamarck would have published so exciting an idea promptly, pursuing it speculatively as far as imagination would take them, Darwin felt that it was "like confessing murder" to record privately his doubts about the stability of species and to confide them to a few selected friends. To his way of thinking, caution was the soul of science. His imagination undoubtedly soared high and wide in unguarded moments, extending the evolutionary hypothesis to man himself, but he soon checked his speculative fancy with the rein of factual evidence.

With respect to my far distant work on species [he wrote a friend in the year following completion of his private essay], I must have expressed myself with singular inaccuracy if I led you to suppose that I meant to say that my conclusions were inevitable. They have become so, after years of weighing puzzles, to myself *alone;* but in my wildest day-dream, I never expect more than to be able to show that there are two sides to the question of the immutability of species, i. e. whether species are *directly* created or by intermediate laws (as with the life and death of individuals). I did not approach the subject on the side of the difficulty in determining what are species and what varieties, but...from such facts as the relationship between the living and extinct mammifers in South America, and between those living on the Continent and on adjoining islands, such as the Galapagos. It occurred to me that a collection of all such analogous facts would throw light either for or against the view of related species being co-descendants from a common stock. A long searching amongst agricultural and horticultural books and people makes me believe (I well know how absurdly presumptuous this must appear) that I see the way in which new varieties become exquisitely adapted to the external conditions of life and to other surrounding beings. I am a bold man to lay myself open to being thought a complete fool, and a most deliberate one. From the nature of the grounds which make me believe that species are mutable in form, these grounds cannot be restricted to the closest-allied species; but how far they extend I cannot tell, as my reasons fall away by degrees, when applied to species more and more remote from each other. Pray do not think that I am so blind as not to see that there are numerous immense difficulties in my notions but they appear to me less than on the common view (11).

The essay of 1844 began, as the *Origin of Species* was to begin fifteen years later, with a discussion of variation in domestic stocks, for Darwin, like Maupertuis and Wells before him, conceived the origin of new types in nature by analogy to the formation of domestic breeds by artificial selection. Like these writers and Buffon, Darwin began with the fact of hereditary variation in organic beings. It was variation which supplied the novelties constituting the raw materials of organic change. Garden plants, horses, cattle, and the like presented no problem in this respect; their tendency to vary had been noticed by many observers. Darwin had only to gather the facts by patient reading of

261

horticultural and agricultural journals, by hobnobbing with gardeners and breeders, and by experimenting himself with artificial selection. To determine which variations were hereditary was not so easy, however. In this matter Darwin, like so many of his predecessors, assumed the transmissibility of acquired characters.

> Thus, in animals [he noted], the size and vigour of body, fatness, period of maturity, habits of body or consensual movements, habits of mind and temper, are modified or acquired during the life of the individual, and become inherited. There is reason to believe that when long exercise has given to certain muscles great development, or disuse has lessened them, that such development is also inherited. Food and climate will occasionally produce changes in the colour and texture of the external coverings of animals; and certain unknown conditions affect the horns of cattle in parts of Abyssinia; but whether these peculiarities, thus acquired during individual lives, have been inherited, I do not know. It appears certain that malconformation and lameness in horses, produced by too much work on hard roads, — that affections of the eyes in this animal probably caused by bad ventilation, — that tendencies towards many diseases in man, such as gout, caused by the course of life and ultimately producing changes of structure, and that many other diseases produced by unknown agencies, such as goitre, and the idiotcy [sic] resulting from it, all become hereditary (12).

On the whole, however, Darwin placed less emphasis on variations arising in the course of life than on differences appearing at birth or soon after. These differences, whether slight (as in the case of eye color) or very pronounced (as in the case of rumpless fowls and tailless sheep), were generally hereditary. They could scarcely be attributed to the direct influence of the environment. Nevertheless environment seemed to have something to do with them, since domestication itself was a change in environment and seemed to excite variability in the plants and animals domesticated, presumably by some kind of influence on their reproductive systems. The variability of plants and animals in a state of nature was a more difficult matter to prove, or so it seemed to Darwin. Buffon had simply assumed the variability of all the productions of nature. Or, he had inferred it as a corollary of his conception of visible nature as a kaleidoscope of changing effects produced by the operation of the laws, elements, and forces which constituted primary nature. Darwin preferred not to take anything for granted; he wanted solid empirical grounds for asserting the natural variability of wild stocks. The proverbial expression that no two animals or plants were born absolutely alike seemed to Darwin "much truer when applied to those under domestication, than to those in a state of nature." The problem was further complicated by the inability of naturalists to agree whether certain forms constituted varieties sprung from slightly different parent forms or independent species dating from the creation. Some natural species were known to be more variable than others, however, and it seemed likely that this variability of certain traits in wild stocks was of the same character as the more general variability of domestic races. Since the evidence concerning variation in nature was scanty and dubious, Darwin had to content himself with an argument from analogy. Since all wild stocks showed a tendency to vary when domesticated, they must possess a potential variability in the wild state. The problem, therefore, was to show how conditions might arise in nature *analogous* to the conditions which seemed to excite variability under domestication.

> Domestication [he wrote] seems to resolve itself into a change from the natural conditions of the species (generally perhaps including an increase of food); if this be so, organisms in a state of nature must *occasionally,* in the course of ages be exposed to analogous influences; for geology clearly shows that many places must, in the course of time, become exposed to the widest range of climatic and other influences; and if such places be isolated, so that new and better adapted organic beings cannot freely emigrate, the old inhabitants will be exposed to new influences, probably far more varied, than man applies under the form of domestication.... Whatever might be the result of these slow geological changes, we may feel sure, from the means of dissemination common in a lesser or greater degree to every organism taken conjointly with the changes of geology, which are steadily (and sometimes suddenly, as when an isthmus at last separates) in progress, that occasionally organisms must suddenly be introduced into new regions, where, if the conditions of existence are not so foreign as to cause its extermination, it will often

263

be propagated under circumstances still more closely analogous to those of domestication; and therefore we expect will evince a tendency to vary. It appears to me quite *inexplicable* if this has never happened; but it can happen very rarely (pp. 83–84).

So far there was little in Darwin's argument that Buffon would not have assented to. But at this point, where Buffon had inferred the variability of organic beings from the variability of environmental conditions, assuming (at least in some passages) that new races and perhaps new species might arise by the continued influence of changed conditions, Darwin recognized the necessity for a selective agency to establish new types. It would do this by choosing among the variations resulting from the environmentally induced plasticity of the organism. Although every part of the organism would tend to vary under such conditions, it would vary "in no determinate way," that is, it would vary without continuous direction and without reference to the needs of the organism, "and therefore *without selection* the free crossing of these small variations (together with the tendency to reversion to the original form) would constantly be counteracting this unsettling effect of the extraneous conditions on the reproductive system." Since variation itself was not adaptive (Lamarck to the contrary notwithstanding), it could produce no trend of variation unless by the selective influence of some outside agency.

Here again the analogy of domestic breeds provided the clue. Scientific breeding of plants and animals had accomplished wonders in Darwin's own day. Less systematic selection had been practiced since the dawn of history; in the course of time it had produced the various kinds of cereals, sheep, dogs, and other domestic plants and animals known to man. But — and here Darwin hit upon the same idea which had suggested itself to Maupertuis and William Wells — might not there be a selection in nature comparable to the selection exercised by man on domestic stocks? In the unpublished essay of 1844, Darwin personified the selective power of the environment. In language oddly reminiscent of Buffon's contrast between the works of man and the works of nature, Darwin asked his imaginary reader to suppose the existence of:

264

a Being with penetration sufficient to perceive differences in the outer and innermost organization quite imperceptible to man, and with forethought extending over future centuries to watch with unerring care and select for any object the offspring of an organism produced under the foregoing circumstances; I can see no conceivable reason why he could not form a new race (or several were he to separate the stock of the original organism and work on several islands) adapted to new ends. As we assume his discrimination, and his forethought, and his steadiness of object, to be incomparably greater than those qualities in man, so we may suppose the beauty and complications of the adaptations of the new races and their differences from the original stock to be greater than in the domestic races produced by man's agency.... With time enough, such a Being might rationally (without some unknown law opposed him) aim at almost any result. ...Seeing what blind capricious man has actually effected by selection during the few last years, and what in a ruder state he has probably effected without any systematic plan during the last few thousand years, he will be a bold person who will positively put limits to what the supposed Being could effect during whole geological periods (pp. 85–87).

A striking conception, this idea of a Master Breeder infinitely wise and patient, with infinite time at his disposal, who, carefully selecting from among the variations in nature those which suited his purposes, molded organic nature to his own wise ends. Such a Being could be little less than God Himself. But the Creator, Darwin observed, seemed to prefer to govern the universe by "secondary means." Adaptive selection, if it did indeed occur in nature, must be the work of the system of nature itself. It must be *natural* selection. What, then, was nature's method of adapting organic beings to their places in her economy? It was, said Darwin, the simple method of endowing every creature with a capacity to breed far in excess of the capacity of the environment to provide food and living space and to select by a process of elimination those which happened to have some slight advantage in the ensuing struggle for existence.

It is the doctrine of Malthus applied in most cases with tenfold force....Yearly more are bred than can survive; the smallest grain in the balance, in the long run, must tell on which death shall fall, and which shall survive. Let this work of selection, on the one hand, and death on the other, go on for a thousand generations; who would pretend to affirm that it would produce no effect, when we remember what in a few years [Robert] Bakewell effected in cattle and [Lord] Western in sheep, by this identical principle of selection (pp. 88, 91).

There it was in all its beautiful simplicity, the hypothesis of random variation, struggle for existence, and adaptive "selection" by the environment. The next problem was to show how this system of processes might form new species by selecting among the progeny of an already existing species. Could natural selection produce races of plants and animals which bred true and which, like natural species, seldom produced fertile offspring when crossed? As to the first point, the production of true-breeding races, Darwin resorted again to the analogy of artificial selection. If breeders were able by selection to form new races which bred true, was there not every reason to believe that races formed by natural selection would breed even truer? The "trueness" of a race seemed to depend on (1) steady selection with one end in view, (2) preventing crosses with individuals of other stocks, and (3) maintaining the new race in conditions suited to it. But would not natural selection be more efficient in all these respects than human selection?

> Man selects chiefly by the eye, and is not able to perceive the course of every vessel and nerve, or the form of the bones. ...He has bad judgment, is capricious, he does not, or his successors do not, wish to select for the same exact end for hundreds of generations. He cannot always suit the selected form to the properest conditions; nor does he keep those conditions uniform: he selects that which is useful to him, not that best adapted to those conditions in which each variety is placed by him....He seldom allows the most vigorous males to struggle for themselves and propagate, but picks out such as he possesses, or such as he prefers, and not necessarily those best adapted to the existing conditions....He often grudges to destroy an individual which departs considerably from the required type. He often begins his selection by a form or sport considerably departing from the parent form. Very differently does the natural law of selection act; the varieties selected differ only slightly from the parent forms; the conditions are constant for long periods and change

The Rock-pigeon

English Pouter

English Fantail

African Owl

English Carrier

Short-faced English Tumbler

FIG. 9.4 — Artificial selection illustrated with reference to various breeds of pigeons, as in Darwin's **The Variation of Animals and Plants under Domestication.** The common rock pigeon, parent form of all the other breeds, is shown at top center. "The kinds examined by me," Darwin writes, "form eleven races, which include several sub-races; and even these latter present differences that would certainly have been thought of specific value if observed in a state of nature."

slowly; rarely can there be a cross; the selection is rigid and unfailing, and continued through many generations; a selection can *never be made* without the form be *better* adapted to the conditions than the parent form; the selecting power goes on without caprice, and steadily for thousands of years adapting the form to these conditions. The selecting power is not deceived by external appearances, it tries the being during its whole life; and if less well (?) adapted than its *congeners,* without fail it is destroyed; every part of its structure is thus scrutinized and proved good towards the place in nature which it occupies.... How incomparably 'truer' then would a race produced by the above rigid, steady, natural means of selection, excellently trained and perfectly adapted to its conditions, free from stains of blood or crosses, and continued during thousands of years be compared with one produced by the feeble, capricious, misdirected and ill-adapted selection of man (pp. 94–95).

But what of the sterility criterion? Here the analogy to artificially selected domestic stocks seemed to fail, for these stocks were not sterile when crossed, as was generally true of species found in nature. Darwin's answer to this difficulty was to question the assumption that crosses between natural species were invariably sterile and, by producing examples of fertile species crosses in nature, to suggest that the ordinary sterility of hybrids depended on complex causes which were yet to be discovered. Since races produced by natural selection would undoubtedly be "truer" than those produced by man, there was every reason to expect that naturally formed races would be less capable of crossing successfully. The truly astonishing fact, said Darwin, was not that crosses between natural species were unusual but rather that they should occur at all and that, when they did occur, the effects on the resulting progeny were indistinguishable from those produced by crosses between races known to be descended from a common stock. In some cases of repeated crossing between species the characters of one parent species were totally absorbed and obliterated by those of the other parent species. "Marvellous that one act of creation should absorb another or even several acts of creation!" Was it not more probable that what naturalists called species were simply strongly marked races which, having undergone selective adaptation in special environments, had lost the capacity to interbreed with closely related types?

268

Having shown how races indistinguishable from species might arise in nature, Darwin proceeded at once to attack a major difficulty still lurking in the mind of his imaginary reader. If this could be shown to be less formidable than it appeared at first glance, the reader's mind would then be open to conviction that descent with modification, far from being merely possible, had actually taken place. The difficulty, long a stumbling block in Darwin's own thinking, was that of conceiving how natural selection could have produced the marvellous instincts of animals and complicated organs like the eye, so perfectly adapted to their function. In the case of instincts, the experience of breeders provided a clue. Certain instincts of domestic animals seemed to have been highly developed by man, either by rigorous training inducing habits which subsequently became instinctual or, more usually, by selection of those individuals which happened to be born with the desired tendencies.

> In the same manner as peculiarities of corporeal structure slowly acquired or lost during mature life..., as well as congenital peculiarities, are transmitted; so it appears to be with the mind. The inherited paces in the horse have no doubt been acquired by compulsion during the lives of the parents: and temper and tameness may be modified in a breed by the treatment which the individuals receive. Knowing that a pig has been taught to point, one would suppose that this quality in pointer-dogs was the simple result of habit, but some facts, with respect to the occasional appearance of a similar quality in other dogs, would make one suspect that it originally appeared in a less perfect degree, 'by chance,' that is from a congenital tendency in the parent of the breed of pointers (pp. 115–16).

Why, then, could not natural selection develop instinct in the same ways? It was known that the nest-building powers of birds varied enormously from species to species and, more important, among individuals of the same species. Was it not likely, then, that those individuals of a species whose nest-building habits gave their eggs a better chance of hatching would eventually outnumber and supplant those individuals in whom the instinct was less well developed? So with any other instinct. As long as it was variable and the variation hereditary, there was no theoretical limit to the complication it might undergo by natural selection during thousands of ages. "Once grant that dispositions, tastes, actions or habits can be slightly modified, either by slight congenital differences or by the force of external circumstances, and that such slight modifications can be rendered inheritable, — a proposition which no one can reject, — and it will be difficult to put any limit to the complexity and wonder of the tastes and habits which may *possibly* be thus acquired" (pp. 127–28).

But if complicated instincts could be produced gradually by natural selection, why not complicated organs as well? Here again the key to nature's secret lay in the gradation observable in organs performing the same general function in various plants and animals. If the visual faculty, for example, were represented in nature only by the highly developed eye of the mammal, it would be folly to imagine that it had been produced by natural selection. But if, as was actually the case, the faculty was found in varying degrees in various classes of the animal kingdom, associated with organic structures of varying complexity, it was at least possible that the more complex organs of vision had been developed from simpler forms by natural selection. If so, nature should exhibit in the case of each such faculty a graduated series of structures rising from the simplest to the most complex without substantial discontinuity. In the case of the eye the most that could be shown was "a multitude of different forms, more or less simple, not graduating into each other, but separated by sudden gaps or intervals." But when it was recalled that innumerable races of creatures with visual organs had perished, leaving few or no traces of their existence, it became evident that the gaps in the known series of visual organs were not conclusive evidence that a continuous series did not, in retrospect, exist. With this argument Darwin passed over from the case for the abstract possibility of species formation by natural selection to the evidence that it had actually occurred.

Even at this point, however, Darwin did not present the facts without first discussing the a priori probabilities as to how the Creator would proceed in making species. Like Buffon, Lamarck, and his own grandfather, he was fully convinced that God chose to accomplish his purposes by secondary causes. "It is in every case more conformable with what we know of the government of this earth, that the Creator should have imposed only general laws." If, then, it could be shown that the known facts of natural history were intelligible on the assumption of common descent with modification, and if the same facts were arbitrary and unconnected on the assumption of special creations, there was every reason to adopt the evolutionary view. "That all the organisms of this world have been produced on a scheme is certain from their general affinities," declared Darwin, "and if this scheme can be shown to be the same with that which would result from allied organic beings descending from common stocks, it becomes highly improbable that they have been separately created by individual acts of the will of a Creator. For as well it might be said that, although the planets move in courses conformably to the law of gravity, yet we ought to attribute the course of each planet to the individual act of the will of the Creator" (pp. 133–34). A more just analogy would have been to Immanuel Kant's nebular hypothesis. In propounding it, Kant had professed to enjoy "the pleasure without having recourse to arbitrary hypotheses, of seeing a well-ordered whole produced under the regulation of the established laws of motion, and this whole looks so like that system of the world which we have before our eyes, that I cannot refuse to identify it with it" (See Chap. 2, p. 30).

The facts most difficult to square with an evolutionary view of nature were those of the fossil record, where discontinuity rather than continuity seemed to prevail. But Darwin undertook to show that the discontinuity resulted from the necessary imperfection of the geological record. Since certain kinds of geological process were unfavorable to the preservation of organic remains, the formations arising in epochs marked by those kinds of process would present a blank in the history of organic forms. Yet there was nothing to support Cuvier's idea of the sudden extinction of whole floras and faunas by sudden upheavals. On the contrary, said Darwin, the facts seemed to show that the duration of any species was proportionate to its geographical range; this was to be expected if extinction was a gradual phenomenon produced by local disturbances in the equilibrium of nature rather than by wholesale transformations of the earth's crust. Some species of tortoises seemed to have survived unchanged throughout millions of years. This fact was fatal to the supposition of wholesale obliteration of succeeding fauna, but it presented no great difficulty for Darwin. For, unlike Lamarck, he renounced completely the notion of an

271

inherent tendency toward change and development in living matter and accepted the extinction of species as a matter of course. To Lamarck extinction seemed improbable because he believed organisms had the capacity to undergo change in response to the demands of the environment. For Darwin, however, extinction of those types which did not happen to fit the requirements of their situation was the *vera causa*, the true means, by which nature produced new forms. Extinction was the ultimate penalty for maladaptation in the never-ending struggle for existence. It was not, as Cuvier and other catastrophists had conceived it, a dramatic event heralding the creation of a brave new world. On the contrary, it was a common occurrence in nature's ample scheme of reckoning, no more portentous than the death of an individual. "If the rule is that organisms become extinct by becoming rarer and rarer, we ought not to view their extinction, even in the case of the larger quadrupeds, as anything wonderful and out of the common course of events." Like Lyell, Darwin applied the uniformitarian principle to the extinction of species. But he went beyond Lyell in seeing that selective extinction implied the creation of new species. Buffon had hinted at natural selection, but had found his *vera causa* in the direct influence of climate and diet on organic forms. Darwin saw that change-with-adaptation required a steady selection of those changes, whatever their cause, which met the conditions of existence. The birth and the death of species were but two aspects of the same natural process.

Paleontological difficulties out of the way, Darwin turned to the more enjoyable task of marshalling the evidence favoring the theory of descent with modification. He began with the evidence which had first led him to doubt the fixity of species, namely, the facts concerning the geographic distribution of organic forms, living and extinct. Just as Buffon had undertaken to compare the quadrupeds of the New World and the Old, so Darwin began by analyzing the geographic distribution of the mammals of the world.

> If we divide the land [area of the world] into two divisions, according to the amount of difference [among mammals], and disregarding the numbers of the terrestrial mammifers inhabiting them, we shall have first Australia including New Guinea; and secondly the rest of the world: if we make a

three-fold division, we shall have Australia, S. America, and the rest of the world; I must observe that North America is in some respects neutral land, from possessing some S. American forms, but I believe it is more closely allied (as it certainly is in its birds, plants and shells) with Europe. If our division had been four-fold, we should have had Australia, S. America, Madagascar (though inhabited by few mammifers) and the remaining land: if five-fold, Africa, especially the southern eastern parts, would have to be separated from the remainder of the world (p. 152).

These facts, Darwin emphasized, were difficult to explain on the time-honored assumption that the different species had been separately created in the various regions of the world, each form being wisely adapted to the environment it was to inhabit. If that were the case, one would expect desert animals to be much alike throughout the world, tropical forms to be created on another pattern, and so on. But this was not true. Although in some ways the tropical species of South America and those of Africa resembled each other superficially, the basic similarities of anatomical structure were to be found among South American animals as a whole as contrasted to the quite different group of African animals. Generally speaking, organic beings seemed to resemble each other structurally more in proportion to how close they were geographically than to how similar their environments were. Geographic barriers, rather than the requirements of diverse habitats, seemed to be the effective causes of organic diversity. Thus, each of the five regions distinguished by its mammalian species was separated from the rest of the world by formidable geographic obstacles to the migration of mammals, yet each contained within itself an astonishing variety of environmental conditions. Again, the productions of oceanic islands resembled those of the neighboring continent much more than those of islands similar in geologic structure but situated near other continents. The mammals of the Canary and Cape de Verde Islands resembled those of Europe and Africa; those of the Galapagos were South American in general character. The same law held true among mammals recently extinct. A tri-partite division by mammalian species of the land area of the world in the late Tertiary period would yield three regions — Australia and its dependent islands; South America; and Europe-Asia-Africa — "almost as distinct as at the present day, and intimately related in each division to the existing forms in that division." Still farther back in geologic time the pattern would change, but so would the pattern of world geography which dictated the arrangement of organic beings.

These were the laws of geographic distribution. On the creationist hypothesis they made no sense whatever; they were simply arbitrary facts to be ascribed to the arbitrary will of the Creator. Why should New Zealand, with its great diversity of environmental conditions, have only four or five hundred flowering plants while the Cape of Good Hope, characterized by the monotony of its scenery, swarmed with plant species? Why should the Galapagos Islands abound with terrestrial reptiles while many equally large islands in the Pacific had few or none? Why were the islands of the open ocean destitute of native quadrupeds? Why should the living mammals in every region be clearly related to the recently extinct mammals of the same region? The creationist could only say that God willed it so. But for Darwin, as for Buffon a century earlier, the discovery of uniformities in the way the facts of nature presented themselves to the observer was but the groundwork for a larger scientific enterprise. It was, said Darwin, "absolutely opposed to every analogy, drawn from the laws imposed by the Creator on inorganic matter, that facts, when connected, should be considered as ultimate and not the direct consequences of more general laws." It remained, therefore, to show how the laws of geographical distribution might be rendered intelligible, one might almost say inevitable, on the theory of descent with gradual modification.

The theory that allied species had descended from common stocks was, in general, the hypothesis by which Buffon had attempted to account for the similarity-with-a-difference which he found in comparing the quadrupeds of the Old World and the New. But he imagined that modification had been produced by the direct influence of new environmental conditions on those quadrupeds which had wandered from the eastern hemisphere to the western. Usually, though not always, Buffon viewed the alteration thus produced as a degeneration of the original stock into smaller, less vigorous forms. Darwin, on the contrary, attributed little modifying power to the direct influence of environmental conditions. The role of the environment, as he conceived it, was twofold: first, to stimulate variation in organisms exposed to new and changing conditions, and, second, to provide conditions of relative isolation in which newly arising variations might flourish because of their superior adaptation for survival and because of the difficulty of crossing with parent forms. The Galapagos Islands were plainly in Darwin's mind as he developed an elaborate hypothetical model of geologic change accompanied by the evolution and extinction of species.

Let us now take the simplest natural case of an islet upheaved by the volcanic or subterranean forces in a deep sea, at such a distance from other land that only a few organic beings at rare intervals are transported to it, whether borne by the sea (like the seeds of plants to coral-reefs), or by hurricanes, or by floods, or on rafts, or in roots of large trees, or the germs of one plant or animal attached to or in the stomach of some other animal, or by the intervention (in most cases the most probable means) of other islands since sunk or destroyed. . . . Let this island go on slowly, century after century, rising foot by foot; and in the course of time we shall have instead (of) a small mass of rock, lowland and highland, moist woods and dry sandy spots, various soils, marshes, streams and pools: under water on the sea shore, instead of a rocky steeply shelving coast we shall have in some parts bays with mud, sandy beaches and rocky shoals. . . . It is impossible that the first few transported organisms could be perfectly adapted to all these stations; and it will be a chance if those successively transported will be so adapted. . . . Now as the first transported and any occasional successive visitants spread or tended to spread over the growing island, they would undoubtedly be exposed through several generations to new and varying conditions: it might also easily happen that some of the species *on an average* might obtain an increase of food, or food of a more nourishing quality. According then to every analogy with what we have seen takes place in every country, with nearly every organic being under domestication, we might expect that some of the inhabitants of the island would 'sport,' or have their organization rendered in some degree plastic. . . . We might therefore expect on our island that although very many slight variations were of no use to the plastic individuals, yet that occasionally in the course of a century an individual might be born of which the structure or constitution in some slight degree would allow it better to fill up some office in the insular economy and to struggle against other species. The struggle for existence would go on annually selecting such individuals until a new race or species was formed. Either few or all the first visitants to the island might become modified, according as the physical conditions of the island and those resulting from the kind and number of other transported species were different from those of the parent country — according to the difficulties offered to fresh immigration — and according to the length of time since the first inhabitants were introduced. It is obvious that whatever was the country, generally the nearest from which the first tenants were transported, they would show an affinity, even if all had become modified, to the natives of that country... (pp. 184–87).

Darwin next extended his argument to the case of several volcanic islands, and from these to a continent formed by the conjunction of several such islands. That continents had been formed in this manner and had subsequently undergone many periods of uplift and subsidence, most geologists agreed.

During the sinking of a continent and the probable generally accompanying changes of climate the effect would be little, *except* on the numerical proportions and in the extinction (from the lessening of rivers, the drying of marshes and the conversion of high-lands into low &c.) of some or of many of the species. As soon however as the continent became divided into many isolated portions or islands, preventing free immigration from one part to another, the effect of climatic and other changes on the species would be greater. But let the now broken continent, forming isolated islands, begin to rise and new stations thus to be formed, exactly as in the first case of the upheaved volcanic islet, and we shall have equally favourable conditions for the modification of old forms, that is the formation of new races or species. Let the islands become reunited into a continent; and then the new and old forms would all spread, as far as barriers, the means of transportal, and the preoccupation of the land by other species, would permit. Some of the new species or races would probably become extinct, and some perhaps would cross and blend together. We should thus have a multitude of forms, adapted to all kinds of slightly different stations, and to diverse groups of either antagonist or food-serving species. The oftener these oscillations of level had taken place (and therefore generally the older the land) the greater the number of species [which] would tend to be formed.

The inhabitants of a continent being thus derived in the first stage from the same original parents, and subsequently from the inhabitants of one wide area, since often broken up and reunited, all would be obviously related together and the inhabitants of the most *dissimilar* stations on the same continent would be more closely allied than the inhabitants of two very *similar* stations on two of the main divisions of the world (pp. 189–90).

Thus the facts of the geographic distribution of species, both living and fossil, were explained "as a simple consequence of specific forms being mutable and of their being adapted by natural selection to diverse ends, conjoined with their powers of dispersal, and the geologico-geographical changes now in slow progress and which undoubtedly have taken place." The *record* of these organic transformations was bound to be fragmentary, Darwin explained, since the periods of rapid organic change (during continental elevation) were the very periods in which few fossiliferous strata could be formed. Only during periods of subsidence would these strata be formed abundantly, hence they would be "the tomb, not of transitional forms, but of those either becoming extinct or remaining unmodified."

Darwin now turned from the facts of geographic distribution to questions related to the "natural system" of classification, the discovery of which Linnaeus had set as the ultimate goal of the natural historian. Thus far, said Darwin, the problem had proved insoluble. Early classifiers had proceeded on the assumption that the most important parts for classificatory purposes were those related to the organism's way of life. But it had soon become apparent that superficial resemblances between animals adapted for swimming or flying, such as a fish and a dolphin or a bat and a bird, afforded no safe clue to organic affinity. Lamarck and others had suggested that comparison of organs indispensable to survival was the key to the problem. But it was well known to systematists that, although comparisons of this kind were useful in some cases, in other cases characters of no great physiological importance, such as the nature of the body covering, were far more serviceable. Most systematists seemed to guide themselves by resemblances among those parts or organs which varied least in a given group. Unfortunately, however, parts or organs which varied little in one group varied widely in others. Everyone talked about "the plan of the Creator," but no two naturalists could agree what "the plan" was.

Suppose, however, that allied species and genera were descended from common ancestors. Would not the doubts and difficulties which had beset the idea of a natural system of classification vanish overnight? Once again Darwin appealed to the analogy of domestic breeds. In classifying the various kinds of cattle, sheep, dogs, cabbages, and the like, systematists encountered the same kinds of problems involved in distinguishing wild forms, but the problems were rendered soluble, at least in principle, by the knowlege that the various forms were related to each other by descent. Who could doubt that a genealogical classification of domestic varieties would be the true and "natural" system of classification? And if, as the previous argument had shown, species were but strongly marked races formed by natural selection in the struggle for existence, would not the genealogical be the natural system in the realm of nature too?

> Let us suppose for example that a species spreads and arrives at six or more different regions...exposed to different conditions, and with stations slightly different, not fully occupied with other species, so that six different races or species were formed by selection, each best fitted to its new habits and station....The races or new species supposed to be formed would be closely related to each other; and would either form a new genus or sub-genus, or would rank (probably forming a slightly different section) in the genus to which the parent species belonged. In the course of ages, and during the contingent physical changes, it is probable that some of the six new species would be destroyed; but the same advantage, whatever it may have been (whether mere tendency to vary, or some peculiarity of organization, power of mind, or means of distribution), which in the parent-species and in its six selected and changed species-offspring, caused them to prevail over other antagonist species, would generally tend to preserve some or many of them for a long period. If then, two or three of the six species were preserved, they in their turn would, during continued changes, give rise to as many small groups of species: if the parents of these small groups were closely similar, the new species would form one great genus, barely perhaps divisible into two or three sections: but if the parents were considerably unlike, their species-offspring would, from inheriting most of the peculi-

arities of their parent-stocks, form either two or more sub-genera of (if the course of selection tended in different ways) genera. And lastly species descending from different species of the newly formed genera would form new genera, and such genera collectively would form a family (pp. 208–10).

If this were the way in which natural classes, orders, families, genera, and species had actually been formed, Darwin continued, one would expect to find in nature that "unity of type" Cuvier and Candolle had discerned. One would also expect to find the metamorphoses described by Goethe, Geoffroy Saint-Hilaire and others, in which the sepals, petals, stamens, and pistils of flowers appeared as metamorphosed leaves and the vertebrate skull as a group of metamorphosed and fused vertebrae. Indeed, metamorphosis ("change of form") would no longer be a figure of speech, as it had been for all naturalists who assumed the fixity of species. Instead, it would be a term describing actual transformations which had taken place in nature as organic forms were molded by forces analogous to those involved in the evolution of domestic breeds. The facts of embryology, too, would take on new meaning. It was a curious fact, totally unintelligible in the creationist hypothesis, that the embryo of a fish resembled the embryo of a bird, reptile, or mammal more than it resembled its own parents, the resemblance being most striking in the earliest stages of embryonic development. Likewise, the larvae of the various orders of insects bore more resemblance to adult forms of the lower articulated animals than to their own parents, and the embryo of a jellyfish looked more like a polyp than like an adult jellyfish. But it was to be expected that if the various forms of vertebrates were related by descent and if natural selection operated chiefly on characters appearing in the adult organism, vertebrates would resemble each other more in the embryonic state than when fully mature, since divergence would have occurred largely in characters appearing in the adult stage. In that case, it was clear why embryological characters had proved so useful to systematists in determining the affinities of organic beings. More important still, the study of embryonic development promised eventually to disclose the order in which the various kinds of animals had appeared on earth. For if, as was probably the case, the embryo of the parent stock of the vertebrate class was itself similar to the adult animal of that stock (dissimilarity between embryo and adult having developed only gradually in the course of time), "the embryos of the existing vertebrata will shadow forth the full-grown structure of some of those forms of this great class which existed at the earlier periods of the earth's history: and accordingly, animals with a fishlike structure ought to have preceded birds and mammals; and of fish, that higher organized division with the vertebrae extending into one division of the tail ought to have preceded the equal-tailed, because the embryos of the latter have an unequal tail; and of Crustacea, entomostraca ought to have preceded the ordinary crabs and barnacles — polypes ought to have preceded jelly-fish, and infusorial animalcules to have existed before both" (p. 230).

One last kind of evidence remained to be considered: the evidence afforded by so-called "rudimentary" and "abortive" organs of plants and animals. Such were the teeth formed in the jaws of whales and rhinoceri which never emerged to serve for biting; the wings of insects never destined to be used in flight; the petals reduced to mere scales in some plants; "useless organs" of the kind Buffon had cited to prove that some things in nature had no purpose, and which Candolle had likened to empty plates placed on the banquet table of creation to preserve the symmetry of the whole display. Like his grandfather, Erasmus, Darwin was quick to see the evolutionary significance of these regularly formed but nonfunctional organs. But what his grandfather had viewed as early and imperfect attempts of nature in her steady striving onward and upward, Darwin conceived as variations produced either by disuse or by chance and preserved in succeeding generations either because they did not hurt the organism's chances of survival or because they had become adapted to positive uses quite different from those which their appearance suggested.

> In as far then as it is admitted as probable that the effects of disuse (together with occasional true and sudden abortions during the embryonic period) would cause a part to be less developed, and finally to become abortive and useless; then during the infinitely numerous changes of habits in the many descendants from a common stock, we might fairly have expected that cases of organs becom(ing) abortive would have been numerous.... Again, by gradual selection, we can see

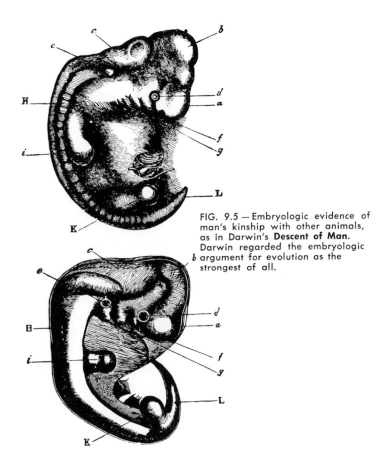

FIG. 9.5 — Embryologic evidence of man's kinship with other animals, as in Darwin's **Descent of Man.** Darwin regarded the embryologic argument for evolution as the strongest of all.

Upper figure human embryo, from Ecker. Lower figure that of a dog, from Bischoff.

a. Fore-brain, cerebral hemispheres, etc.	f. First visceral arch.
b. Mid-brain, corpora quadrigemina.	g. Second visceral arch.
c. Hind-brain, cerebellum, medulla oblongata.	H. Vertebral columns and muscles in process of development.
d. Eye.	i. Anterior ⎱ extremities.
e. Ear.	K. Posterior ⎰
	L. Tail or os coccyx.

how an organ rendered abortive in its primary use might be converted to other purposes; a duck's wing might come to serve for a fin, as does that of the penguin; an abortive bone might come to serve, by the slow increment and change of place in the muscular fibres, as a fulcrum for a new series of muscles; the pistil of the marigold might become abortive as a reproductive part, but be continued in its function of

281

sweeping the pollen out of the anthers; for if in this latter respect the abortion had not been checked by selection, the species must have become extinct from the pollen remaining enclosed in the capsules of the anthers (pp. 237–38).

Here again, the "abortion" which was only metaphorical in the creationist view of nature was an actual historical modification of organic form on the theory of descent with selective modification.

Darwin had reached the end of his evidence. It remained only to recapitulate the entire argument for his theory and to indicate the lengths to which it might be pushed. The chief argument against it, he declared, was the impossibility of demonstrating the intermediate steps by which the supposed changes of organic form had taken place. "The mind cannot grasp the full meaning of the term of a million or hundred million years, and cannot consequently add up and perceive the full effects of small successive variations accumulated during almost infinitely many generations." But the same difficulty applied in geology. There, the cumulative effect of everyday processes operating throughout millions of years had gradually been recognized and made the basis of a great new science. As in the case of geology, the strongest argument for saying that the structures of nature had been modified gradually by the relentless operation of ordinary processes throughout eons of time was that innumerable facts were made meaningful which in any other hypothesis remained disconnected and unintelligible.

> Shall we [Darwin asked] then allow that the three distinct species of rhinoceros which separately inhabit Java and Sumatra and the neighbouring mainland of Malacca were created, male and female, out of the inorganic materials of these countries? Without any adequate cause, as far as our reason serves, shall we say that they were merely, from living near each other, created very like each other, so as to form a section of the genus dissimilar from the African section, some of the species of which section inhabit very similar and some very dissimilar stations? Shall we say that without any apparent cause they were created on the same generic type with the ancient woolly rhinoceros of Siberia and of the other species which formerly inhabited the same main division of the world: that they were created, less and less closely related, but still with interbranching affinities, with all the other living and extinct mammalia? That without any apparent ade-

quate cause their short necks should contain the same number of vertebrae with the giraffe; that their thick legs should be built on the same plan with those of the antelope, of the mouse, of the hand of the monkey, of the wing of the bat, and of the fin of the porpoise. That in each of these species the second bone of their leg should show clear traces of two bones having been soldered and united into one; that the complicated bones of their head should become intelligible on the supposition of their having been formed of three expanded vertebrae; that in the jaws of each when dissected young there should exist small teeth which never come to the surface. That in possessing these useless abortive teeth, and in other characters, these three rhinoceroses in their embryonic state should much more closely resemble other mammalia than they do when mature. And lastly, that in a still earlier period of life, their arteries should run and branch as in a fish, to carry the blood to gills which do not exist.... For my own part I could no more admit that the planets move in their courses, and that a stone falls to the ground, not through the intervention of the secondary and appointed law of gravity, but from the direct volition of the Creator (pp. 249–51).

But if the theory of common descent among organic beings were admitted, how far might it be carried? Might all living creatures, man included, be derived from some lowly one-celled animal? On this question Darwin was cautious. It was true, he conceded, that the arguments favoring common descent became weaker as the species compared became more and more widely separated in the classificatory system. But if the unity of type manifested in the great divisions of nature's productions was in fact a product of descent with modification, a common origin must be assigned to all the members of the same class, thus reducing the number of prototypes to less than ten. Scientific evidence might take one so far if it were stretched to the limit. But if evidence would carry the argument no farther, scientific imagination, stirred to its depths by a new sense of the range and amplitude of nature's operations and reaching out to fathom the ultimate implications of the general principles discovered in specific facts, might soar to a breathtaking and awe-inspiring view of the consequences and corollaries locked up in the premises of the theory:

We cease to be astonished that a group of animals should have been formed to lay their eggs in the bowels and flesh of other sensitive beings; that some animals should live by and even delight in cruelty; that animals should be led away by false instinct; that annually there should be an incalculable waste of the pollen, eggs and immature beings; for we see in all this the inevitable consequences of one great law, of the multiplication of organic beings not created immutable. From death, famine, and the struggle for existence, we see that the most exalted end which we are capable of conceiving, namely, the creation of the higher animals, has directly proceeded.... There is a [simple] grandeur in this view of life with its several powers of growth, reproduction and of sensation, having been originally breathed into matter under a few forms, perhaps into only one, and that whilst this planet has gone cycling onwards according to the fixed laws of gravity and whilst land and water have gone on replacing each other — that from so simple an origin, through the selection of infinitesimal varieties, endless forms most beautiful and most wonderful have been evolved (pp. 254–55).

A great intellectual revolution had taken place since the day nearly two centuries ago when Robert Boyle had suggested that the origin of forms and qualities in the physical world might be explained in terms of the local motions of ultimate particles of matter in a lawbound system of nature. Boyle was thinking of chemical substances. He was arguing that they became what they were, not because of some indwelling form acting on the potentiality of matter, but through a "concourse of accidents" in the local motions of atoms. This way of regarding the structures of nature was gradually extended to much larger structures. In astronomy it produced Laplace's nebular hypothesis and Kant's theory of cosmic evolution. In geology it resulted in the uniformitarian doctrine of Hutton and Lyell. In biology it inspired the speculations of Buffon, Erasmus Darwin, and Lamarck, reaching full fruition in Darwin's *Origin of Species,* which was indeed an essay on the origin of forms and qualities.

If Darwin's theory was but the consummation of an intellectual movement which had taken its rise in the seventeenth century, Darwin himself showed little awareness of the fact. His point of view was stubbornly, even naively, nonhistorical. To Lyell he acknowledged a profound debt for opening his eyes to the scale of nature's operations on the surface of the globe. But with respect to his theory of species formation he recognized no precursors.

With respect to books on this subject [he wrote the great botanist Joseph Hooker] I do not know of any systematical one, except Lamarck's, which is veritable rubbish; but there are plenty, as Lyell, Pritchard, &c., on the view of immutability.... Isidore G. St. Hilaire has written some good essays tending towards the mutability side, in the 'Suites à Buffon,' entitled 'Zoolog. Generale.' Is it not strange that the author, of such a book as the 'Animaux sans Vertèbres' [Lamarck], should have written that insects, which never see their eggs, should *will*...to be of particular forms, so as to become attached to particular objects. The other, common (especially Germanic) notion is hardly less absurd, viz. that climate, food, &c. should make a Pediculus [louse] formed to climb hair, or wood-pecker, to climb trees. I believe all these absurd views arise, from no one having, as far as I know, approached the subject on the side of variation under domestication, and having studied all that is known about domestication (13).

Darwin's attitude toward Lamarck never changed. He was annoyed when Lyell used the expression "Lamarck's views improved by yours," as if the theory of natural selection were simply a modification of Lamarck's development hypothesis. "If this is your deliberate opinion there is nothing to be done about it," Darwin replied, "but it does not seem so to me. Plato, Buffon, my grandfather before Lamarck, and others, propounded the *obvious* views that if species were not created separately they must have descended from other species, and I can see nothing else in common between the 'Origin' and Lamarck. I believe this way of putting the case is very injurious to its acceptance, as it implies necessary progression, and closely connects Wallace's and my views with what I consider, after two deliberate readings, as a wretched book, and one from which (I well remember my surprise) I gained nothing" (14). This comment shows clearly that in Darwin's mind the difference between his theory and Lamarck's did *not* concern the question of the hereditary transmissibility of characters acquired during the life of an organism. As has been said, Darwin accepted this premise. Darwin objected to Lamarck on the grounds, first, that he attributed organic change to the striving of the organisms concerned; second, that he assumed a necessary progression, or upward tendency, of organic change; and, third, that he presented no substantial evidence in support of his theory. These criticisms were not without some justification. Lamarck did assume a psychological response of the organism to changes in its environment and attributed the observable progression in organic forms to the adaptive character of this response. As to whether progression was necessary, Lamarck was of two minds, as we have seen. At times he wrote of an innate tendency toward organic complication; at other times he insisted that there was no inherent direction or purpose in nature, that local circumstances determined entirely whether change would take place and in what direction. The evidence which Lamarck produced in support of his views was admittedly sketchy and unsatisfactory, and this deficiency, taken in conjunction with differences in theoretical approach, unquestionably weighed heavily in Darwin's evaluation of Lamarck. For Darwin, as he himself confessed to Lyell, was "always searching books for facts" with which to bolster his own theory of evolution. Even so, his judgment of Lamarck was too severe, for Lamarck redefined the whole science which he named *biology,* redefined it in terms of a thoroughgoing uniformitarianism. If his intuition of the possibilities inherent in this new conception outran his ability to realize them in a scientifically convincing way, he nonetheless deserved great credit for his vision and courage. Darwin was naive in supposing that any fool could postulate an evolutionary view of nature. Little did he realize the extent to which the scientific enterprise was conditioned by pre-existing conceptions of nature. Little did he suspect how important it had been for his own intellectual development that, at the very time when his suspicions concerning the stability of species were aroused, he should have had at his disposal, both in Lyell's second volume and in the works of Lamarck himself, a full-scale discussion of the idea not only that species were mutable but that the whole array of living forms had arisen from the simplest beginnings by the slow working of natural processes. If Buffon had had such a discussion available when he first began to question the fixity of species he might well have anticipated Darwin's entire theory, for his thought about the nature and mechanisms of organic change was much closer to Darwin's than was Lamarck's.

It is ironic that the one writer who more than any other before Darwin "approached the subject on the side of variation under domestication...having studied all that is known about domestication" should have remained unread by Darwin. Yet this seems to have been the case. Although Darwin refers to Geoffroy Saint-Hilaire's contribution to the *Suites à Buffon,* he seems never to have had the curiosity to go back and read Buffon himself. Thus, when he sent his theory of inheritance, which he named *pangenesis,* to Huxley for criticism, he was astonished to be told that Buffon had anticipated him by a hundred years. "It would have annoyed me extremely to have republished Buffon's views, which I did not know of," Darwin wrote to Huxley in 1865, "but I will get the book; and if I have strength I will also read Bonnet." Darwin satisfied himself that his own views on heredity were different in important respects from Buffon's, but he does not seem to have read further to discover Buffon's interest in artificial selection, even though Buffon's work and Robert Chambers' *Vestiges of the Natural History of Creation* had repeatedly been coupled with his own book by Richard Owen, "one of my chief enemies (the sole one who has annoyed me)." Darwin's refusal to regard Chambers as a precursor was largely justified, not only because Chambers' approach was strongly Lamarckian but also because Darwin had outlined his own views privately before Chambers published. But Buffon could not have been so lightly dismissed if Darwin had taken the trouble to read him with the same attention he paid to the *Vestiges.* "I am not likely to take a low view of Darwin's position in the history of science," Thomas Huxley wrote to a correspondent in 1882, "but I am disposed to think that Buffon and Lamarck would run him hard in both genius and fertility. In breadth of view and in extent of knowledge these two men were giants, though we are apt to forget their services" (15). Darwin himself conceded more than a little to his predecessors when he wrote: "Whether the naturalist believes in the views given by Lamarck, by Geoffroy St. Hilaire, by the author of the 'Vestiges,' by Mr. Wallace and myself, or in any other such view, signifies extremely little in comparison with the admission that species have descended from other species, and have not been created immutable; for he who admits this as a great truth has a wide field opened to him for further inquiry" (16).

This was the essential point, the idea of descent with modification, but it was Darwin, and Darwin alone, who made this idea scientifically respectable by showing that natural selection was a process which could form new species and by amassing a vast store of carefully weighed and sifted evidence indicating that species had in fact been so formed. What Copernicus was to Aristarchus, what Harvey was to Realdus Columbus, Cesalpino, and others who suggested but never demonstrated the circulation of the blood, Darwin was to Buffon, Lamarck, and the other pioneers of evolutionary biology.

It was this work of collecting, sifting, and pondering the evidence bearing on his theory which absorbed Darwin's energies in the fruitful years between the writing of the unpublished essay of 1844 and the publication of the *Origin of Species* in 1859. Owing to ill health, his working day was a short one, and many precious hours during eight of these years were consumed in dissecting, describing, and classifying barnacles. It was laborious, frustrating work, but, like everything Darwin did, it was grist to the species mill. "How painfully (to me) true is your remark," he wrote Joseph Hooker, "that no one has hardly a right to examine the question of species who has not minutely described many." Two things in particular became clear to Darwin as he toiled over his barnacles. One was that variation in nature was much more common than he had supposed. When the same organ was minutely compared in many barnacles of the same species, slight variations were always discovered. These were the despair of the systematist, but they were most welcome to a speculative naturalist whose theory of evolution presupposed a high degree of natural variability in organisms. The second discovery was equally welcome: in many cases it was impossible to draw the line between species and varieties with any degree of assurance. "Certainly I have felt it humiliating," Darwin confessed to Hooker, "discussing and doubting, and examining over and over again, when in my own mind the only doubt has been whether the form varied today or yesterday" (17). Whatever the case, Darwin emerged from the ordeal a trained naturalist competent to pass judgment on the species question.

When the last barnacle had been dissected, classified, and described, Darwin returned with renewed zest to the problems connected with his theory of species formation, devising ever new methods of testing its adequacy. "I suppose that I am a very slow thinker," he wrote to Lyell in 1859, "for you would be surprised at the number of years it took me to see clearly what some of the problems were which had to be solved, such as the necessity of the principle of divergence of character, the extinction of intermediate varieties, on a continuous area with graduated conditions; the double problem of sterile first crosses and sterile hybrids, &c. &c. Looking back, I think it was more difficult to see what the problems were than to solve them..." (18). Facts, facts, facts were Darwin's passion, but the facts were always viewed in relation to his grand hypothesis.

"No subject gives me so much trouble and doubt and difficulty as the means of dispersal of the same species of terrestrial productions on the oceanic islands," he confided to a friend, W. D. Fox, in October, 1856. For more than a year he had been experimenting on this question. First he tried soaking plant seeds in salt water for longer and longer periods of time to test the possibility that vegetation could be established on ocean islands by floating seeds. The seeds survived the immersion, but they refused to float. Darwin then fed soaked seeds to the fish in the aquarium of the Zoological Society of London, but while he was imagining that these fish had been swallowed by a heron and had subsequently been regurgitated on a distant island, the fish "ejected vehemently, and with disgust equal to my own, *all* the seeds from their mouths." But Darwin did not give up easily. "I find," he reported to Hooker somewhat later, "fish will greedily eat seeds of aquatic grasses, and that millet-seed put into fish and given to a stork, and then voided, will germinate.... But I am not going to give up the floating yet: in the first place I must try fresh seeds, though of course it seems far more probable that they will sink; and secondly, as a last resource, I must believe in the pod or even whole plant or branch being washed into the sea; with floods and slips and earthquakes, this must continually be happening, and if kept wet, I fancy the pods, &c. &c., would not open and shed their seeds. Do try your Mimosa seed at Kew" (19).

Lizard eggs were also set afloat in sea water in Darwin's basement, and schoolboys in neighborhoods where lizards were common were offered two shillings a dozen to collect specimens for the experiment. The geographic distribution of fresh-water mollusks proved an even knottier problem until Darwin discovered that they might be transported from place to place on the feet of aquatic birds — "when first hatched they are very active, and I have had thirty or forty crawl on a dead duck's foot; and they cannot be jerked off, and will live fifteen and even twenty-four hours out of water."

Meanwhile other theoretical difficulties were being put to practical test. To his friend Fox, Darwin wrote for assistance in determining whether the young of domestic breeds differed from each other as much as from their parents, a point closely related to his theory of embryonic resemblances. He described the progress of his experiments in pigeon breeding also: "I have found my careful work at pigeons really invaluable, as enlightening me on many points on variation under domestication. The curious old literature, by which I can trace the gradual changes in the breeds of pigeons has been extraordinarily useful to me. I have just had pigeons and fowls *alive* from Gambia! Rabbits and ducks I am attending to pretty carefully, but less so than pigeons." With another friend, a former governess in the Darwin family, Darwin was making a collection of all the plants growing in a field which had remained uncultivated for fifteen years. These would be compared with plants collected in an adjoining cultivated field "just for the fun of seeing what plants have survived or died out." Then there was Darwin's "weed garden," three by two feet square, in which he marked each seedling as it appeared and kept count of those which were destroyed by slugs or other agents.

Some problems could not be approached experimentally, especially those concerning the range, distribution, and relative numbers of species and genera. Here again Darwin enlisted the aid of naturalist friends and acquaintances, searching their publications for facts, pumping them with questions by letter or in person, setting them problems to solve and projects to work on. In botanical matters he leaned heavily on Joseph Dalton Hooker and the American botanist Asa Gray.

Hooker had first met Darwin in 1839 — "a rather tall and rather broad-shouldered man [Hooker recalled] with a slight stoop, an agreeable and animated expression when talking, beetle brows, and a hollow but mellow voice." The encounter was full of interest for Hooker; he had already heard of Darwin from Lyell and had seen some of the proof sheets of Darwin's account of the voyage of the *Beagle*. From reading these, he derived inspiration and practical instruction for his own adventure as naturalist on Sir James Ross's expedition to the

Antarctic. No sooner had Hooker returned from that expedition than Darwin wrote him, soliciting his cooperation in correlating the flora of Tierra del Fuego with those of the Cordillera and of Europe and offering his own Fuegian, Patagonian, and Galapagos collections for Hooker's inspection. Personal visits to Darwin's home at Down soon followed.

A more hospitable and more attractive home under every point of view could not be imagined [Hooker wrote] — of Society there were most often Dr. Falconer, Edward Forbes, Professor Bell, and Mr. Waterhouse — there were long walks, romps with the children on hands and knees, music that haunts me still. Darwin's own hearty manner, hollow laugh, and thorough enjoyment of home life with friends; strolls with him all together, and interviews with us one by one in his study to discuss questions in any branch of biological or physical knowledge that we had followed; and which I at any rate always left with the feeling that I had imparted nothing and carried away more than I could stagger under. Latterly, as his health became more seriously affected, I was for days and weeks the only visitor, bringing my work with me and enjoying his society as opportunity offered. It was an established rule that he every day pumped me, as he called it, for half an hour or so after breakfast in his study, when he first brought out a heap of slips with questions botanical, geographical, &c., for me to answer, and concluded by telling me of the progress he had made in his own work, asking my opinion on various points. I saw no more of him till about noon, when I heard his mellow ringing voice calling my name under my window — this was to join him in his daily forenoon walk round the sandwalk (20).

Hooker was an inexhaustible source of information to Darwin. He was also a keen and stubborn critic on whom Darwin could try his unorthodox theories with full assurance that nothing slipshod or questionable would pass muster. "You cannot imagine how pleased I am that the notion of Natural Selection has acted as a purgative on your bowels of immutability," Darwin wrote his friend in 1858. To convert Hooker would be to convert the whole scientific community in the long run. Converted he was. In 1859, a few months before the publication of the *Origin of Species,* Hooker came to Darwin's support in the "Introductory Essay" of his *Flora of Tasmania.* "The mutual relation of the plants of each great botanical province, and, in fact, of the world generally [he wrote], is just such as would have resulted if variation had gone on operating throughout indefinite periods, in the same manner as we see it act in a limited number of centuries, so as gradually to give rise in the course of time, to the most widely divergent forms." Hooker had enlisted under Darwin's banner.

Second only to Hooker as a botanical mentor and critic was Asa Gray, professor of natural history at Harvard University. Darwin had met Gray at Kew Gardens, Hooker's workshop, during one of Gray's visits to Europe. Soon after, he wrote Gray asking for information concerning American alpine plants. He was collecting facts about variation, he explained, "and when I find that any general remark seems to hold good amongst animals, I try to test it in Plants." In his essay of 1844 Darwin had noted that some alpine regions in widely separated parts of the world (North America and India, for example) contained species identical with each other and with those of Arctic regions. He had attempted to explain this phenomenon by the gradual retreat of glaciers: the Arctic species in the once-glaciated regions had retreated up the mountain slopes of those regions at the same time that they had moved northward before the advancing warmer climate. Gray was in an excellent position to help test this theory. Not only did he have full command of American flora but he had also worked through the specimens collected on Lt. Charles Wilkes' expedition to the Antarctic, various Pacific Islands, and the northwest coast of America. There were other matters which Darwin urged Gray to investigate. It would be useful, Darwin suggested, to divide the American flora into three groups: first, species found in the Old World as well as the New; second, species found only in America but belonging to genera found in the Old World; third, species belonging to genera confined to the New World.

Gray was naturally curious to know what theoretical ax Darwin was grinding on the whetstone of hard fact. In a letter written July 20th, 1856, Darwin divulged his secret.

> To be brief [he wrote], I *assume* that species arise like our domestic varieties with *much* extinction; and then test this hypothesis by comparison with as many general and pretty well-

established propositions as I can find made out, — in geographical distribution, geological history, affinities, &c., &c. And it seems to me that, *supposing* that such hypothesis were to explain such general propositions, we ought, in accordance with the common way of following all sciences, to admit it till some better hypothesis be found out. For to my mind to say that species were created so and so is no scientific explanation, only a reverent way of saying it is so and so.... But as an honest man, I must tell you that I have come to the heterodox conclusion that there are no such things as independently created species — that species are only strongly defined varieties. I know that this will make you despise me. I do not much underrate the many *huge* difficulties on this view, but yet it seems to me to explain too much, otherwise inexplicable, to be false (21).

A year later Darwin sent Gray a fuller exposition of his views. By this time he had begun, at Lyell's urgent insistence, to draw up an exposition of his theory of species formation, and the subject was much on his mind. His views were still substantially those he had recorded privately in 1844, but the letter to Gray showed some new developments and emphases. He was more than ever convinced that the physical environment was important only insofar as changes in it stimulated variability in the creatures inhabiting it. These changes were the chief cause of variability, though some variation probably took place from other causes. In the work of natural selection, however, the relation of organisms to other organisms was of much greater importance than climate and geography. In the competition among living creatures any small advantage conferred by some slight variation tipped the balance in favor of some at the expense of others. In this struggle for existence a gradual divergence of character was inevitable. Some of the varying offspring of the parent stocks would fit into the diverse niches available in the economy of nature, supplant the parent forms and win out over their brothers and sisters. "This I believe to be the origin of the classification and affinities of organic beings at all times; for organic beings always *seem* to branch and subbranch like the limbs of a tree from a common trunk, the flourishing and diverging twigs destroying the less vigorous — the dead and lost branches rudely representing extinct genera and families" (22).

While Darwin was taking Asa Gray into his confidence he was discovering also a close congeniality of theoretical views with an English naturalist engaged in exploring the zoology of the Malay Archipelago. This man was Alfred Russel Wallace, whose article in 1855 "On the Law That Has Regulated the Introduction of New Species" had stirred Lyell to warn Darwin that he must publish his theory or find himself anticipated by others. Almost two years elapsed before Wallace and Darwin began to correspond, Wallace taking the initiative. It was plain to Darwin from the beginning that Wallace's views were much like his own and that Wallace's researches would be most valuable to him in developing his theory. But he was totally unprepared for the blow which fell in June 1858, when he received from Wallace a manuscript expounding a theory of species modification by natural selection. Wallace requested Darwin to send it to Lyell to be considered for publication. The story of Darwin's dismayed but honorable reaction, his appeal to Hooker and Lyell to adjudicate the matter, their decision to have both Wallace's manuscript and certain excerpts from Darwin's essay of 1844 and his letter to Asa Gray in 1857 read before the Linnaean Society and published in their *Journal*, has been told too often to require repetition. These events had a galvanic effect on Darwin. Stirred to the depths, he abandoned the ponderous volume he had been composing and set out to make an "abstract" of it. It was this "abstract" which was published in November, 1859, under the title *On the Origin of Species by Means of Natural Selection, or the Preservation of Favoured Races in the Struggle for Life.*

The public reaction to the *Origin of Species* and the controversies to which it gave rise are by now a familiar story. Less familiar but no less significant and interesting is the discussion which Darwin carried on with his own supporters, especially Lyell and Gray, in his correspondence with them after the publication of the *Origin*. Lyell's attitude toward Darwin's theory is especially interesting, for Lyell had been among the first to befriend and encourage Darwin when he returned from the voyage of the *Beagle* laden with scientific booty and eager to find naturalists who would help him to examine it. Like Hooker, Lyell had been informed of Darwin's biological heresy shortly after Darwin committed it to paper; like Hooker, he had contributed both facts and criticisms to its subsequent development. It was Lyell who kept after Darwin to publish his views. Yet all the while Lyell had had reservations about his protegé's theory. Writing to Hooker in July, 1856, Darwin expressed surprise that such a "wonderful man" as Lyell should argue in one breath that it made no difference whether species were mutable or not and assert, in the next, that belief in their mutability would multiply specific names unnecessarily and contradict the evidence that the various types of organisms had radiated from specific geographic centers.

In 1859, the year in which the *Origin* was published, Lyell was still struggling to escape being "perverted," as he expressed it to Darwin. His objection seems to have been less to the idea of organic evolution itself (though he had tried that and found it wanting in 1832) than to Darwin's heavy emphasis on natural selection as the mechanism of evolution. Even Hooker thought Darwin too strong on this point: "You certainly make a hobby of Natural Selection, and probably ride it too hard; that is a necessity of your case. If the improvement of the creation-by-variation doctrine is conceivable, it will be by unburthening your theory of Natural Selection, which at first sight seems overstrained — i. e., to account for too much. I think, too, that some of your difficulties which you override by Natural Selection may give way before other explanations" (23).

Darwin's response to these criticisms was a respectful but unyielding reaffirmation of faith in natural selection. For years he had wrestled with the problem of explaining the marvellous adaptations of structure and function in nature on evolutionary assumptions, and it was the analogy of artificial selection which had finally given him his clue. How else could these adaptations have been produced if not by natural selection? He did not contend, however, that natural selection was the *only* process which could modify organic forms. "Can Sir Wyville Thomson name any one who has said that the evolution of species depends only on Natural Selection?" he demanded of the editor of *Nature* in 1880 in response to an article published in that magazine. "As far as concerns myself, I believe that no one has brought forward so many observations on the effects of use and disuse of parts as I have done in my *Variation of Animals and Plants under Domesti-*

cation; and these observations were made for this special object. I have likewise there adduced a considerable body of facts, showing the direct action of external conditions on organisms; though no doubt since my books were published much has been learnt on this head" (24).

As this declaration late in life shows, Darwin gradually accorded a somewhat larger role to agencies other than natural selection in the evolutionary process. Perhaps this was because Lord Kelvin's calculation of the age of the earth, granting a maximum of two hundred million years, seemed to restrict the time available for organic development. At the same time, a British engineer named Fleeming Jenkin had shown mathematically that "sports," or sudden large variations, of the kind on which Darwin had placed some reliance, could not be preserved if Darwin's theory of blending inheritance was correct. Impressed by Jenkin's argument, Darwin fell back on slight variations, or "individual differences," as the raw materials of evolution. These he conceived to occur in some degree in all organisms, but he was never willing to concede that large-scale variation could take place in nature without the stimulus of changes in the conditions of existence.

> You speak of 'an inherent tendency to vary wholly independent of physical conditions!' [he wrote Hooker] This is a very simple way of putting the case. . . but two great classes of facts make me think that all variability is due to change in the conditions of life; firstly, that there is more variability and more monstrosities (and these graduate into each other) under unnatural domestic conditions than under nature; and secondly, that changed conditions effect in an especial manner the reproductive organs — those organs which are to produce a new being. But why one seedling out of thousands presents some new character transcends the wildest power of conjecture (25).

In any case, Darwin wrote the botanist George Bentham, the argument for natural selection was grounded on general considerations rather than on absolute proofs. "When we descend to details, we can prove that no one species has changed [i. e., we cannot prove that a single species has changed]; nor can we prove that the supposed changes are beneficial, which is the groundwork of the theory" (26).

The question whether variation necessarily constituted "improvement" had been raised by Lyell in his correspondence with Darwin. Lyell had noted Darwin's pronounced tendency to identify change-by-natural-selection with "improvement" in the *Origin of Species*. "It may metaphorically be said," Darwin had written, "that natural selection is daily and hourly scrutinising, throughout the world, every variation, even the slightest; rejecting that which is bad, preserving and adding up all that is good; silently and insensibly working, *whenever and wherever opportunity offers*, at the improvement of each organic being in relation to its organic and inorganic conditions of life" (27). To this way of writing Lyell made strenuous objection. Could mere random variation accompanied by the extinction of those organisms which happened to vary in an unfortunate direction produce a trend toward "improvement" in nature? What of those species which had remained substantially unchanged throughout long geologic epochs? What of those which had retrogressed rather than progressed? If in fact there had been progress in nature, as the fossil record seemed to indicate, must not this progress depend on some "principle of improvement," some "power of adaptation," independent of and superior to mere variation-with-selection?

The question posed by Lyell was by no means new to Darwin. For years he had wrestled with the problem Buffon had propounded but had never solved: Why should change in nature move in any particular direction, and what justification was there for describing the changes observed as "degeneration" or "improvement." Were not these expressions relative to human convenience? Were they not devoid of meaning when applied to nature herself? Rejecting Lamarck's solution, which assumed a tendency toward progression and a power of adaptive response in living matter, Darwin had found in the principle of natural selection a means by which change might become directional. But on what grounds could the selected line of variation be called "improvement"? On the ground, answered Darwin, that the selected variations favored the survival of the organisms exhibiting them. But could not a variation which improved an organism's chances of survival in a given environment constitute "retrogression" when viewed against the background of organic development as a whole? Could not a line of variation which had proved advantageous to some organism at one stage of evolutionary history prove its undoing and lead to extinction in a later stage? But if mere survival was not an adequate criterion of biological progress, what other criterion was there? Naturalists talked about "higher" and "lower" forms of life, but what did they mean by these expressions? Darwin was not sure.

> With respect to 'highness' and 'lowness' [he had written Hooker in 1854], my ideas are only eclectic and not very clear. It appears to me that an unavoidable wish to compare all animals with men, as supreme, causes some confusion; and I think that nothing besides some such vague comparison is intended, or perhaps is even possible, when the question is whether two kingdoms such as the Articulata or Mollusca are the highest. Within the same kingdom I am inclined to think that 'highest' usually means that form which has undergone most 'morphological differentiation' from the common embryo or archetype of the class; but then every now and then one is bothered (as Milne Edwards has remarked) by 'retrograde development,' i.e., the mature animal having fewer and less important organs than its own embryo. The specialisation of parts to different functions, or 'the division of physiological labour' of Milne Edwards exactly agrees (and to my mind is the best definition, when it can be applied) with what you state is your idea in regard to plants. I do not think zoologists agree in any definite ideas on this subject; and my ideas are not clearer than those of my brethren (28).

In the end, Darwin worked out a conception of "competitive highness" which seemed to suffice for the discussion of natural selection. In the general struggle for existence those forms which had had to compete with the greatest variety of other plants and animals would have risen highest in the scale of organization by natural selection. This meant that the forms inhabiting large areas (continental or oceanic) would develop faster and further than those confined to small areas where competition was less intense. Hence it was clear why the most primitive aquatic forms among living species were invariably fresh-water forms; in the more strenuous competition of the ocean they had lost out, surviving only where they had managed to adapt themselves to fresh-water existence. It was equally clear why European and Indian plants and animals spread rapidly when introduced into Australia and New Zealand, whereas flora and fauna from those regions had little success when introduced into Europe.

On our theory of Natural Selection [Darwin wrote Hooker in 1858], if the organisms of any area belonging to the Eocene or Secondary periods were put into competition with those now existing in the same area (or probably in any part of the world) they (i.e. the old ones) would be beaten hollow and be exterminated; if the theory be true, this must be so. In the same manner, I believe, a greater number of the productions of Asia, the largest territory in the world, would beat those of Australia, than conversely. So it seems to be between Europe and North America....But this sort of highness (I wish I could invent some expression, and must try to do so) is different from highness in the common acceptation of the word. It might be connected with degradation of organisation; thus the blind degraded worm-like snake *(Typhlops)* might supplant the true earthworm. Here then would be degradation in the class, but certainly increase in the scale of organisation in the general inhabitants of the country....I do not see how this 'competitive highness' can be tested in any way by us. And this is a comfort to me when mentally comparing the Silurian and Recent organisms. Not that I doubt a long course of 'competitive highness' will ultimately make the organisation higher in every sense of the word; but it seems most difficult to test it (29).

Thus Darwin was ready for Lyell's objection when it was raised shortly before the publication of the *Origin of Species.*

When you contrast natural selection and 'improvement,' you seem always to overlook (for I do not see how you can deny) that every step in the natural selection of each species implies improvement in that species in relation to its conditions of life. No modification can be selected without it be an improvement or advantage. Improvement implies, I suppose, each form obtaining many parts or organs, all excellently adapted for their functions. As each species is improved, and as the number of forms will have increased, if we look to the whole course of time, the organic condition of life for other forms will become more complex, and there will be a necessity for other forms to become improved, or they will be exterminated; and I can see no limit to this process of improvement, without the intervention of any other and direct principle of improvement. All this seems to me quite compatible with certain forms fitted for simple conditions, remaining unaltered, or being degraded.

If I have a second edition, I will reiterate 'Natural Selection,' and, as a general consequence, Natural Improvement (30).

It appears, then, that Darwin endeavoured to define improvement or "highness" in terms of fitness for survival. But fitness was to be judged not only in terms of the competition which the organism faced in its actual situation but also in terms of its chances of survival in competition with other organisms of a similar kind in other parts of the world. The highest form of a given kind would be that which was capable of competing successfully with any rival form, past or present, and the capacity so to succeed was presumed to be the cumulative product of innumerable past competitions. But what of forms which were not in competition with each other to any considerable degree? How could their relative highness or lowness be judged? Was an ape higher than a turnip? If so, the criterion of highness was obviously not one of ability to survive a competition between the two forms. As Darwin sensed, there was an a priori conviction that man was the highest form of life on earth and that his rank was not dependent on mere ability to survive. Natural selection might be the *cause* of the progress in nature revealed by the fossil record, but capacity to survive was to prove an inadequate criterion of the degree of progressiveness of the organisms concerned.

The difficulty inherent in attempting to rid biology of normative concepts incapable of definition in purely biological terms became even more evident when Darwin and others tried to find a substitute for the term *natural selection*. Asa Gray and Alfred Russel Wallace objected to the expression because it seemed to imply an intelligent agent selecting according to pre-established standards. "I have been so repeatedly struck by the utter inability of numbers of intelligent persons to see clearly, or at all, the self-acting and necessary effects of Natural Selection," Wallace wrote Darwin, "that I . . . wish to suggest to you the possibility of entirely avoiding this source of misconception in your great work (if not now too late), and also in any future editions of the Origin, . . . by adopting Spencer's term (which he generally uses in preference to Natural Selection) — viz., 'survival of the fittest' " (31). Darwin obligingly introduced the Spencerian phrase alongside his own in later editions of the *Origin*, but "survival of the fittest" was to mislead the public far more grievously than "natural selection" ever had. By 1890 Huxley was lamenting the "unlucky substitution" of Spencer's phrase for Darwin's. "Fittest" seemed to imply some kind of moral excellence. Even Darwin and Spencer, although they protested that the fittest was not always the best, continually used the word in a way implying value judgments. They did so because in their heart of hearts they believed that the processes of nature operated, however slowly and sporadically, to produce ever higher forms of existence. As naturalists they tried to define "improvement," "fitness," "highness," and the like in biological terms, but their use of the terms was subtly colored by the indomitable optimism of their age. The nineteenth century believed in progress, but it was not very careful to define what it meant by progress.

Lyell and Asa Gray believed in progress too, but they were reluctant to grant that it could result from nothing more than chance variation and the elimination of organisms which did not happen to vary in a fortunate way. There must be some principle of improvement, some tendency toward progress, over and beyond the processes of variation and elimination. Sir John Herschel, no less eminent an astronomer than his father had been, was of the same opinion. To him it seemed that Darwin had reduced evolution to "the law of higglety-pigglety." Venturing into the realm of biology in his *Physical Geography of the Globe,* Herschel declared:

> We can no more accept the principle of arbitrary and casual variation and natural selection as a sufficient account, *per se,* of the past and present organic world, than we can receive the Laputan method of composing books. . .as a sufficient one of Shakespeare and the *Principia.* Equally in either case an intelligence, guided by a purpose, must be continually in action to bias the directions of the steps of change — to regulate their amount, to limit their divergence, and to continue them in a definite course. We do not believe that Mr. Darwin means to deny the necessity of such intelligent direction. But it does not, so far as we can see, enter into the formula of this law, and without it we are unable to conceive how far the law can have led to the results (32) .

Darwin was not impressed by this criticism, nor by the similar arguments of Gray and Lyell. To say that the Creator controlled the course of variation in such a way as to provide needed novelties of form and structure at appropriate moments in the history of nature was to scuttle Darwin's whole theory. What need would there be for natural selection if a higher intelligence were directing the course of variation? And if those variations which met the requirements of changing conditions were providentially supplied, were those which did *not* meet them also divinely ordained?

> If you say that God ordained that at some time and place a dozen slight variations should arise [Darwin wrote Lyell], and that one of them alone should be preserved in the struggle for life and the other eleven should perish in the first or few first generations, then the saying seems to me mere verbiage. It comes to merely saying that everything that is, is ordained.... Why should you or I speak of variation as having been ordained and guided, more than does an astronomer, in discussing the fall of a meteoric stone?... Would you have him say that its fall at some particular place and time was 'ordained and guided without doubt by an intelligent cause on a preconceived and definite plan'? Would you not call this theological pedantry or display? I believe it is not [considered] pedantry in the case of species, simply because their formation has hitherto been viewed as beyond law; in fact, this branch of science is still with most people under its theological phase of development (33).

Much the same argument served Darwin in his running controversy with Asa Gray on the subject of design in nature. Darwin took the matter seriously enough to make his position public in the closing pages of *The Variation of Animals and Plants under Domestication*. Likening the work of a breeder to that of an architect erecting a building with miscellaneous fragments of stone found at the base of a cliff, he extended the analogy to the process of natural selection and argued that the variations selected, like the fragments chosen by the builder, were produced by the laws of nature without reference to their possible use for specific ends. Professor Gray to the contrary notwithstanding, they could not be regarded as having been preordained for those ends, except in the sense that everything in nature was an inevitable consequence of the laws which the Creator had imposed on matter. "It is foolish to touch such subjects," Darwin wrote privately to Hooker, "but there have been so many allusions to what I think about the part which God has played in the formation of organic beings, that I thought it shabby to evade the question..." (34).

Gray was not satisfied with Darwin's position, and Darwin himself was anything but happy with it.

With respect to the theological view of the question [he wrote Gray]. This is always painful to me. I am bewildered. I had no intention to write atheistically. But I own that I cannot see as plainly as others do, and as I should wish to do, evidence of design and beneficence on all sides of us. There seems to me too much misery in the world. I cannot persuade myself that a beneficent and omnipotent God would have designedly created the Ichneumonidae with the express intention of their feeding within the living bodies of Caterpillars, or that a cat would play with mice. Not believing this, I see no necessity in the belief that the eye was expressly designed. On the other hand, I cannot anyhow be contented to view this wonderful universe, and especially the nature of man, and to conclude that everything is the result of brute force. I am inclined to look at everything as resulting from designed laws, with the details, whether good or bad, left to the working out of what we may call chance (35).

Thus Darwin discovered, as Laplace and others had before him, that the conception of nature as a law-bound system of matter in motion led straight to stoicism. "The old argument of design in nature, as given by Paley, which formerly seemed to me so conclusive, fails, now that the law of natural selection has been discovered," Darwin wrote in his autobiography. "There seems to be no more design in the variability of organic beings and in the action of natural selection, than in the course which the wind blows. Everything in nature is the result of fixed laws." Immanuel Kant, too, had proclaimed the universal reign of law in the evolution as well as in the regulation of the great structures of nature. Kant had inferred the existence of a wise and benevolent Creator "just because nature even in chaos cannot proceed otherwise than regularly and according to order." But Darwin could not take so cheerful a view of the operations of natural law in the organic realm nor console himself, as did many of his contemporaries, with the thought that natural selection was simply God's way of insuring perpetual progress. Presumably the laws of nature implied a lawgiver, but what kind of lawgiver would achieve the adaptation of structure to function by proliferating millions of variations at random, leaving it to the environment to eliminate those which did not happen to fit? What kind of lawgiver would permit the enormous amount of suffering evident in nature? Law governed the operations of the world-machine, but the details, "whether good or bad," seemed left to chance.

So it went, around and around, in Darwin's head — law and chance, chance and law. In the static view of nature, chance and change had been the opposites of design and permanence. The forms of the species were products of intelligent design and possessed an immutability appropriate to their divine origin. Varieties, on the other hand, were products of time and circumstance, of *chance,* not in the sense that they were uncaused or independent of natural law, but rather in the sense that they were not part of the original plan of creation. But now, in the evolutionary view, change was everywhere, and everything was either law or chance depending on which way one looked at it. Adaptation of structure to function was an outcome of chance in the sense of not being specifically provided for in a preconceived plan for the economy of nature, but it was certainly not chance in the sense of being uncaused or spontaneous. There was no room for genuine chance in Darwin's view of nature. Everything, he asserted repeatedly, was the result of fixed laws. Hence, when critics charged Darwin with eliminating purpose from nature and enthroning chance in its place, Huxley rushed to his defense with the argument that so-called "chance variations" were in reality the result of unknown natural laws. To arrive at a teleological, or purposive, view of evolution, Huxley declared, one had only to suppose "that the original plan was sketched out — that the purpose was foreshadowed in the molecular arrangements out of which the animals have come."

> The teleology which supposes that the eye, such as we see it in man, or one of the higher vertebrata, [Huxley wrote], was made with the precise structure it exhibits, for the purpose of enabling the animal which possesses it to see, has undoubtedly received its death blow. Nevertheless, it is necessary to remember that there is a wider teleology which is not touched by the doctrine of Evolution, but is actually based upon the fundamental proposition of Evolution. This proposition is that the whole world, living and not living, is the result of the mutual interaction, according to definite laws, of the forces possessed by the molecules of which the primitive nebulosity of the universe was composed. If this be true, it is no less certain that the existing world lay potentially in the cosmic vapour, and that a sufficient intelligence could, from a knowledge of the properties of the molecules of that vapour, have predicted, say the state of the fauna of Britain in 1869, with

as much certainty as one can say what will happen to the vapour of the breath on a cold winter's day... (36).

Yet Huxley and Darwin repeatedly admitted that they could see no evidence of purpose in nature. Writing to William Graham about that writer's *Creed of Science,* Darwin questioned the argument that the existence of natural laws implied purpose in nature. Likewise, Huxley assured a correspondent that he could see "no trace" of moral purpose in nature. Later he took violent exception to Herbert Spencer's attempt to derive ethics from the laws of evolutionary biology. But what was a teleological view of nature which denied purpose, or *telos,* in nature? The old words had taken on new meanings. Confusion was rampant.

Oddly enough, it was precisely the element of chance variation, taking chance not simply as the reverse aspect of law but as its genuine opposite, which appealed to the American pragmatists Charles Peirce and William James as a means of deliverance from the mechanical determinism of nineteenth-century physics and chemistry — "the block universe eternal and without a history," as William James described it.

Peirce was generally sympathetic with Darwinian evolutionary thought and excited by it. Still, his interpretation of the theory of natural selection was wholly at variance with the Darwin-Huxley idea of nature as a law-bound system of matter in motion. "In biology," wrote Peirce, "that tremendous upheaval caused in 1860 by Darwin's theory of fortuitous variations was but the consequence of a theorem in probabilities, namely, the theorem that if very many similar things are subject to very many slight fortuitous variations, as much in one direction as in the opposite direction, which when they aggregate a sufficient effect upon any one of those things in one direction must eliminate it from nature, while there is no corresponding effect of an aggregate of variations in other directions, the result must, in the long run, be to produce a change of the average character of the class of things in the latter direction" (37). Peirce then went on to substitute an evolutionary conception of natural law for the idea that the laws of nature were rigid patterns of behavior imposed on matter by the Creator. He envisaged the cosmic process as a gradual growth of "concrete reasonableness" in the universe at large.

A truly evolutionary philosophy of nature [he declared], would suppose that in the beginning — infinitely remote — there was a chaos of unpersonalized feeling, which being without connection or regularity would properly be without existence. This feeling, sporting here and there in pure arbitrariness, would have started the germ of a generalizing tendency. Its other sportings would be evanescent, but this would have a growing virtue. Thus, the tendency to habit would be started; and from this, with the other principles of evolution, all the regularities of the universe would be evolved. At any time, however, an element of pure chance survives and will remain until the world becomes an absolutely perfect, rational, and symmetrical system, in which mind is at last crystallized in the infinitely distant future (38).

Likewise, William James, Henri Bergson, A. N. Whitehead and others, each in his own way, found in the idea of organic evolution the key to a new philosophy of nature in which spontaneity, novelty, creativity, and purpose had a place — a place denied them earlier in the mechanical cosmology inherited from the seventeenth century.

Whitehead, for example, pointed out that Newton's concept of the atom (a hard, impenetrable bit of matter endowed with certain properties by God "in the beginning") was, in reality, a hangover of the static view of nature. The Newtonian atom was a permanent, wisely-designed structure which participated in the world of change but remained unaffected in its essential nature by that participation. A genuinely evolutionary concept of the atom, said Whitehead, would treat it as an organism rather than as a bit of inert stuff. The atom revealed by twentieth-century physics had an organic character. The parts were mutually dependent and mutually subservient to the functioning of the whole. The whole was in equilibrium with its environment, and this equilibrium was essential to its continuance. The atom's environment included the presence of other organisms like itself. Physics studies these smaller organisms, biology the larger ones; nothing was truly "inorganic" in Whitehead's philosophy of nature. There were living things and nonliving things, but the line between them was hard to draw, the difference consisting primarily in the complexity of organization. Thus the ultimate reality was not the "local motions" of particles of inert, valueless "matter," but the cosmic process in which organisms emerged as real events in nature, each "prehending," or grasping, surrounding events into the unity of its own being at the same time that it was prehended by them. Evolution was the development of complex organisms from simple ones, evolutionary theory "nothing else than the analysis of the conditions for the formation and survival of various types of organisms" (39). The highest organisms were those in which the rest of the universe was most fully prehended. Judged by this criterion, man was the highest terrestrial organism, but the universe was not meaningless or purposeless apart from man. Every organism was itself an achievement and enjoyment of value, an individual expression of the underlying creative activity.

Darwin foresaw the future better than he knew when, in 1837, he scribbled in one of his worn notebooks that his theory would affect the whole of metaphysics. On the one hand, it gave the death blow to traditional natural theology by drawing out the ultimate implications of the Newtonian cosmology, exhibiting the adaptation of structure to function in the organic world as a necessary outcome of random variation, struggle for existence, and natural selection. It was this consequence of his theory which impressed and disturbed Darwin. To the end of his days he remained a prisoner in the rigidly deterministic system he had discerned in all the operations of nature, organic as well as inorganic.

For others, however, the theory of evolution was a means of escape from the gloomy confines of that system. Seizing on the elements of chance and development in Darwin's theory, these writers attacked the Newtonian cosmology at its roots and opened nature once more to those aspects of value, purpose, and novelty which Newton and his contemporaries had eliminated from nature except insofar as they thought to find them in the original design of the Creator. The Darwinian revolution in biology was soon followed by an equally spectacular revolution in physics and cosmology. The atom of Niels Bohr was as different from Newton's atom as the "process philosophy" of Whitehead and others was from the "mechanical philosophy" of Robert Boyle and his contemporaries.

*...in fine, he [Lamarck] renounces his belief in the high
genealogy of his species, and looks forward, as if in compensation, to the future perfectibility of man in his physical,
intellectual, and moral attributes.*

LYELL, Principles of Geology, 1832

*I am sorry to say that I have no 'con-
solatory view' as to [the] dignity of
man: I am content that man will
probably advance, and care not much
whether we are looked at as mere
savages in a remotely distant future.
Many thanks for your last note —
Yours affectionately, C. Darwin*

*I have noted in a Manchester news-
paper a rather good squib, showing
that I have proved 'might is right,'
and therefore that Napoleon is right
and every cheating tradesman is also
right —*

DARWIN (to Lyell, May 4, 1860)

10.

Darwin and Adam

Y<small>OU ASK WHETHER</small> I shall discuss 'man,' " Darwin wrote to Alfred Russel Wallace late in 1857. "I think I shall avoid the whole subject, as so surrounded with prejudices; though I fully admit that it is the highest and most interesting problem for the naturalist" (1). True to his resolve, Darwin said nothing about man in the *Origin of Species* except to note that: "Light will be thrown on the origin of man and his history."

Actually, it was not Darwin, but Lyell, Huxley, and Wallace who first attempted to throw light on man's origin and early history in the decade following the publication of the *Origin*. Herbert Spencer already had led the way in his *Principles of Psychology*, published in 1855. Converted to Lamarck's "development hypothesis" fifteen years earlier by reading Lyell's account of it, Spencer attempted in this work to establish psychology as a branch of evolutionary biology. Mental processes, said Spencer, were means by which organisms became adapted to their environments and these means had undergone a progressive evolution. He argued first the similarity of mental processes and life processes in general. Then he proceeded to trace the various ways in which organisms took account of their environment, from the simplest organic responses of one-celled creatures to the highest human thought processes. The progressive development of the modes of interaction between organism and environment corresponded, he said, with the progressive complication of the nervous system. This was natural, since both had developed in response to changes in the environment. Hence, not only reflex and instinctive actions but the very forms of thought, the Kantian categories of space, time, causality, and the like, had been produced by the interaction of living creatures with the conditions of existence.

> The doctrine that the connections among our ideas are determined by experience [wrote Spencer] must, in consistency, be extended not only to all the connections established by the accumulated experiences of every individual, but to all those established by the accumulated experiences of every race. The abstract law of Intelligence being, that the strength of the tendency which the antecedent of any psychical change has to be followed by its consequent, is proportionate to the persistency of the union between the external things they symbolize; it becomes the resulting law of all concrete intelligences, that the strength of the tendency for such consequent to follow its antecedent, is, other things being equal, proportionate to the number of times it has thus followed in experience. The harmony of the inner tendencies and the outer persistencies, is, in all its complications, explicable on the single principle that the outer persistencies produce the inner tendencies (2).

Darwin had not read Spencer's *Psychology* when he wrote the *Origin of Species,* but he knew of it and suspected, as he wrote Lyell, it had "a bearing on Psychology as we should look at it." However, the first treatises on man by the members of Darwin's circle were concerned with man's body rather than his mind. In 1863 Thomas Huxley published *Man's Place in Nature* and Lyell brought out his *Geological Evidences of the Antiquity of Man.* In the first of the three essays forming his contribution, Huxley reviewed the long controversy concerning the zoological classification of man, studying and weighing the evidence accumulated since Andrew Battell's account of the *pongo* and the *engeco* in 1625. In the second essay he examined the relevant anatomic and embryologic evidence and handed down a verdict favorable to Linnaeus' classification of man in the same order with the apes. Darwin's theory was then introduced as a possible explanation of the similarities and differences between man and ape. Huxley accepted the theory but explained that it would not be established beyond question until breeds incapable of generating fertile offspring when crossed with other breeds derived from the same stock had been produced by artificial selection. "But even leaving Mr. Darwin's views aside," Huxley added, "the whole analogy of natural operations furnishes so complete

310

and crushing an argument against the intervention of any but what are termed secondary causes, in the production of all the phenomena of the universe; that, in view of the intimate relations between Man and the rest of the living world, and between the forces exerted by the latter and all other forces, I can see no excuse for doubting that all are coordinated terms of Nature's great progression, from the formless to the formed — from the inorganic to the organic — from blind force to conscious intellect and will" (3).

In his third essay Huxley turned to the fossil evidence of man's early history. Two discoveries in particular excited his interest. The first of these was the Engis skull, found by Professor Philippe Schmerling in the valley of the Meuse River in Belgium alongside human artifacts and the remains of the woolly mammoth and an extinct rhinoceros. The second was the Neanderthal skull, discovered in 1857 in a limestone cave in the Neander Valley near the city of Düsseldorf. Leaving to Lyell the question of the geologic antiquity of the sites where these skulls had been found, Huxley addressed himself to the anatomic issue: "Can either be shown to fill up or diminish, to any appreciable extent, the structural interval which exists between Man and the man-like apes?" In Huxley's opinion the "facial angle" described by the Dutch anatomist Petrus Camper a century earlier could not provide a criterion for settling this question. In every case the facial angle was a resultant of many complex circumstances, "not the expression of one definite organic relation of the parts of the skull." Such a relationship Huxley found in the orientation of the various parts of the skull to the *basicranial axis,* or line drawn through the base of the skull, "on which the bones of the sides and roof of the cranial cavity, and of the face, may be said to revolve downwards and forwards or backwards, according to their position." Using this relationship as a basis of comparison, he concluded that neither the Engis nor the Neanderthal skull fell far enough outside the normal range of variation exhibited in the skulls of the various races of man to warrant regarding it as a "missing link" intermediate between man and apes. The Neanderthal skull, though more apelike than any other human skull yet known, had a capacity equal to that of Polynesian and Hottentot skulls. It was, said Huxley, "the extreme term of a series leading gradually from it to the highest and best developed of human crania."

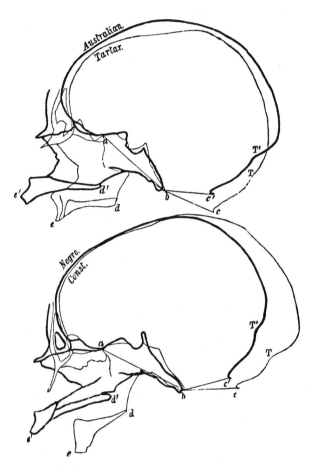

FIG. 10.1 — Huxley's method of comparing skulls, as illustrated in **Man's Place in Nature** (1863). The diagram shows sections of the skull bisected from top to bottom. In each half of the figure two such sections, one from a "round," straight-jawed, the other from a "long," jutting-jawed skull, are superimposed so that their **basicranial axes** (line ab) coincide. The other parts of the skull outline (TT', cc', etc.) are then seen to incline more or less forward or backward with reference to this base line. In the upper half of the figure a straight-jawed Australian skull is compared with a prognathous Tartar skull. In the lower half a straight-jawed Negro skull is compared with a prognathous skull of uncertain race, found in a cemetery near Constantinople. Huxley concludes from these comparisons "that the prognathous skulls, so far as their jaws are concerned, do really differ from the orthognathous in much the same way as, though to a far less degree than, the skulls of the lower mammals differ from those of Man."

Where, then, [asked Huxley] must we look for primaeval Man? Was the oldest *Homo sapiens* pliocene or miocene, or yet more ancient? In still older strata do the fossilized bones of an ape more anthropoid, or a Man more pithecoid, than any yet known await the researches of some unborn paleontologist? Time will show. But, in the meanwhile, if any form of the doctrine of progressive development is correct, we must extend by long epochs the most liberal estimate that has yet been made of the antiquity of Man (4).

Lyell came to a similar conclusion in his *Antiquity of Man* after a much fuller discussion of the geologic and paleontologic evidence. In the first nineteen chapters, Lyell described in detail the relics of early human history unearthed in various parts of the world, from the peat bogs of Denmark and the pile foundations of ancient Swiss lake-dwellings to the cave deposits of western Europe and the curious constructions of the mound builders in North America. The Danish and Swiss remains he assigned to a period four to seven thousand years old. The inhabitants of the Belgian caverns were contemporaries of now-extinct pachyderms, bears, lions, and hyenas, yet they belonged to the species *homo sapiens,* as Huxley had shown. Their inconceivably slow progress in the arts and crafts, attested by the remains found in successive geological strata, showed how bestial their condition must have been. Like Lord Monboddo a century earlier, Lyell discovered truth in the Roman poet's naturalistic picture of man's earliest progenitors — "a dumb and filthy herd, they fought for acorns and lurking-places with their nails and fists."

In the last chapters of his book Lyell turned once more to the transmutation hypothesis. He had examined and rejected it in his *Principles of Geology* thirty years earlier, but much had happened since then, and Lyell was by no means sure that his original judgment had been sound. To Darwin's great distress he insisted on treating Lamarck as the original proponent of the evolutionary hypothesis, relegating Darwin's views to the status of "recent modifications of the Lamarckian theory of progressive development and transmutation." This was entirely natural, for it was Lamarck who had first presented the evolutionary challenge to Lyell. To the very end, moreover, Lamarck's version of evolution was in many ways more palatable to Lyell than Darwin's mechanistic conception. Once again Lyell rehearsed the essentials of Lamarck's theory and his own early objections to it. He had never subscribed to Cuvier's theory of successive creations interrupted by violent catastrophes, he declared. He had always emphasized the uniformity of nature's operations on the surface of the globe, had always explained the apparent discontinuities in the fossil series as resulting from the imperfection of the geological record. In that respect he had always been on the side of the transmutationists.

Lyell then took up the Darwin-Wallace theory of evolution by natural selection, assigning proper credit to Wallace and also to Joseph Dalton Hooker for his application of the theory to the plant kingdom. Point by point he summarized Darwin's leading arguments in favor of the theory, but in so detached and noncommittal a way as to cause Darwin no little pain. One strong advantage of the theory, Lyell conceded, was that it postulated no necessary progression in nature. "It will account equally well for what is called degradation, or a retrograde movement towards a simpler structure, and does not require Lamarck's continual creation of monads; for this was a necessary part of his system, in order to explain how, after the progressive power had been at work for myriads of ages, there were as many beings of the simplest structure in existence as ever" (5) . Darwin and Hooker had shown, moreover, that the character of the flora and fauna of any given region depended more on geographical barriers to immigration than on climatic conditions. Isolation, rather than some "creative plan," seemed to be the key to the peculiar productions of Australia and New Zealand. Yet, when he had said everything for natural selection that he could honestly say, Lyell found something else at work in the development of organic life, something which defied and transcended scientific explanation — "a law of development of so high an order as to stand nearly in the same relation as the Deity himself to man's finite understanding, a law capable of adding new and powerful causes, such as the moral and intellectual faculties of the human race, to a system of nature which had gone on for millions of years without the intervention of any analogous cause." "If we confound 'Variation' or 'Natural Selection' with such creational laws," he declared, "we deify secondary causes or immeasurably exaggerate their influence" (6) .

Thus Lyell refused, in the end, to accept natural selection as an adequate cause of organic evolution or even to affirm that evolution had probably occurred. It was evident that he had been largely "perverted" on the latter issue, however, for his last chapter, dealing with man's place in nature, was largely devoted to answering hypothetical objections to the transmutation hypothesis. The absence of intermediate fossil forms between man and his supposed apelike ancestors was not conclusive, Lyell declared. Negative evidence of this sort had too often been proved valueless by subsequent discoveries. Richard Owen's attempt to prove a decisive difference between the human brain and the brains of apes had likewise fallen short of the mark. Yet it was possible to distinguish man from the apes by his moral, intellectual, and spiritual nature and his "power of progessive and improvable reason." It might have been Rousseau or Lord Monboddo rather than the Archbishop of Canterbury speaking in the passage which Lyell quoted in support of this last point:

> Animals are born what they are intended to remain. Nature has bestowed upon them a certain rank, and limited the extent of their capacity by an impassible decree. Man she has empowered and obliged to become the artificer of his own rank in the scale of beings by the peculiar gift of improvable reason (7).

Even if it should prove that man differed from the higher animals in these respects only in degree, even if it should turn out that he was in fact descended from an apelike progenitor, it was still possible, Lyell insisted, that a sudden jump forward, involving a special exertion of creative power, had occurred when a manlike creature first arose. Asa Gray had shown that evolution, even by natural selection, was not incompatible with natural theology. Indeed, said Lyell, "the supposed introduction into the earth at successive geological periods of life — sensation — instinct — the intelligence of the higher mammalia bordering on reason — and lastly the improvable reason of Man himself, presents us with a picture of the ever-increasing dominion of mind over matter."

The tendency to place human evolution in a different class from prehuman evolution was also prominent in Alfred Russel Wallace's writings on the origin of human races. In an article published in the *Journal* of the Anthropological Society of London in 1864, Wallace undertook to show that the theory of natural selection could bring to an end the long controversy between those who regarded all the races of man as varieties of one species and those who considered each race a separate species. The monogenists had carried the day in the eighteenth century, but the development of archaeology in the nineteenth century had strengthened the case of the polygenists by proving that the races of man had been as distinct from each other several thousand years ago as they were now. Early Egyptian sculptures and paintings showed Negroes and Semites with the same characteristic features familiar in modern times. Unable to deny these facts, the monogenists had availed themselves of the new evidences of the antiquity of man, arguing that many thousand years had been required for the differentiation of the races.

That man was "not a recent introduction into the earth" Wallace quite agreed. "We can with tolerable certainty affirm," he asserted, "that man must have inhabited the earth a thousand centuries ago, but we cannot assert that he positively did not exist. . .for a period of a hundred thousand centuries. We know positively that he was contemporaneous with many now extinct animals, and has survived changes of the earth's surface fifty or a hundred times greater than any that have occurred during the historical period; but we cannot place any definite limit to the number of species he may have outlived, or to the amount of terrestrial change he may have witnessed" (8). Nevertheless, said Wallace, the burden of proof lay on the monogenists. They must either produce archaeological evidence showing that differences among races dwindled as investigation reached back earlier and earlier into human history, or else, explain convincingly how the various races might have become differentiated at an extremely early epoch and yet have remained substantially unchanged since that time. It was this latter paradox which Wallace proposed to elucidate by the theory of natural selection.

In brief, Wallace argued that natural selection could have acted on man's body in any marked degree only during the period before man acquired the intellectual capacities which made him truly man. So long as man's ancestors depended on mere animal strength, agility, and cunning to make their way in the world, their bodies must have been subject, like those of other animals, to the winnowing action of natural selection. Wallace's protoman, unlike Rousseau's, did not possess the ability to perfect himself.

> By a powerful effort of the imagination [Wallace wrote], it is just possible to perceive him at that early epoch existing as a single homogeneous race without the faculty of speech, and probably inhabiting some tropical region.... He must have been even then a dominant race, spreading widely over the warmer regions of the earth as it then existed, and, in agreement with what we see in the case of other dominant species, gradually becoming modified in accordance with local conditions. As he ranged farther from his original home, and became exposed to greater extremes of climate, to greater changes of food, and had to contend with new enemies, organic and inorganic, useful variations in his constitution would be selected and rendered permanent, and would, on the principle of 'correlation of growth,' be accompanied by corresponding external physical changes. Thus arose those striking and special modifications which still distinguish the chief races of mankind (9).

At the same time this creature's mental powers would be sharpened by natural selection. Eventually these would develop to the point where he could invent tools, fashion clothing, lay snares — in short, adapt to the environment by other means than hereditary variation. At that point, said Wallace, natural selection would cease to have much influence on his bodily form. From then on, his success or failure in the struggle for survival would depend on mental and moral qualities rather than on physical factors. The various races of man, already formed by natural selection in the period before man became man, would henceforth continue with very little physical modification except insofar as the development of intellectual capacity was reflected in the shape and size of the cranium. In the mental and moral sphere, however, there would be a severe competition resulting in the spread of the best endowed races and the gradual extinction of the less gifted ones. In this competition some races would "advance and become improved merely by the harsh discipline of a sterile soil and inclement seasons," while others, inhabiting tropical regions, would stagnate from lack of environmental challenge.

It is [declared Wallace] the same great law of *'the preservation of favoured races in the struggle for life,'* which leads to the inevitable extinction of all those low and mentally undeveloped populations with which Europeans come in contact.... The intellectual and moral, as well as the physical qualities of the European are superior; the same power and capacities which have made him rise in a few centuries from the condition of the wandering savage with a scanty and stationary population to his present state of culture and advancement, with a greater average longevity, a greater average strength, and a capacity of more rapid increase, — enable him when in contact with the savage man, to conquer in the struggle for existence, and to increase at his expense, just as the more favourable increase at the expense of the less favourable varieties in the animal and vegetable kingdoms, just as the weeds of Europe overrun North America and Australia, extinguishing native productions by the inherent vigour of their organization, and by their greater capacity for existence and multiplication (10).

If this argument were sound, Wallace continued, both the monogenists and the polygenists could lay claim to a share of the truth. If by "man" was meant a creature possessing the human form but lacking the mental and moral qualities now associated with that form, then the races of "man" could probably be said to be varieties of a single species differentiated from a common stock by variation and natural selection. If, however, man's anthropoid ancestor became "man" only when he acquired the capacity to improve his condition by taking thought, the polygenists could claim an antiquity for the races of man equal to that of man himself. In Wallace's opinion, the "true grandeur and dignity of man" lay in his unique ability to transcend the law of natural selection which ruled the fates of all lower animals.

From the moment when the first skin was used as a covering, when the first rude spear was formed to assist in the chase, the first seed sown or shoot planted, a grand revolution was effected in nature, a revolution which in all the previous ages of the earth's history had had no parallel, for a being had arisen who was no longer necessarily subject to change with the changing universe — a being who was in some degree superior to nature, inasmuch, as he knew how to control and regulate her action, and could keep himself in harmony with her, not by a change in body, but by an advance of mind.... Man has not only escaped 'natural selection' himself, but he actually is able to take away some of that power from nature which, before his appearance, she universally exercised. We can anticipate the time when the earth will produce only cultivated plants and domestic animals; when man's selection shall have supplanted 'natural selection'; and when the ocean will be the only domain in which that power can be exerted, which for countless cycles of ages ruled supreme over all the earth (11).

Looking to the future, Wallace painted a picture of progressive cultural advance issuing from the steady predominance of "the more intellectual and moral" races over the "lower and more degraded" races in the conflict of cultures.

While his external form will probably ever remain unchanged, except in the development of that perfect beauty which results from a healthy and well organized body, refined and ennobled by the highest intellectual faculties and sympathetic emotions, his mental constitution may continue to advance and improve till the world is again inhabited by a single homogeneous race, no individual of which will be inferior to the noblest specimens of existing humanity. Each one will then work out his happiness in relation to that of his fellows; perfect freedom of action will be maintained, since the well balanced moral faculties will never permit any one to transgress on the equal freedom of others; restrictive laws will not be wanted, for each man will be guided by the best of laws; a thorough appreciation of the rights, and a perfect sympathy with the feelings, of all about him; compulsory government will have died away as unnecessary (for every man will know how to govern himself), and will be replaced by voluntary associations for all beneficial public purposes; the passions and animal propensities will be restrained within those limits which most conduce to happiness; and mankind will have at length discovered that it was only required of them to develope the capacities of their higher nature, in order to convert this earth, which had so long been the theatre of their unbridled passions, and the scene of unimaginable misery, into as bright a paradise as ever haunted the dreams of seer or poet (12).

Wallace was the author of this purple passage, but the ideas were derived from Herbert Spencer, the self-educated philosopher who applied the idea of natural selection to the evolution of human races several years before Darwin and Wallace first published their views. A devout believer in free, private enterprise unhampered by government regulation, Spencer had written his *Social Statics* in 1850 to show that the laissez faire policy in political and social matters was in keeping with nature's "stern discipline" for accomplishing progress in the biological realm. Just as nature insured the survival of the fittest races by subjecting all to a harsh struggle for existence, so society should compel its members to develop self-reliance, thrift, foresight, and industry by exposing them to the rigors of economic competition. By this policy the elevation of man from his original savage condition, in which he could be governed only by force and fear, to the perfect society, in which every individual would be free to do as he pleased but none would wish to harm any other, would be greatly accelerated. The discipline of economic competition, seconding the stronger discipline of racial conflict, would develop a higher breed of men capable of living without government. So ran Spencer's argument in *Social Statics,* the book which suggested to Wallace the general idea of his article and some of its particular applications (13).

Darwin was pleased with Huxley's views on man, but he was disturbed by those of Lyell and Wallace. His alarm increased when Wallace, reviewing the tenth edition of Lyell's *Principles of Geology* in 1869, asserted that neither natural selection nor the more general theory of evolution could explain the origin of conscious life or the moral and intellectual nature of man. It could not even explain such purely physical features as the human brain and hand, the organs of speech, man's smooth and hairless skin, or his erect posture and the beautiful symmetry of his body. Darwin's anxiety lest arguments of this kind should undermine confidence in the principle of natural selection may have goaded him to publish his own views, but he had other motives as well. He had intended to include a chapter on man in his *Variation of Animals and Plants under Domestication* but had found the subject too large for so brief a treatment. Exhausted from his labors on that lengthy treatise, he decided to "amuse" himself by writing a "short essay" on human origins. "I was partly led to do this by having been taunted that I concealed my views, but chiefly from the interest which I had long taken in the subject," he wrote to Alphonse de Candolle in 1868. "Now this essay has branched out into some collateral subjects, and I suppose will take more than a year to complete" (14). As usual, the work dragged on longer than Darwin expected. It was 1871 before *The Descent of Man* issued from the press. Darwin's views on the touchiest, most challenging aspect of evolutionary biology were no longer subject to doubt.

"The sole object of this work," Darwin announced, "is to consider, firstly, whether man, like every other species, is descended from some pre-existing form; secondly, the manner of his development; and thirdly, the value of the differences between the so-called races of man." On man's anatomic and embryologic resemblance to other animals Darwin elaborated but little, referring his readers to the writings of Huxley and others for further details. To these two types of evidence bearing on the origin of man's body he added a third: the various rudimentary organs, such as the vermiform appendix, the muscles which enabled some individuals to wiggle their ears, the projecting point of the ear (inherited from some ancestor with pointed ears?), the hair scattered over the human body, and the *os coccyx,* or tail bone. To Darwin it was no scientific explanation of these anatomic similarities and vestiges to say that they formed a harmonious design in the mind of the Creator. They could be explained much less metaphysically and far more simply by assuming common descent with selective modification. To take any other view was, in Darwin's opinion, "to admit that our own structure, and that of all the animals around us, is a mere snare laid to entrap our judgment."

But what about man's mental, moral, and spiritual faculties, those aspects of human nature which naturalists from the time of Tyson and Linnaeus had regarded as raising man immeasurably above the level of the brute creation? Once more Darwin appealed to the principle of gradation. If it could be shown that the difference between man and other animals in these respects was one of degree and not one of kind, it could then be argued that the more highly developed forms of intellect, esthetic sensitivity, group spirit, and the like had evolved from lower forms in the course of time. So saying, Darwin proceeded to marshal the evidence he had been collecting for many years. A close observer of animals, he had been impressed by the range of their capacity for feeling and learning. In varying degrees he had found them capable, like man, of pleasure and pain, happiness and misery, terror and shame, playfulness and boredom, courage and devotion, pride and jealousy. On the intellectual and esthetic side the higher animals exhibited a sense of wonder and curiosity, a capacity to learn by experience, an ability to communicate with each other by cries and sounds, and strong esthetic preferences. These qualities in animals should be compared, said Darwin, not with the highest manifestations of human art and intellect but with the attitudes, emotions, and thought processes of the crudest savages. He who had seen the naked Fuegians gathering limpets and mussels in the cold rain or squatting in their wretched shelters "conversing" in hoarse grunts, he who had seen one of their chiefs dash his own child against the rocks for dropping a basket of sea-urchins, would not be inclined to exaggerate the difference between the lowest man and the highest animals or reject as preposterous the suggestion that both might be descended from common ancestors.

Having prepared his reader to entertain the possibility that man's ancestry was brutish, Darwin proceeded to investigate the genealogy of man more fully. Like Huxley, he traced man's line of descent to the progenitor of the Old World monkeys, possibly some lemur-like creature from whom both the Old World and the New World monkeys had sprung. Among the Old World monkeys most naturalists distinguished the anthropoid apes as a subgroup. If this was correct (Darwin was inclined to think it was), man must have developed from some extinct member of this subgroup.

> The early progenitors of man must have been once covered with hair, both sexes having beards; their ears were probably pointed, and capable of movement; and their bodies were provided with a tail, having the proper muscles.... The foot was then prehensile, judging from the condition of the great toe in the foetus; and our progenitors, no doubt, were arboreal in their habits, and frequented some warm, forest-clad land. The males had great canine teeth, which served them as formidable weapons (15).

This early forerunner of man and the anthropoid apes probably bore no close resemblance to any existing monkey or ape, Darwin

322

warned. Africa was probably its habitat, since the gorilla and the chimpanzee, man's nearest relatives, lived there now. The discovery of an extinct ape called *Dryopithecus* showed that the higher apes had diverged from the lower apes by the time of the Upper Miocene, but the transitional forms between man and his apelike ancestor were yet to be discovered. That they would be discovered in due time Darwin had no doubt, but he did not live to see his prediction confirmed by numerous discoveries in Europe, Asia, and Africa.

Darwin had given man a pedigree "of prodigious length, but not, it may be said, of noble quality." He next undertook to show the manner of man's development from his apelike ancestor. The principles of organic evolution expounded in the *Origin of Species* were as applicable to man as to other animals, he declared. The variability of the human constitution was well known, though its causes were still uncertain. The tendency of human populations to press continually on the food supply had been noted by Malthus long ago. It followed that there must have been a struggle for existence from time to time in the long course of human history, especially at that remote period "before man had arrived at the dignity of manhood."

> Our early semi-human progenitors would not have practiced infanticide or polyandry; for the instincts of the lower animals are never so perverted as to lead them regularly to destroy their own offspring, or to be quite devoid of jealousy. There would have been no prudential restraint from marriage, and the sexes would have freely united at an early age. Hence the progenitors of man would have tended to increase rapidly; but checks of some kind, either periodical or constant, must have kept down their numbers, even more severely than with existing savages....Beneficial variations of all kinds will thus, either occasionally or habitually, have been preserved, and injurious eliminated (16).

Darwin conceded that the most beneficial variations would be those in man's intellectual faculties and social instincts, but he was far from agreeing with Wallace that natural selection could act only on man's mind after he had become a toolmaker, or that it was incapable of giving rise to man's erect position, dextrous hands, smooth skin, facial aspect, and the like. The gradual attainment of an erect posture would give man's progenitor an immense advantage in the struggle for life by freeing his hands and arms for uses other than locomotion, Darwin explained. As man began to fashion tools, those individuals would survive who possessed the best coordination of mind and muscle; the hand would undergo a progressive refinement. At the same time the jaws and teeth would diminish in size, being no longer required for attack and defense. Smaller jaw muscles would require less prominent skull processes for their attachment. The progressive development of the brain would also influence the shape of the skull. The face and cranium would take on an increasingly human aspect. To what extent these changes had resulted from natural selection, to what extent from the inherited effects of use and disuse or "the action of one part on another," Darwin found it hard to say. In the *Origin of Species* he had placed overwhelming emphasis on natural selection, perhaps because, as he himself admitted, the influence of natural theology had led him to assume that every part and organ must either have, or once have had, a use. Now, however, he was prepared to admit that structures neither beneficial nor harmful in the struggle for existence might undergo a progressive development from the continued action of the unknown causes governing their production and inheritance. In any case, Darwin had shown to his own satisfaction that many distinctively human features which Wallace and others had declared beyond the power of natural selection to produce had probably originated by that very means.

There was, said Darwin, another type of selection which must have played an important role not only in differentiating man from his anthropoid cousins but also in producing varieties in the human stock itself. This was sexual selection, or the long run genetic effect of esthetic preference in the choice of mates. James Prichard had laid great emphasis on sexual selection as a cause of racial diversity, and Darwin was closely familiar with Prichard's ideas. But Darwin investigated the subject far more thoroughly than Prichard had, extending his researches beyond man to the entire animal kingdom. The whole of Part Two of *The Descent of Man,* comprising a treatise within the larger treatise, was devoted to sexual selection in animals other than man, ranging from crustaceans to monkeys, baboons, and apes. From the last of these it was but a step, both logically and biologically, to man.

In the human species, said Darwin, sexual selection had probably produced its greatest effects at a very early period, "when man had only just attained to the rank of manhood." Judging from the present social habits of man, it was probable that he originally lived in small communities, "each with a single wife, or if powerful with several, whom he jealously guarded against all other men."

> During primordial times there would be no early betrothals, for this implies foresight. Nor would women be valued merely as useful slaves or beasts of burden. Both sexes, if the females as well as the males were permitted to exert any choice, would choose their partners not for mental charms, or property, or social position, but almost solely from external appearance. All the adults would marry or pair, and all the offspring, as far as that was possible, would be reared; so that the struggle for existence would be periodically excessively severe. Thus during these times all the conditions for sexual selection would have been more favourable than at a later period, when man had advanced in his intellectual powers but had retrograded in his instincts (17).

That the law of battle in the acquisition of choice females prevailed in those days was apparent, said Darwin, both from the practices of savage peoples and from the occasional appearance in modern man of canine teeth projecting above the other teeth, with traces of an open space for receiving the canines from the opposite jaw. The greater size and strength of man as compared to woman, his greater courage and pugnacity, his higher powers of imagination and reason — all these were undoubtedly due in large part to the struggles he had endured both in securing a mate and in the battle for life generally. These qualities would be transmitted chiefly to the male progeny, Darwin thought, but, as in mammals generally, there would be a tendency to transmit them to both sexes in some degree. Had this not been so, man's superiority over woman would have become even more pronounced than was actually the case. "Woman seems to differ from man in mental disposition, chiefly in her greater tenderness and less selfishness," Darwin declared. "It is generally admitted that with women the power of intuition, of rapid perception, and perhaps of imitation, are more strongly marked than in man; but some, at least, of these faculties are

325

characteristic of the lower races, and therefore of a past and lower state of civilization" (18). The softer, higher voice of the female and her beardlessness were probably also acquired through the advantage they conferred in the competition for male notice. But the males did not do all the choosing. The male beard and other masculine features had probably been developed or accentuated by the attraction they exerted on the most eligible females. Music and the dance probably originated in the efforts of both sexes to attract each other.

Sexual selection also provided Darwin with a clue to the origin of human races. Natural selection might explain man's gradual triumph over other anthropoid creatures, but the characteristic traits of the various human races, such as skin color, hair type, skull form, and the like, seemed to confer little or no advantage in the struggle for survival. In the competition for mates, however, they had undoubtedly played an important role, especially at the dawn of human history.

> The strongest and most vigorous men — those who could best defend and hunt for their families, who were provided with the best weapons and possessed the most property, such as a large number of dogs or other animals, — would succeed in rearing a greater average number of offspring than the weaker and poorer members of the same tribes. There can, also, be no doubt that such men would generally be able to select the more attractive women. At present the chiefs of nearly every tribe throughout the world succeed in obtaining more than one wife.... We have seen that each race has its own style of beauty, and we know that it is natural to man to admire each characteristic point in his domestic animals, dress, ornaments, and personal appearance, when carried a little beyond the average. If then the several foregoing propositions be admitted..., it would be an inexplicable circumstance if the selection of the more attractive women by the more powerful men of each tribe, who would rear on an average a greater number of children, did not after the lapse of many generations somewhat modify the character of the tribe (19).

Exactly what weight should be assigned to sexual selection as compared to natural selection, the inherited effects of use and disuse, the long-continued action of external conditions, crossing, and the like in the formation of races it was impossible to say, Darwin conceded. But since sexual selection had undoubtedly played an important role in developing special traits in other animals, it seemed likely that it had been effective with respect to man's racial peculiarities, especially those which had no apparent survival value. In any case, there could be little doubt that the various races of man had been developed from a common stock at a very early period. Whether these races were called species or subspecies or varieties made relatively little difference as long as their common descent from one anthropoid stem was accepted. The old controversy between monogenists and polygenists deserved a speedy burial now that species were seen to be nothing more than strongly marked varieties.

When Darwin turned from tracing the development of man's body to discussing his mental and cultural evolution, he took his place in the ranks of those who, like Herbert Spencer, aspired to erect a social science on biological foundations. "You will readily believe how much interested I am in observing that you apply to moral and social questions analogous views to those which I have used in regard to the modifications of species," Darwin wrote to a correspondent in 1869. It had not occurred to him, he added, that his principles could be applied so widely. This statement showed a becoming modesty, but it was scarcely true. His friend Joseph Hooker had often proposed to him "that morals and politics would be very interesting if discussed like any branch of natural history," and Darwin had agreed heartily. His *Descent of Man* showed clearly that he was a moral as well as a natural philosopher, a social as well as a natural scientist.

Like most social theorists of the nineteenth century, Darwin assumed that history was the record of man's progress from savagery to civilization. "To believe that man was aboriginally civilised and then suffered utter degradation in so many regions, is to take a pitiably low view of human nature," he wrote. "It is apparently a truer and more cheerful view that progress has been much more general than retrogression; that man has risen, though by slow and interrupted steps, from a lowly condition to the highest standard as yet attained by him in knowledge, morals and religion" (20). The problem, therefore, was to describe the steps by which man had progressed to his present condition and to explain the causes of the whole development.

Darwin's theory of social progress was similar in many respects to

Herbert Spencer's. It was Spencer who had first proposed that the evolution of man and his institutions should be viewed as a simple extension of cosmic and organic evolution, continuous with them and subject to the same general laws. In his essay entitled "Progress: Its Law and Cause," published in 1857, Spencer had discerned in the whole universe a progressive development from homogeneity to heterogeneity. Progress seemed written into the structure of things. It was "not an accident, not a thing within human control, but a beneficent necessity" (21). In human history it had come about primarily through a competition of individuals and races. Those best adapted to the changing requirements of the environment had prevailed over those less well adapted, thus setting the stage for new advances in human progress.

Darwin's reaction to Spencer's theorizing was mixed. On the one hand, he was highly suspicious of Spencer's intellectual methods. "My mind," he wrote Spencer's American disciple John Fiske, "is so fixed by the inductive method, that I cannot appreciate deductive reasoning." On the other hand, there can be little question that Darwin shared Spencer's belief in necessary, if somewhat sporadic, improvement in both nature and history and regarded natural selection as the chief engine of progress in both realms. "I cannot explain why," he wrote to Lyell in 1860, "but to me it would be an infinite satisfaction to believe that mankind will progress to such a pitch that we should [look] back at [ourselves] as mere Barbarians." That natural selection had played an important role in past progress Darwin was firmly convinced. It was natural selection which had produced man's erect posture, bipedal gait, and manual dexterity. Above all, it had acted to develop man's mental and social character, for these attributes had been decisive in the struggle for survival. "I suppose that you do not doubt that the intellectual powers are as important for the welfare of each being as corporeal structure," he wrote Lyell in 1859; "if so, I can see no difficulty in the most intellectual individuals of a species being continually selected; and the intellect of the new species thus improved, aided probably by effects of inherited mental exercise. I look at this process as now going on with the races of man; the less intellectual races being exterminated" (22).

Man had not become truly human until his mind had developed

328

to a certain point, Darwin continued. At that time man had not yet spread over the whole globe, and the races of man had not yet arisen. Although early man's intellect was probably little inferior to that of the lowest living savages, his tools were limited to the spear, the club, and perhaps the canoe. He knew the use of fire, but of language he had only the rudest beginnings, probably derived from musical sounds uttered in courtship during the mating season. As the "half-art and half-instinct" of language developed further, it stimulated the development of the brain and of the social sentiments. These, in turn, brought about group progress through imitation of the inventions and discoveries of the most gifted members of the group.

> If the invention were an important one, the tribe would increase in number, spread, and supplant other tribes. In a tribe thus rendered more numerous there would always be a rather greater chance of the birth of other superior and inventive members. If such men left children to inherit their mental superiority, the chance of the birth of still more ingenious members would be somewhat better, and in a very small tribe decidedly better. Even if they left no children, the tribe would still include their blood-relations; and it has been ascertained by agriculturists that by preserving and breeding from the family of an animal, which when slaughtered was found to be valuable, the desired character has been obtained (23).

Meanwhile, social solidarity and common morality developed by a similar process of natural selection. To Darwin it seemed likely that any animal possessing strong social instincts would acquire a moral sense if its intellectual powers became developed to the point where it was conscious of a conflict between its immediate impulses and its enduring social instincts.

> In the same manner as various animals have some sense of beauty, though they admire widely different objects, so they might have a sense of right and wrong, though led by it to follow widely different lines of conduct. If, for instance, to take an extreme case, men were reared under precisely the same conditions as hive-bees, there can hardly be a doubt that our unmarried females would, like the worker bees, think it a sacred duty to kill their brothers, and mothers

329

would strive to kill their fertile daughters; and no one would think of interfering. Nevertheless, the bee, or any other social animal, would gain in our supposed case, as it appears to me, some feeling of right or wrong, or a conscience. For each individual would have an inward sense of possessing certain stronger or more enduring instincts, and others less strong or enduring; so that there would often be a struggle as to which impulse should be followed; and satisfaction, dissatisfaction, or even misery would be felt, as past impressions were compared during their incessant passage through the mind. In this case an inward monitor would tell the animal that it would have been better to have followed the one impulse rather than the other (24).

There was no question, Darwin continued, that man had strong social instincts, probably acquired at a very early period through the advantage they conferred on the tribes possessing them. When to these instincts were added growing powers of reflection on past events and impressions, a moral sense would inevitably emerge. The individual would feel that he *ought* to obey the dictates of group sentiment; feelings of shame and guilt would follow if he did not. Thus he would acquire a conscience, whose influence would increase as the development of language strengthened group cohesion and the habit of obeying the wishes of the group was reinforced by instruction and example.

Every advance in morality and social solidarity would have survival value for the group in which it occurred, Darwin added. "A tribe including many members who, from possessing a high degree of the spirit of patriotism, fidelity, obedience, courage, and sympathy, were always ready to aid one another, and to sacrifice themselves for the common good, would be victorious over most other tribes; and this would be natural selection. At all times throughout the world tribes have supplanted other tribes; and as morality is one important element in their success, the standard of morality and the number of well-endowed men will thus everywhere tend to rise and increase" (25). Thus, for Darwin as for Spencer, human progress depended on the rise and spread of ever superior breeds of men. The most striking examples of this process were those which presented themselves wherever Western civilization impinged on savage cultures. In Tasmania, Darwin had found the few remaining aboriginal inhabitants confined under guard on a promontory of the island. In New South Wales the aborigines wandered free, and Darwin had had the good fortune to come across a band of them.

> In their own arts they are admirable [he noted in his diary]; a cap being fixed at thirty yards distance, they transfixed it with the spear delivered by the throwing stick, with the rapidity of an arrow from the bow of a practised archer; in tracking animals & men they show most wonderful sagacity & I heard many of their remarks, which manifested considerable acuteness.... Their numbers have rapidly decreased; during my whole ride with the exception of some boys brought up in the houses, I saw only one other party.... The decrease in numbers must be owing to the drinking of Spirits, the European diseases, even the milder ones of which such as the Measles, are very destructive, & the gradual extinction of the wild animals. It is said, that from the wandering life of these people, great numbers of their children die in very early infancy (26).

The problem of the causes of the decline and extinction of savage races continued to fascinate Darwin for the rest of his life. After long reflection and study he came to the conclusion that barbarous peoples declined in fertility when exposed to civilizing influences, like wild animals in a zoo. This decrease in fertility, combined with the "poor constitution" of the children, was a major cause of their progressive extinction. The problem, though difficult, was no more so than that presented by the extinction of other higher animals. "The New Zealander," Darwin wrote in *The Descent of Man,*

> seems conscious of this parallelism, for he compares his future fate with that of the native rat now almost exterminated by the European rat. Though the difficulty is great to our imagination, and really great, if we wish to ascertain the precise causes and their manner of action, it ought not to be so to our reason, as long as we keep steadily in mind that the increase of each species and each race is constantly checked in various ways; so that if any new check, even a slight one, be superadded, the race will surely decrease in number; and decreasing numbers will sooner or later lead to extinction; the end, in most cases, being promptly determined by the inroads of conquering tribes (27).

The decline of savage nations was a somber theme, but there was consolation in the reflection that their extinction signified the upward progress of mankind as a whole through the triumph of the higher over the lower varieties of the human species. Anchored safely in the cove at Sydney, Australia, Darwin had seen in the busy commercial life of this flourishing outpost of empire "a most magnificent testimony to the power of the British nation." A few decades of British industry had effected more progress in this remote country than centuries of Spanish rule had accomplished in South America. "My first feeling," Darwin wrote in his diary, "was to congratulate myself that I was born an Englishman."

In general, progress seemed to result from the competition of individuals, tribes, and races. To specify the exact causes of progress or lack of progress in any given historical situation was no easy matter, however. It was too early, Darwin thought, to attempt to explain why some barbarous peoples had progressed to a civilized state while others had remained sunk in savagery. The causes of the rise and decline of civilized nations were equally mysterious, but Darwin was sure that natural selection had played its part. The human body seemed to have been developed and strengthened, rather than debilitated, by civilization. As for the mind, it was obvious that the intellectually superior members of each class in society had the best chance of succeeding in their occupations and hence of rearing more children than the less gifted members of the same class. Unfortunately, however, there were innumerable factors which inhibited the beneficent action of natural selection in civilized societies.

> With savages [Darwin wrote], the weak in body or mind are soon eliminated; and those that survive commonly exhibit a vigorous state of health. We civilised men, on the other hand, do our utmost to check the process of elimination; we build asylums for the imbecile, the maimed, and the sick; we institute poor-laws; and our medical men exert their utmost skill to save the life of every one to the last moment. There is reason to believe that vaccination has preserved thousands, who from a weak constitution would formerly have succumbed to small-pox. Thus the weak members of civilised societies propagate their kind. No one who has attended to the breeding of domestic animals will doubt that this must be highly injurious to the race of man. It is surprising how soon a want of care, or care wrongly directed, leads to the degeneration of a domestic race; but excepting

in the case of man himself, hardly any one is so ignorant as to allow his worst animals to breed (28).

Thus Darwin joined Spencer, Wallace, Francis Galton and other prophets of doom in warning against policies which might endanger social progress by diminishing the competitive struggle which was its basic prerequisite.

> Man, like every other animal, has no doubt advanced to his present high condition through a struggle for existence consequent on his rapid multiplication [Darwin warned]; and if he is to advance still higher, it is to be feared that he must remain subject to a severe struggle. Otherwise he would sink into indolence, and the more gifted men would not be more successful in the battle of life than the less gifted. Hence our natural rate of increase, though leading to many and obvious evils, must not be greatly diminished by any means. There should be open competition for all men; and the most able should not be prevented by laws or customs from succeeding best and rearing the largest number of offspring (29).

In such passages, Darwin seemed to subscribe heart and soul to Spencer's "every - man - for - himself - and - the - Devil - take - the - hindmost" social philosophy. But he was not consistent in this view. Deep in his character there was a warm humanitarianism and a strong holdover of the Christian ethic in which he had been trained. Although he thought the moral sense had originated by natural selection of those tribes in whom the social instinct was strongest, he recognized that this primitive ethic had been gradually developed into a "higher morality" through the effects of habit, rational reflection, religious instruction, and the like. By these means, Darwin explained, man's sympathies had gradually reached out beyond the narrow confines of his own tribe, "extending to men of all races, to the imbecile, maimed, and other useless members of society, and finally to the lower animals." Not "the survival of the fittest" but "do unto others as you would have them do unto you" had come to be regarded as the true maxim of human conduct. Nor was moral progress at an end. "Looking to future generations," Darwin prophesied, "there is no cause to fear that the social instincts will grow weaker, and we may expect that virtuous habits will grow stronger, becoming perhaps fixed by inheritance. In this case the struggle between our higher and lower impulses will be less severe, and virtue will be triumphant" (30).

A happy prospect indeed, but there were serious contradictions in Darwin's account of man's advance toward moral perfection. On the one hand he subscribed to the idea that human progress resulted from a competitive struggle in which superior individuals, tribes, and races triumphed over inferior ones and drove them to the wall. "Progress" in this view of things tended to have a purely biological meaning. It consisted in the continual improvement of the hereditary endowment of the human stock. Darwin assumed, of course, that this superior endowment would manifest itself in superior achievements in art, letters, science, and morals; otherwise it would not be progress. Who could doubt that an increase in intellectual and physical vigor would produce moral and cultural improvement? This was the way progress took place.

At the same time, however, Darwin believed in a "higher morality" which taught men to respect and love their fellow men as human beings, to preserve the poor, the imbecile, and the maimed regardless of their fitness to succeed in the battle of life. Such a morality, if taken seriously, would soften or eliminate the competitive struggle and thus destroy the very basis of social progress. But how could an ethic which removed the conditions essential for future progress be regarded as a "higher" ethic? Was it not, as the German philosopher Nietzsche soon suggested, a herd morality designed to delude the gifted few and prevent them from asserting their rightful dominion over the stupid many?

Herbert Spencer had become involved in a similar dilemma. So long as he viewed human history as a slow but steady progress from the militant, regimented societies of primitive times to the peaceful, voluntary associations of the future — a progress brought about through individual and racial competition — he could regard British society of his own day as the highest development in the historical process. But when it became apparent, as it did toward the end of the nineteenth century, that industrial societies might regress toward militancy or move toward regulating business, Spencer's criterion of progress broke down. If change did not automatically constitute progress, who was to say which changes were progressive and which were not? If the militant German empire were to challenge the supremacy of industrial Britain and defeat her in combat, would not victory in battle prove the superiority of the victor's culture? And if in Britain itself, as well as in other western democracies, the growing public demand for government regulation of business were to restrict freedom of private enterprise, who was to say whether this latest development constituted progress or retrogression?

The difficulty was that biology afforded no criteria for judging the progress of a creature like man. Survival was a precondition of progress, but it did not insure progress or define its essence. In the last analysis natural selection meant not the survival of the fittest but the survival of those who survived. Survival was a brute fact, not a moral victory. Herbert Spencer never grasped this truth, but Thomas Huxley was in firm possession of it by the time of his famous Romanes lecture in 1893. Denouncing the "fanatical individualism" of those who prated about the "ethics of evolution," Huxley called for a clear recognition that the law of the jungle was no adequate law for human beings. "Let us understand, once for all," he declared, "that the ethical progress of society depends, not on imitating the cosmic process, still less in running away from it, but in combating it" (31).

The admission that nature provided no clue to man's duty and destiny left Huxley in a precarious position with respect to the basis of moral judgments. The commandments of Scripture no longer possessed divine sanction in his eyes. Yet, strangely enough, the moral vision of the Hebrew prophets, in particular Micah's injunction to do justly, love mercy, and walk humbly with God, commended itself to him as the highest expression of man's duty. Stripping the prophet's message of the religious beliefs which were inextricably associated with it in the prophet's mind, Huxley in effect fell back on a kind of moral intuitionism (32).

Darwin had less faith in the intuitions of the human mind. His social ideal, like Spencer's, was derived to a considerable extent from the utilitarianism of Adam Smith and his school, but it was strongly influenced by his Christian upbringing and his innate humanitarianism. His emphasis on religious instruction as a factor in the develop-

3 3 5

ment of a "higher morality" was rather incongruous in view of his early rejection of the Bible's claim to constitute a divine revelation and his subsequent loss of faith in the argument for God's existence from the wise design of nature. As an anthropologist, Darwin believed that religion had been born in the fears and misconceptions of primitive men. "The idea of a universal and beneficent Creator does not seem to arise in the mind of man, until he has been elevated by long-continued culture," he declared. Once this idea had arisen, he conceded, it became "a potent influence on the advance of morality." Yet the latest advances in science, to which Darwin himself had contributed mightily, seemed to undermine belief in such a Creator. In discovering the secret of man's lowly origin Darwin had lost confidence in the power of human reason and intuition to penetrate the riddle of the universe. He had, he confessed, an "inward conviction" that the universe was not the result of mere chance. "But then," he added, "with me the horrid doubt always arises whether the convictions of man's mind, which has been developed from the mind of the lower animals, are of any value or at all trustworthy. Would any one trust in the convictions of a monkey's mind, if there are any convictions in such a mind?" (33).

Here, indeed, was agnosticism — an agnosticism which trusted in the power of science to discover the origin of stars and planets, mountains and species, morality and religion, but which to all the deepest questions of the human spirit returned an *Ignoro,* followed by an *Ignorabo.* This desperate situation was only partially relieved by a compensating belief in human progress. Progress, Darwin agreed with Hooker, was "painfully slow." Worse yet, it was menaced by an awful threat — "the certainty of the sun some day cooling and we all freezing. To think of millions of years, with every continent swarming with good and enlightened men, all ending in this, and with probably no fresh start until this our planetary system has been again converted into red-hot gas. *Sic transit gloria mundi,* with a vengeance..." (34).

There was an even more terrifying prospect which Darwin never glimpsed, namely, the possibility that man might perish from the face of the earth, not by some natural catastrophe, but by his own hand, because the progress of his intellect had outrun the progress of human sympathy and understanding. That this possibility did not occur to Darwin is highly significant. It shows how implicitly he assumed that scientific and technological progress would be accompanied by moral and cultural progress. Rousseau and Lord Monboddo had been less naive. Rousseau, in particular, had sensed the morally ambiguous character of human progress. He had seen that man's ability to "perfect" himself, that is to improve his condition by extending his control over nature, did not necessarily make for human happiness and social justice. If it was true that "man makes himself" (as Lord Monboddo, anticipating the modern prehistorian V. Gordon Childe, asserted), it was also true that man was not in control of the process by which his nature was remolded. Man's increasing mastery of nature did not imply his increasing mastery of himself. To the very end man remained the creature as well as the creator of human progress. Every step upward in the knowledge and control of nature augmented his power for evil along with his power for good. Whether one expected this growing power to be used for good rather than evil depended in the last analysis on one's estimate of human nature.

Certainly Darwin had done little to raise man's estimate of his own nature. The creature whom the Psalmist viewed as "little lower than the angels" Darwin showed to be but little higher than the brutes. If man were in fact descended from the animals, what reason was there to believe that his "better instincts" or his reason would prevail over his self-regarding instincts and aggressive passions? Sigmund Freud, exploring the realm of the unconscious, could see none. In Freud's view, the progressive imposition of social controls over man's "lower nature," a process which Darwin viewed with high hope, was the source of neurosis and hence of antisocial behavior. As for "reason," it was reduced to the role of seeking to uncover the hidden conflict between the demands of society and the demands of man's biological nature in the hope that the "ego," struggling to maintain a precarious balance between the claims of the "super-ego" and those of the "id," would be able to get along in the world with a minimum of pain and frustration. That reason could discover in human nature a standard of right conduct, a clue to the good life, Freud held out no hope. Religion was an escape from reality, a kind of mass neurosis. Science could describe the eternal conflict between the erotic and the aggressive instincts, but it could offer no way of escape from the conflict either for the individual or for society.

Not all who accepted the practical view of human reason implied in Darwin's account of man's origin drew such gloomy conclusions. The "instrumentalists," led by the American philosopher John Dewey, stressed the plasticity of human nature and the power of human reason to remold it by intelligent planning of the social environment. From their "instrumental" view of reason arose the "adjustment philosophy" which has had so great a vogue in psychology and education. The neo-Darwinians — biologists, like Sir Julian Huxley, who reformulated the theory of natural selection in the light of brilliant new discoveries concerning the mechanism of heredity — took an even bolder line. Could not man, through his knowledge of hereditary processes, take control of evolution itself? "Judged by any reasonable criteria," declares the eminent geneticist Theodosius Dobzhansky, "man represents the highest, most progressive, and most successful product of organic evolution.... Most remarkable of all, he is now in the process of acquiring knowledge which may permit him, if he so chooses, to control his own evolution. He may yet become 'business manager for the cosmic process of evolution,' a role which Julian Huxley has ascribed to him, perhaps prematurely" (35).

"If he so chooses" — ah, there was the rub. A free, intelligent agent, man is in a position to plan all sorts of things, himself included. But who is to plan the planners? Who will prevent them from using the powers entrusted to them to establish a tyranny over man's mind and body? Who is to restrain man from choosing the evil which he would not do in place of the good which he would do? Is man in truth a kind of Prometheus unbound, ready and able to assume control of his own and cosmic destiny? Or is he, as the Bible represents him, a God-like creature who, having denied his creatureliness and arrogated to himself the role of Creator, contemplates his own handiwork with fear and trembling lest he reap the wages of sin, namely, death? The events of the twentieth century bear tragic witness to the realism of the Biblical portrait of man. The planned society looks less inviting in its grim reality than it did when still a dream. The conflict of nations and races, far from raising mankind to ever higher levels of virtue, freedom, and culture, threatens to accomplish the destruction of the human race. Science and technology, which were to have led the way to a bright new future, have become increasingly preoccupied with devising new and more dreadful weapons of obliteration. The historical Adam is dead, a casualty of scientific progress, but the Adam in whom all men die lives on, the creature and the creator of history, a moral being whose every intellectual triumph is at once a temptation to evil and a power for good.

Notes to Chapter 1

1. In 1673 Ray published an account of this tour entitled *Observations Topographical, Moral, & Physiological; Made in a Journey through Part of the Low-Countries, Germany, Italy, and France; with a Catalogue of Plants not Native of England, Found Spontaneously Growing in Those Parts, and Their Virtues*. The second edition, issued at London in 1738, was used.

2. John Ray, "Preface" to *Synopsis Methodica Stirpium Britannicarum...* (London: 1690), translated in Charles E. Raven, *John Ray Naturalist: His Life and Works* (Cambridge: 1942), p. 251. This is an excellent and very comprehensive biography of Ray.

3. John Ray, *The Wisdom of God Manifested in the Works of the Creation...* (3rd ed.; London: 1701), "Preface" (not paged).

4. *Ibid.*, pp. 219, 222-23.

5. *Ibid.*, p. 246. See also in this connection Richard S. Westfall's excellent study *Science and Religion in Seventeenth-Century England* (New Haven: 1958).

6. *Ibid.*, p. 44.

7. *Ibid.*, p. 65.

8. *Ibid.*, p. 56.

9. Robert Boyle, *The Origin of Forms and Qualities*, in *The Philosophical Works of the Honourable Robert Boyle...*, Peter Shaw, ed., (2nd ed.; London: 1738), I, 214-15.

10. Robert Boyle, *The Excellence and Grounds of the Mechanical Philosophy*, in *Works*, p. 187.

11. Isaac Newton, *Opticks...*, (4th ed.; London: 1730), p. 376.

Notes to Chapter 2

1. Galileo Galilei, *The Systeme of the World: in Four Dialogues*...in Thomas Salusbury, ed., *Mathematical Collections and Translations*... (London: 1661), I, 25.

2. William Whiston, *A New Theory of the Earth*... (London: 1696). Whiston was chaplain to the Bishop of Norwich at the time he published his first work. He succeeded Newton in the Lucasian Professorship in 1703 but was dismissed from it in 1710 for heterodoxy. He continued to write and lecture on astronomical and Biblical topics, aiming always at proving the concordance of science and Scripture. A severe critic of Newton's writings on chronology and Biblical prophecy, he failed election to the Royal Society largely because of Newton's opposition. He died in 1752. Three years later, the sixth edition of his *New Theory* appeared. For an account of him see his own *Memoirs of the Life and Writings of Mr. William Whiston*... (London: 1749). Katherine B. Collier, *Cosmogonies of our Fathers, Some Theories of the Seventeenth and Eighteenth Centuries* (New York: 1934), summarizes a great many cosmogonical speculations of Whiston's day. Pierre Busco, *Les Cosmogonies modernes et la théorie de la connaissance* (Paris: 1924), provides a stimulating analysis of the philosophical and epistemological presuppositions of the main cosmogonists from Descartes on. Charles Wolf, *Les Hypothèses cosmogoniques,* begins with Kant and includes a full translation of Kant's *Universal Natural History and Theory of the Heavens*. See also Hervé Faye, *Sur l'Origine du monde. Théories cosmogoniques des anciens et des modernes* (2nd ed.; Paris: 1885) Agnes M. Clerke, *Modern Cosmogonies* (London: 1905); Hector Macpherson, *Modern Cosmologies: A Historical Sketch of Researches and Theories Concerning the Structure of the Universe* (London: 1929).

3. Halley's two papers, with an explanation of his reluctance to publish them, were subsequently published in the *Philosophical Transactions*, XXXIII (1724–1725), 118–25, "at the Desire of a late Committee of the Society, who were pleased to think them not unworthy of the Press."

4. Whiston, *New Theory*, p. 26. Whiston conceived his work as an improvement on Burnet's *Sacred Theory of the Earth* (London: 1680), in which the formation of the earth was explained on Cartesian principles. For an account of Burnet's theory see page 39 ff.

5. *Ibid.*, "Discourse," p. 3. Whiston's *New Theory* begins with a lengthy "Discourse Concerning the Nature, Style, and Extent of the Mosaic History of the Creation," devoted to proving his contention concerning the scope of the Biblical narrative. The "Discourse" is paged separately.

6. *Ibid.*, pp. 59–60. See also p. 40: "Every unbyass'd Mind would easily allow, that like Effects had like Causes; and that Bodies of the same general Nature, Uses, and Motions, were to be deriv'd from the same Originals; and consequently, that the Sun and the fixed Stars had one, as the Earth, and the other Planets another sort of Formation. If therefore any free Considerer found that one of the latter sort, that Planet which we Inhabit, was deriv'd from a Chaos; by a parity of Reason he would suppose, every one of the other to be so deriv'd also; I mean each from its peculiar Chaos."

7. *Ibid.*, pp. 145–46. Whiston notes also the geologic evidences of the Noachian Deluge and refers his readers to Prof. John Woodward's *Essay Towards a Natural History of the Earth*. See page 53.

8. *Ibid.*, pp. 115–16.

9. Letter from John Locke, Oates, England, to William Molyneaux, February 22, 1696–7, quoted in Whiston, *Memoirs*, p. 44.

Chapter 2. THE INCONSTANT HEAVENS

10. Letter from Sir Isaac Newton to Dr. Thomas Burnet, no date, quoted in Sir David Brewster, *Memoirs of the Life, Writings, and Discoveries of Sir Isaac Newton* (Edinburgh and London: 1855), II, 448–49, 452–53. The spelling has been modernized somewhat. Brewster comments (page 454): "There is no signature to this letter, but the whole is distinctly written in Sir Isaac's hand, and almost without any corrections or interlineations, which is very unusual in his manuscripts." See also Whiston, *Memoirs*, page 43: "This Book [Whiston's] was shewed in MS. to Dr. Bentley, and to Sir Christopher Wren, but chiefly laid before Sir *Isaac Newton* himself, on whose Principles it depended, and who well approved of it."

11. Georges Louis Leclerc, Comte de Buffon, *A Natural History. General and Particular*...William Smellie, tr., (3rd ed.; London: 1791), I, 108. The first three volumes of the *Histoire naturelle*, published in 1749, dealt with the solar system, the earth, and man. Before his death Buffon had finished the quadrupeds and birds and had begun preparation of a volume on fishes.

12. *Ibid.*, pp. 77, 81.

13. See page 73 ff.

14. Thomas Wright, *An Original Theory or New Hypothesis of the Universe, Founded upon the Laws of Nature, and Solving by Mathematical Principles the General Phaenomena of the Visible Creation; and Particularly the Via Lactea*...(London: 1750). Of humble origin, Wright began as a clock and instrument maker and eventually acquired a considerable proficiency as a tutor, lecturer, and writer on mathematics, navigation, and astronomy. In 1742 he was offered, but declined, a professorship in the Imperial Academy of St. Petersburg. His *Original Theory* attracted some notice on the Continent but seems to have been ignored in England. In the United States it found a belated champion in the eccentric but able naturalist C. S. Rafinesque, who undertook to reprint it at Philadelphia in 1837 under the title: *The Universe and the Stars, or the Theory of the Visible and Invisible Creation.* Said Rafinesque: "We have not yet found him quoted any where, and a PHILOSOPHER AND ASTRONOMER, equal to Plato, Copernic, Newton and Herschell [sic], was to this day nearly unknown, until we found his work, and determined at once to restore him to life and fame." Actually, the scientific world paid little attention to Wright until Augustus De Morgan rehabilitated him in an article in *The Philosophical Magazine* in 1848 ("An Account of the Speculations of Thomas Wright of Durham," *The London, Edinburgh and Dublin Philosophical Magazine and Journal of Science*, XXXII [January–June, 1848], 241 ff.). F. A. Paneth, "Thomas Wright of Durham," *Endeavour*, IX (1950), 117–24, contains a good account of Wright with reproductions of some of the plates of the *Original Theory* and a bibliography of writings by Wright and about him. Edward Hughes, "The Early Journal of Thomas Wright of Durham," *Ann. Sci.*, VII (March 28, 1951), 1–24, reproduces Wright's private journal and describes some Wright manuscripts brought to light in connection with the bicentennial celebration of the publication of the *Original Theory.*

15. Thomas Wright, *Original Theory*, pp. 51–52.

16. *Ibid.*, pp. 80–81. Conceptions similar to Wright's were developed independently by the Alsatian physicist Johann Heinrich Lambert (1728–1777), in his *Kosmologische Briefe über die Einrichtung des Weltbaues*, published at Augsburg in 1761. Lambert assumed the stability and perfect contrivance of the present order of the universe and made no attempt to account for its origin. Wright offered no theory of cosmic evolution either, but he did visualize the dissolution of worlds within the present system of nature. Thus, p. 76: "In

this great Celestial Creation, the Catastrophe of a World, such as ours, or even the total Dissolution of a System of Worlds, may possibly be no more to the great Author of Nature, than the most common Accident in Life with us, and in all Probability such final and general Doom-Days may be as frequent there, as even Birth-Days, or Mortality with us upon the Earth."

17. Immanuel Kant, *Universal Natural History and Theory of the Heavens...*, in William Hastie, ed. and tr., *Kant's Cosmogony...* (Glasgow: 1900), p. 23. For a full account of the genesis, reception, and importance of Kant's theory of the heavens, see the "Translator's Introduction" to *Kant's Cosmogony.*

18. *Ibid.,* p. 145.

19. *Ibid.,* p. 26.

20. Letter from William Herschel to James Hutton, no date given, quoted in Constance A. Lubbock, ed., *The Herschel Chronicle. The Life Story of William and His Sister Caroline Herschel* (New York and Cambridge: 1933), p. 59. This is a very interesting account of the Herschels.

21. William Herschel, "On the Proper Motion of the Sun and Solar System; with an Account of Several Changes That Have Happened among the Fixed Stars Since the Time of Mr. Flamsteed," reprinted from *Philos. Trans. Roy. Soc. London,* LXXIII (1783) in *The Scientific Papers of Sir William Herschel Including Early Papers Hitherto Unpublished* (London: 1912), I, 108–30.

22. William Herschel, "Account of Some Observations Tending to Investigate the Construction of the Heavens," *Scientific Papers,* I, 158. The article appeared originally in the *Philosophical Transactions* for 1784.

23. William Herschel, "On the Construction of the Heavens," *Scientific Papers,* I, 259. Original article in the *Philosophical Transactions* for 1785.

24. *Ibid.,* p. 225. See also Herschel's "Catalogue of a Second Thousand of New Nebulae and Clusters of Stars; with a Few Introductory Remarks on the Construction of the Heavens," *Scientific Papers,* I, 329–69.

25. William Herschel, "On Nebulous Stars, Properly so Called," *Scientific Papers,* I, 421–22, originally published in the *Philosophical Transactions* for 1791. Herschel is quoting from his journal for Nov. 13, 1790. Remarking on the phenomenon, he observes: "If the point be a generating star, the further accumulation of the already condensed, luminous matter, may complete it in time."

26. William Herschel, "Astronomical Observations Relating to the Sidereal Part of the Heavens...," *Scientific Papers,* II, 541. See also his "Astronomical Observations relating to the Construction of the Heavens," in the same volume, 459–97. These two papers appeared in the *Philosophical Transactions;* the first in 1814, the second in 1811.

27. William Herschel, "On the Nature and Construction of the Sun and Fixed Stars," *Scientific Papers,* I, 484; original in the *Philosophical Transactions* for 1795. For Herschel's conversation with Napoleon and Laplace, see Lubbock, *Herschel Chronicle,* p. 310.

28. Review of Herschel's "Catalogue of a Second Thousand of New Nebulae and Clusters of Stars..." in *The Monthly Review,* II (London: 1790), 158–59. See also Lubbock, *Herschel Chronicle,* p. 196–98 on the reception of Herschel's theories.

29. Pierre-Simon Laplace, *Exposition du système du monde* (Paris: 1796), II, 296–304. Laplace expanded the hypothesis in the 3rd edition in 1808. In the 5th edition (1824), the last published during his lifetime, the nebular hypothesis was placed in a note at the end of the work, the famous Note VII.

The English translation by the Rev. Henry H. Harte (Dublin: 1830) is based on the 5th French edition.

30. P.-S. Laplace, *The System of the World,* H. H. Harte, tr., (Dublin: 1830), 332–33. In the same note Laplace acknowledges his debt to Herschel.

31. Laplace, *Exposition du système du monde,* II, 311.

Notes to Chapter 3

1. Thomas Burnet, *The Sacred Theory of the Earth...* (7th ed.; London: 1759), I, 66–67. The work was first published in Latin (London: 1681); the first English translation appeared in 1684. Besides the standard histories of geology, see Katherine Collier, *Cosmogonies of Our Fathers. Some Theories of the Seventeenth and Eighteenth Centuries (Columbia University Studies in History, Economics and Public Law,* No. 402), New York: 1934; Carl C. Beringer, *Geschichte der Geologie und des geologischen Weltbildes* (Stuttgart: 1954); Edwin T. Brewster, *This Puzzling Planet...* (Indianapolis: 1928); Ruth Moore, *The Earth We Live On* (New York: 1956); Don C. Allen, *The Legend of Noah* (Urbana, Ill.: 1949).

2. Burnet, *Sacred Theory,* I, 5–6.

3. *Ibid.,* I, 132.

4. *Ibid.,* I, 47–49. Victor Harris, *All Coherence Gone* (Chicago: 1949) discusses the prevalence of the decay-of-nature theme in seventeenth-century England.

5. Cecil Schneer, "The Rise of Historical Geology in the Seventeenth Century," *Isis,* XLV (1954), 256–68, advances an interesting hypothesis concerning the connection between antiquarian researches and the growth of interest in the history of the earth in this period. See also William Henry Fitton, "Notes on the History of English Geology," *The London, Edinburgh and Dublin Philosophical Magazine and Journal of Science,* 3rd Ser., I (1832), 147–60, 268–75, 442–50; II (1833), 37–57. See also Robert Lenoble, *La Géologie au milieu du XVIIᵉ siècle* (Paris: 1954).

6. Hooke's various papers on fossils were published in *The Posthumous Works of Robert Hooke....* Richard Waller, ed., (London: 1705), 279–328. John Ray was equally convinced that the fossils were organic remains. See Charles E. Raven, *John Ray Naturalist* (Cambridge: 1942), Chap. 16.

7. Hooke, *Posthumous Works,* p. 290. Paper read September, 1668.

8. *Ibid.,* p. 327–28. Paper read September 15, 1668.

9. *Ibid.,* p. 321. Paper read September 15, 1668.

10. *Ibid.,* p. 435–36. Paper read May 29, 1689.

11. *Ibid.,* p. 450. Paper read July 25, 1694.

12. Nicolaus Steno, *The Prodromus of Nicolaus Steno's Dissertation Concerning a Solid Body Enclosed by Process of Nature within a Solid: An English Version with an Introduction and Explanatory Notes,* John B. Winter, tr. and ed., (University of Michigan Studies, Humanistic Series, XI, Part II [Ann Arbor: Michigan, 1950]). Steno was born in Copenhagen on January 10, 1638, the son of a goldsmith. He studied at Copenhagen, Leyden, and Paris before accepting an appointment as physician to Grand Duke Ferdinand II in 1665. Ten years later he exchanged a career devoted to science for one dedicated to religion. Converted to Catholicism in 1667, he took Holy Orders in 1675, was appointed Bishop of Titopolis and Apostolic Vicar of North Germany and Scandinavia the following year, and gave himself completely to the Church

until his death in 1686. Oldenburg's translation of the *Prodromus* was reviewed in the *Philos. Trans. Roy. Soc. London,* VI (1671), 2186–87.

13. Steno, *Prodromus,* p. 263.
14. *Ibid.,* p. 269.
15. John Woodward, *An Essay Towards a Natural History of the Earth...with an Account of the Universal Deluge: and of the Effects It Had upon the Earth* (3rd ed.; London: 1723), 167. The first edition appeared in 1695.
16. *Ibid.,* p. 49.
17. *Ibid.,* p. 94.
18. Antonio Lazzaro Moro, *De' Crostacei e degli altri marini corpi che si truovano su' monti* (Venice: 1740); Bernard de Jussieu, "Examen des causes des impressions des plantes marquées sur certaines pierres des environs de Saint-Chaumont dans le Lyonnois," *Hist. Acad. Roy. Scis.,* (1718) (Paris: 1741), pp. 287–97 (Fontenelle's comments pp. 3–6); R. A. Ferchault de Réaumur, "Remarques sur les coquilles fossiles de quelques cantons de la Touraine, & sur les utilités qu'on en tire," *Hist. Acad. Roy. Scis.,* (1720), pp. 400–416 (Fontenelle's comments pp. 5–9).
19. George Louis Leclerc, Comte de Buffon, *A Natural History. General and Particular;...,*William Smellie, tr., (3rd ed.; London: 1791), I, 34.
20. In a posthumous work entitled *Telliamed: ou Entretiens d'un philosophe indien avec un missionaire français...* (1748), Benoit Demaillet (1656–1738) advanced the hypothesis that the continents had been formed in the bosom of a gradually retreating ocean. He attempted to estimate the rate of subsidence of the primitive ocean and arrived at a figure which implied that the earth was much older than generally supposed. He also suggested that land animals, including man, were derived from sea-creatures. Despite the novelty of his views and his rejection of the Deluge as an explanation of terrestrial phenomena, Demaillet was more classical than modern in his view of nature. Eternal recurrence, not evolution, was his theme. Moreover, by casting his speculations in the form of a romance he lessened their appeal to men of science. Whether Buffon drew on him is doubtful, since Buffon's theory of the earth, though not published until 1749, was composed several years earlier.
21. Buffon, *Natural History* Smellie, tr., (3rd ed.) I, 365–66.
22. *Ibid.,* I, 485.
23. *Ibid.,* I, 57–8.
24. *Ibid.,* I, 13.
25. Buffon, *Histoire naturelle, générale et particulière...,* IV (Paris: 1753), xii.
26. Johann Gottlob Lehmann, *Versuch einer Geschichte von Floetz-Gebürgen...* (Berlin: 1756).
27. *Ibid.,* pp. 84–85.
28. Jean Étienne Guettard, "Quelques Montagnes de la France qui ont été des volcans," *Hist. Acad. Roy. Scis.,* (1752), pp. 27–59. The achievements of Guettard, Desmarest, and several others treated in this chapter are vividly set forth in Sir Archibald Geikie's *The Founders of Geology* (London: 1905). See also Karl Alfred von Zittel, *History of Geology and Palaeontology to the End of the Nineteenth Century,* Maria M. Ogilvie-Gordon, tr. (London: 1901) and Frank Dawson Adams, *The Birth and Development of the Geological Sciences* (Baltimore: 1938). The Auvergne country is described with handsome engravings by G. P. Scrope in *The Geology and Extinct Volcanos of Central France,* (2nd ed.; London: 1858).
29. Nicolas Desmarest, "Sur l'Origine & la nature du basalte à grandes colonnes

Chapter 3. A WRECK OF A WORLD

polygones, determineés par l'histoire naturelle de cette pierre, observée en Auvergne," *Hist. Acad. Roy. Scis.,* (1771), pp. 705–75; "Mémoire sur le basalte. Troisième partie, ou l'on traite du basalte des anciens...," *Hist. Acad. Roy. Scis.,* (1773), pp. 599–670.

30. Nicolas Desmarest, "Extrait d'un mémoire sur la détermination de quelques époques de la nature par les produits des volcans...," *Observations sur la physique, sur l'histoire naturelle et sur les arts...,* XIII (1779), 117. The memoir was read before the Royal Academy of Sciences in 1775.

31. Sir William Hamilton, *Observations on Mount Vesuvius, Mount Etna, and Other Volcanos: in a Series of Letters, Addressed to the Royal Society...* (London: 1772), p. 160. Hamilton adds, p. 161: "May not subterraneous fire be considered as the great plough...which Nature makes use of to turn up the bowels of the earth, and afford us fresh fields to work upon, whilst we are exhausting those we are actually in possession of...?" He concludes his last letter with reflections concerning "the great changes our globe suffers, and the probability of its great antiquity."

32. Rudolf Raspe, *An Account of Some German Volcanos and Their Productions ...Published as Supplementary to Sir William Hamilton's Observations on the Italian Volcanos* (London: 1776), pp. 110–11.

33. Giovanni Arduino, *Sagio fisico-mineralogico di lythogonia, e orognosia... tratto dal tomo V. degli Atti della R. Accademia delle Scienze di Siena* [1774], in *Raccolta di memorie chimico-mineralogische, metallurgiche, e orittografische del Signor Giovanni Arduino, e di alcuni suoi amici, tratte dal Giornale d'Italia, &c...* (Venice: 1775), pp. 138–39. The ideas developed in this paper of 1774 were indicated more briefly in 1760. See *Due lettere del Signor Giovanni Arduino sopra varie sue osservazioni naturali,* in Angelo Calogerà, ed., *Nuova raccolta d'opuscoli scientifici e filologici,* VI (Venice: 1760), xcvii–clxxx.

34. Arduino, *Sagio fisico-mineralogico,* p. 201.

35. Abraham Werner, *Kurze Klassification und Beschreibung der verschiedenen gebirgsarten* (Dresden: 1787), p. 25.

36. J. F. d'Aubuisson, *An Account of the Basalts of Saxony, with Observations on the Original of Basalt in General,* P. Neill, tr. (Edinburgh: 1814), pp. 239–40. Says Aubuisson: "The remarkable reserve of Werner, has prevented him hitherto from giving to the world any full exposition of his geological doctrines; but what I have now stated, I know to accord with his sentiments." For another exposition of Werner's doctrines, see Robert Jameson, *System of Mineralogy...* (Edinburgh: 1808). III. Jameson was Werner's most devoted and influential Scotch disciple.

37. Peter Simon Pallas, *Observations sur la formation des montagnes, et les changemens arrivés à notre globe, pour servir à l'histoire naturelle de M. le Comte de Buffon* (Paris: 1782), p. 76. Zittel says that this work was first published by the St. Petersburg Academy in 1777. More famous and substantial than Pallas' work on mountains was Horace Bénédict de Saussure's *Voyages dans les Alpes,* the first volume of which appeared in 1779. Important as he is in the history of geology, Saussure contributed little to the general theory of the earth. Like Desmarest, he shunned speculation and emphasized field work. Insofar as he theorized, he leaned toward Werner's teachings, resorting to earthquakes and violent deluges to explain the presence of granite boulders on the calcareous slopes of Mount Salève and in the crevices of the Jura. In his later volumes he toyed with the idea that subterranean fires had fractured the concentric strata formed in a primeval ocean; on the whole, however, he was inclined to suppose periodic resurgences and disturbances of the ocean, caused perhaps by astronomical events affecting the inclination of the earth's

Chapter 3. A WRECK OF A WORLD

axis. See Vol. I, 150, 156, 184–5; II (1786), 118–19, 339–40, 403–4; III (1796), 107; IV (1796), 431–32.

38. Georges Louis Leclerc, Comte de Buffon, *Les Époques de la nature* (Paris: 1780), I, 3–4. The *Époques* first appeared in the fifth volume of the *Suppléments* to Buffon's *Histoire naturelle*, published in 1778.

39. *Ibid.*, I, 68–69.

40. James Hutton, "Theory of the Earth; or an Investigation of the Laws Observable in the Composition, Dissolution and Restoration of the Land upon the Globe," *Trans. Roy. Soc. Edinburgh*, I, Part II (1788), 209–304. See the *Proc. Roy. Soc. Edinburgh*, LXIII (1948–1949), Part IV, Section B, for a collection of papers about Hutton, commemorating the one hundred fiftieth anniversary of his death. Pages 380–82 reproduce an "Abstract of a Dissertation Read in the Royal Society of Edinburgh upon the Seventh of March, and Fourth of April, MDCCLXXXV, Concerning the System of the Earth, Its Duration, and Stability," which seems to have been the first printed exposition of Hutton's views.

41. Jean Deluc, "Letters to Dr. Hutton," *Monthly Review, or Literary Journal*, II (1790), 206–27, 582–95; III (1791), 573–86; V (1791), 564–85; Richard Kirwan, "Examination of the Supposed Igneous Origin of Stony Substances," *Trans. Roy. Irish Acad.*, V (1793), 51–81. See also Deluc's *An Elementary Treatise on Geology...*, the Rev. Henry de la Fite, tr. (London: 1809), and Charles C. Gillispie, *Genesis and Geology. A Study in the Relations of Scientific Thought, Natural Theology, and Social Opinion in Great Britain, 1790–1850* (Cambridge, Mass.: 1951).

42. James Hutton, *Theory of the Earth, with Proofs and Illustrations* (Edinburgh: 1795), I, 19.

43. *Ibid.*, I, 221–22.

44. *Ibid.*, I, 200.

45. *Ibid.*, II, 547.

46. *Ibid.*, II, 239.

47. *Ibid.*, I, 209.

48. Richard Kirwan, *Geological Essays* (London: 1799).

49. John Playfair, *Illustrations of the Huttonian Theory of the Earth*, in James G. Playfair, ed., *The Works of John Playfair...* (Edinburgh: 1822), I, 137. The same volume contains Playfair's biographical memoir of Hutton.

50. Hutton, *Theory of the Earth*, I, 372–73.

51. Playfair, *Works*, I, 147–48. See also his note on "Prejudices Relating to the Theory of the Earth," I, 497 ff.

52. Nicolas Desmarest, *Encyclopédie méthodique. Géographie physique* (Paris: 1794), I, 732 ff., especially 763.

53. A. Lacroix, *Déodat Dolomieu. Sa Vie aventureuse — sa captivité — ses oeuvres — sa correspondance.* (Paris: 1921), I, 1–li.

54. [John Murray], *A Comparative View of the Huttonian and Neptunian Systems of Geology: in Answer to the Illustrations of the Huttonian Theory of the Earth, by Professor Playfair* (Edinburgh, 1802), p. 255. See also d'Aubuisson, *Basalts of Saxony*, "Translator's Preface."

55. Desmarest, *Encyclopédie Méthodique. Géographie physique*, I, 249–50. For Desmarest's account of Rouelle's teaching, see *Ibid.*, I, 409 ff.

56. Letter from Dolomieu to M. Picot de Lapéyrouse, dated "Rome, end of 1788," quoted in Lacroix, *Dolomieu*, I, 213. To H. B. de Saussure, Paris, April 26,

1792 (pp. 42–43): "You have perhaps found it extraordinary that I do not accept the ideas of those who attribute to our continents an antiquity of more than one hundred thousand years. . ."

57. Letter from Dolomieu to Pierre Picot, Paris, December 19, 1796, quoted in Lacroix, *Dolomieu*, p. 132.

58. Smith's *Delineation of the Strata of England* was not published until 1815, although his ideas and discoveries were used to some extent in the Rev. Joseph Townsend's *The Character of Moses Vindicated*, published in 1813. Smith's *Strata Identified by Organized Fossils* appeared in 1816, his *Stratigraphical System of Organized Fossils* in 1817. See Thomas Sheppard, *William Smith: His Maps and Memoirs* (Hull: 1920).

59. Georges Cuvier and Alexandre Brongniart, *Essai sur la géographie minéralogique des environs de Paris, avec une carte géognostique, et des coupes de terrain* (Paris: 1811), pp. 1–2.

60. *Ibid.*, p. 253.

Notes to Chapter 4

1. John Ray, *Three Physico-Theological Discourses*. . . (3rd ed.; London: 1713), pp. 149–50. The other quotation in the paragraph is from Ray's *Travels Through the Low-Countries, Germany, Italy, and France*. . . (2nd ed.; London: 1738), I, 106.

2. Letter from Joseph Dudley, Roxbury, Mass., to the Rev. Cotton Mather, July 10, 1706, quoted in full in John C. Warren, *The Mastodon Giganteus of North America,* (2nd ed. with additions; Boston: 1855), 196–97.

3. "An Extract of Several Letters from Cotton Mather, D. D. to John Woodward, M. D. and Richard Waller, Esq.; S. R. Secr.," *Philos. Trans. Roy. Soc. London*, XXIX (1714–1716), 63.

4. Sir Hans Sloane, "An Account of Elephants' Teeth and Bones Found Underground," *Philos. Trans. Roy. Soc. London*, XXXV (1727–1728), 468.

5. Sir Hans Sloane, "Of Fossil Teeth and Bones of Elephants. Part the Second," *Philos. Trans. Roy. Soc. London*, XXXV (1727–1728), 498.

6. John P. Breyne, "A Letter from John Phil. Breyne, M. D. F. R. S. to Sir Hans Sloane, Bart. Pres. R. S. with Observations, and a Description of Some Mammoth's Bones Dug up in Siberia, Proving Them to Have Belonged to Elephants," *Philos. Trans. Roy. Soc. London*, XL (1737), 124–38.

7. Georges Louis Leclerc, Comte de Buffon, "Animaux communs aux deux continents," *Histoire naturelle, générale et particulière,*. . .in P. Flourens, ed., *Oeuvres complètes de Buffon* (Paris: 1853–1855), III, 53–54.

8. Jean M. L. Daubenton, "Mémoire sur des os et des dents remarquables par leur grandeur," *Mém. Acad. Roy. Sci. Paris,* (1762), 206–29. Buffon, *Oeuvres,* III, 247–49.

9. Letter from James Wright to John Bartram, August 22, 1762, quoted by G. G. Simpson, "The Discovery of Fossil Vertebrates in North America," *Jour. Paleon.,* XVII (1943), 36.

10. Croghan's journal of the expedition in 1765 is printed in *The Monthly American Journal of Geology and Natural Science,* I (1831), 257–72. A journal of the second expedition was kept by Capt. Harry Gordon. The relevant parts of this journal are quoted in W. R. Jillson's *Big Bone Lick. An Outline of Its History, Geology and Paleontology*. . .Big Bone Lick Association Pub-

lications: No. 1 (Louisville, Ky.: 1936), 16 ff., and in E. M. Kindle, *The Story of the Discovery of Big Bone Lick*, in *The Kentucky Geological Survey*, Ser. VI, XLI (Frankfurt, Ky.: 1931), 198 ff. The most careful and exhaustive discussion of the early fossil discoveries in the Ohio country is contained in two articles by George Gaylord Simpson: "The Beginnings of Vertebrate Paleontology in North America," *Proc. Amer. Philos. Soc.*, LXXXVI (1943), 130–88; "The Discovery of Fossil Vertebrates in North America," *Jour. Paleon.*, XVII (1943), 26–38. The accounts of Jillson and Kindle should be checked against Simpson's findings. The standard biography of Croghan is Albert T. Volwiler's *George Croghan and the Westward Movement, 1741– 1782* (Cleveland: 1926).

11. Peter Collinson, "An Account of Some Very Large Fossil Teeth, Found in North America...," *Philos. Trans. Roy. Soc. London*, LVII (1767), 464–67; "Sequel to the Foregoing Account of the Large Fossil Teeth," *ibid.*, 468–69. According to Collinson, the Shelburne collection included two large tusks, a jaw-bone with two pronged teeth in it, and several separate teeth; Franklin's included four tusks in varying condition, a vertebra, and three large, pronged teeth.

12. William Hunter, "Observations on the Bones Commonly Supposed to Be Elephant Bones, Which Have Been Found near the River Ohio in America," *Philos. Trans. Roy. Soc. London*, LVIII (1769), 45.

13. See Franklin's letter to George Croghan, August 5, 1767, and that to the Abbé Chappe, January 31, 1768, in A. H. Smyth, ed., *The Writings of Benjamin Franklin* (London: 1907), V, 39–40, 92–93.

14. Buffon, *Des Époques de la nature, Oeuvres*, IX, 455–660. For a discussion of the teeth figured by Buffon in the "Notes justificatives," see Osborn, *Proboscidea: A Monograph of the Discovery, Evolution, Migration and Extinction of the Mastodonts and Elephants of the World* (New York: 1942), I, 131 ff. See also L. P. Tolmachoff, "The Carcasses of the Mammoth and the Rhinoceros Found in the Frozen Ground of Siberia," *Trans. Amer. Philos. Soc.*, n. s. XXIII, Part I (1929), 5–74.

15. "A Description of Bones, &c. Found near the Ohio River," *Columbian Magazine*, I (November, 1786), 106. See also Samuel H. Parsons, "Discoveries Made in the Western Country, by General Parsons," *Mem. Amer. Acad. Arts Scis.*, II, Part I (1793), 119–27.

16. Robert Annan, "Account of a Large Animal Found near Hudson's River," *Mem. Amer. Acad. Arts Scis.*, II, Part I (1793), 163–64.

17. Thomas Jefferson, *Notes on Virginia*, in *The Works of Thomas Jefferson*, P. L. Ford, ed. (New York and London: 1904), III, 411.

18. *Ibid.*, III, 427.

19. Thomas Jefferson, "A Memoir on the Discovery of Certain Bones of a Quadruped of the Clawed Kind in the Western Parts of Virginia," *Trans. Amer. Philos. Soc.*, IV (1799), 255–56.

20. Georges Cuvier, "Sur le Mégatherium, autre animal de la famille des paresseux...dont un squelette fossile presque complet est conservé au cabinet royal d'histoire naturelle à Madrid," *Ann. Mus. Natl. Hist. Nat. Paris*, V (1804), 387. The same volume, pp. 358–76, contains Cuvier's "Sur le Mégalonix, animal de la famille des paresseux...dont les ossemens ont été découverts en Virginie, en 1796." For Wistar's account, see *Trans. Amer. Philos. Soc.*, IV (1799), 526–31. Prof. George Gaylord Simpson characterizes it as "a model of cautious, accurate scientific description and inference, an achievement almost incredible in view of the paleontological naiveté of his asso-

Chapter 4. LOST SPECIES

ciates and of the lack of comparative materials." See Simpson, "Beginnings of Vertebrate Paleontology," p. 153, for a thorough discussion of the discovery and interpretation of the remains of the megalonix.

21. George Turner, "Memoir on the Extraneous Fossils Denominated Mammoth Bones; Principally Designed to Show That They Are the Remains of More Than One Species of Non-Descript Animal," *Trans. Amer. Philos. Soc.*, IV (1799), 518. Turner moved to Philadelphia about 1792. He was elected curator of the American Philosophical Society in 1800 but was expelled soon after for defrauding the Society of $500.

22. Nicholas Collin, "Philological View of Some Very Ancient Words in Several Languages," *Trans. Amer. Philos. Soc.*, IV (1799), 506–7.

23. John Drayton, *A View of South Carolina, as Respects Her Natural and Civil Concerns* (Charleston: 1802), p. 46.

24. Georges Cuvier, "Mémoire sur les espèces d'éléphans vivantes et fossiles," *Mém. Inst. Nat. Scis. Arts*, II (1799), 1–22. For a history and bibliography of the nomenclature of the mammoth and the mastodont, see Osborn, *Proboscidea*, I, 165 ff, 1363 ff; II, 1117–24.

25. *Ibid.*, pp. 20–21.

26. James G. Graham, "Further Account of the Fossil Bones in Orange and Ulster Counties: in a Letter from Dr. James G. Graham, one of the Senators of the Middle District, to Dr. Mitchill; dated Shawangunk, Sept. 10, 1800," *Medical Repository*, IV (1801), 212.

27. "Medical and Philosophical News," *Medical Repository*, IV (1801), 419–20. Miller's letter to the editor appears in the same volume, p. 212.

28. Livingston's letter dated Jan. 7, 1801, is quoted by Jefferson in his reply to Wistar, dated Washington, Feb. 3, 1801, printed in Lipscomb, *The Writings of Thomas Jefferson*, X, 196.

29. Rembrandt Peale, *An Historical Disquisition on the Mammoth, or Great American Incognitum*... (London: 1803), 32–33. For another detailed account of the expedition, with quotations from C. W. Peale's diary, see Charles C. Sellers, *Charles Willson Peale* (Philadelphia: 1947), II, 127–37.

30. Concerning the Peales' stay in London, see Sellers, *op. cit.*, II, 167 ff. See also the letter written by C. Roume, private agent of the French Government in Santo Domingo and a close friend of C. W. Peale, published in *The Philosophical Magazine*, XIII (1802), 206–7. Roume was one of those invited to inspect the first skeleton with the members of the American Philosophical Society. In this letter to an unidentified person in France he describes Peale's plans for a European tour. He adds: "I beg you will communicate these details to the National Institute. Get Mr. Peale chosen as a corresponding member. You will not repent it. I have already prevailed on him to send his son first to France; and I have promised that you will give him a good reception, and present him to the Institute, who certainly will not fail to induce the first consul to purchase the skeleton which Mr. Peale jun. will bring with him, either that it may be deposited in the National Museum immediately, or after he has been allowed to exhibit it in different parts of Europe."

31. Peale, *Historical Disquisition*, pp. 90–91. See also Peale's "A Short Account of the Mammoth," *The Philosophical Magazine*, XIV (1802), 162–69.

32. The *Ann. Mus. Natl. Hist. Nat. Paris*, VIII (1806), contain the memoirs referred to: "Sur les Éléphans vivans et fossiles," 1–58, 93–155, 249–69; "Sur le Grand Mastodonte...," 270–312; "Sur Différentes Dents du genre des mastodontes...," 401–20; "Résumé général de l'histoire des ossemens fossiles

351

de pachydermes...," 420–24. Concerning the discovery of a woolly mammoth carcass in 1806 see: Michael Adams, "Some Account of a Journey to the Frozen Sea, and of the Discovery of the Remains of a Mammoth, *"The Philosophical Magazine,* XXIX (1807–1808), 141–53; Tolmachoff, *op. cit.,* pp. 23–24; Osborn, *Proboscidea,* II, 1148; Wilhelm Gottlieb Tilesius von Tilnau, "De skeleto mammonteo Sibirico ad maris glacialis littora anno 1807, effosso, cui praemissae elephantini generis specierum distinctiones," *Mém. Acad. Imp. Sci. St. Petersburg,* V (1812), 406–513; Henry H. Howorth, *The Mammoth and the Flood...* (London: 1887). It should be noted that the Peales were not the only Americans who provided Cuvier and his European colleagues with information and specimens. Jefferson had a keen sense of the importance of this kind of contribution to the progress of science. In 1807 he arranged for Captain William Clark to make a collection of bones at Big Bone Lick in Kentucky. When these specimens arrived at the White House, Jefferson allowed Dr. Caspar Wistar to select some of them for the collections of the American Philosophical Society; he then took a few for his own cabinet at Monticello and sent the rest to the National Institute of Arts and Sciences in Paris. (See Jefferson's letter to Wistar, March 20, 1808, quoted in Jillson, *Big Bone Lick,* pp. 53–54.) Benjamin Smith Barton, professor of natural history and materia medica at the University of Pennsylvania, carried on an extensive correspondence with European naturalists. It was he who kept Cuvier informed of discoveries in America and called his attention to the references to fossil teeth and bones in American writings, such as Drayton's *View of South Carolina.* Some of his letters to Cuvier were published in *The Philadelphia Medical and Physical Journal,* edited by Barton during its brief existence, and in his *Archaeologiae Americanae Telluris collectanea et specimina. Or Collections, with Specimens, for a Series of Memoirs on Certain Extinct Animals and Vegetables of North America. Part First* (Philadelphia: 1814). See also his "Letter to Mr. Lacépède, of Paris, on the Natural History of North America," *The Philosophical Magazine,* XXII (1805), 97–103; 204–11.

33. Georges Cuvier, "Résumé général de l'histoire des ossemens fossiles de pachydermes, des terrains meubles et d'alluvion," *Ann. Mus. Natl. Hist. Nat. Paris,* VIII (1806), 422. See Osborn, "History of the Classification of the Mastodontoidea, Families and Subfamilies," *Proboscidea,* I, Chap. V, for subsequent modifications of Cuvier's classification.

34. Jean Baptiste Pierre Antoine de Monet, Chevalier de Lamarck, "Mémoires sur les fossiles des environs de Paris...," *Ann. Mus. Natl. Hist. Nat. Paris,* I (1802), 302–3.

35. James Parkinson, *Organic Remains of a Former World...* (London: 1804–1811), I, 467. On the formation of coal measures see I, 238 ff.

36. *Ibid.,* III, 455.

37. Georges Cuvier, *Essay on the Theory of the Earth...,* (1st American ed.; New York: 1818), p. 44.

38. *Ibid.,* pp. 172–73.

Notes to Chapter 5

1. See Agnes Arber, "From Medieval Herbalism to the Birth of Modern Botany," in *Science Medicine and History: Essays on the Evolution of Scientific Thought and Medical Practice Written in Honour of Charles Singer,* E. A. Underwood, ed. (London, New York, and Toronto: 1953), I, 323, 329.

2. John Ray's "A Discourse on the Specific Differences of Plants" was not pub-

lished in the *Philosophical Transactions* of the Royal Society but appeared subsequently in Thomas Birch's *History of the Royal Society of London* (London: 1756–57), III, 162–73. The present excerpts are quoted from the reprinted version in Robert W. T. Gunther, ed., *Further Correspondence of John Ray* (London: 1928), pp. 77–83.

3. *Ibid.,* p. 83. Professor Émile Guyenot and others have emphasized Ray's reluctance to dogmatize concerning the fixity of species, but Ray's latest biographer, Charles Raven, concludes that Ray's early doubts on this score had been resolved by the time of the *Historia generalis plantarum.* See C. E. Raven, *John Ray, Naturalist, His Life and Works* (Cambridge, England: 1942), p. 234; Émile Guyenot, *Les Sciences de la vie aux XVIIe et XVIIIe siècles: l'Idée d'évolution* (Paris: 1941), pp. 360–61; Jean Rostand, *L'Évolution des espèces: Histoire des idées des idées transformistes* (Paris: 1932), p. 20; Edmond Perrier, *La Philosophie zoologique avant Darwin* (3rd ed.; Paris: 1896), p. 31; Conway Zirkle, "The Knowledge of Heredity before 1900," in *Genetics in the 20th Century...,* L. C. Dunn, ed. (New York: 1951), pp. 48–49; Zirkle, *The Beginnings of Plant Hybridization* (Philadelphia and London: 1935), p. 76 ff.

   More generally, see J. Rostand, *Esquisse d'une histoire de la biologie* (Paris: 1945); H. F. Osborn, *From the Greeks to Darwin* (New York: 1929); Walter Zimmerman, *Evolution; die Geschichte ihrer Probleme und Erkenntnisse* (Freiburg and Munich: 1953); Loren Eiseley, *Darwin's Century: Evolution and the Men Who Discovered It* (New York: 1958); G. S. Carter, *A Hundred Years of Evolution* (New York: 1957); Edward S. Russell, *Form and Function: A Contribution to the History of Animal Morphology* (London: 1916).

4. C. Linnaeus, "Life of Carl Linnaeus...," entry for July, 1757, as quoted in Richard Pulteney, *A General View of the Writings of Linnaeus,* W. G. Maton, ed., (2nd ed.; London: 1805), p. 549.

5. C. Linnaeus, *Reflections on the Study of Nature,* J. E. Smith, tr. (Dublin: 1786), pp. 10–11; this essay was first published as a preface to Linnaeus' *Museum Suae Regiae Majestatis Adolphi Friderici Regis Suecorum...,* (Stockholm: 1754–1764).

6. "The study of natural history, simple, beautiful, and instructive, consists in the collection, arrangement, and exhibition of the various productions of the earth." C. Linnaeus, *A General System of Nature...Translated from Gmelin's Last Edition...,* William Turton, tr. and ed., (London: 1806), I, 2. See especially H. K. Svenson, "On the Descriptive Method of Linnaeus," *Rhodora,* XLVII (1945), 273–302, 363–88. Linnaeus' idea of a natural system of plant orders is shown in Fig. 5.1, p. 135. The reader will note that the transitional genera between orders are written in fine script on the plate. The orders on the plate, together with the number of genera in each order, the connecting genera, and the orders which they connect, are enumerated below:

I. Palmae 10
   Nipa — Filices
   Hydrocharis — Tripetaloideae
II. Piperitae 10
   Acorus — Tripetaloideae
III. Calamariae 12
   Carex — Gramineae
   Scripus & Schoenus — Tripetaloideae
IV. Graminae 54
   Cenchrus & Cynosurus — Calamariae

V. Tripetaloideae 8
   Butomus — Palmae
   Sagittaria — Ensatae
   Juncus — Calamaria
VI. Ensatae 10
   Ixia — Tripetaloideae
   Crocus — Spathaceae
   Gladiolus — Orchideae
   Iris — Coronariae
VII. Orchideae 11
   Serapis — Ensatae

Chapter 5. FROM MONAD TO MAN

VIII. Scitamineae 13
IX. Spathaceae 12
  Colchicum — Ensatae
  Erythronium — Coronariae
X. Coronariae 20
  Lilium & Martagon — Ensatae
  Tulipa — Sparthaceae
  Amaryllis — Sarmentaceae
XI. Sarmentaceae 21
  Alstroemeria — Coronariae
XII. Oleraceae 36
XIII. Succulentae 29
  Sedum
XIV. Gruinales 14
  Linum — Caryophylleae
XV. Inundatae 10
XVI. Lacking
XVII. Calycanthaceae 17
  Rhexia — Bicornes
XVIII. Bicornes 23
  Kalmia — Calycanthemae
XIX. Hesperideae 19
XX. Rotaceae 14
XXI. Preceae 12
XXII. Caryophylleae 31
  Lychnis — Gruinales
XXIII. Trihalatae 13
XXIV. Corydales 10
  Fumaria — Rhoeadeae
XXV. Putamineae 8
  Capparis — Rhoeadeae
XXVI. Multisiliquae 24
  Trollius — Rhoeadeae
XXVII. Rhoeadeae 6
  Chelidonium — Corydales
  Sanguinaria — Fumarineae
  Podophyllum — Multisiliquae
XXVIII. Luridae 19
  Pedalium & Datura — Personatae
XXIX. Campanaceae 15
  Lobelia — Contortae
XXX. Contortae 25
XXXI. Asperulae 10
XXXII. Papilionaceae 55

XXXIII. Lomentaceae 10
XXXIV. Cucurbitaceae 12
XXXV. Senticosae 12
XXXVI. Pomaceae 10
XXXVII. Columniferae 43
XXXVIII. Tricoccae 35
XXXIX. Siliquosae 31
XL. Personatae 63
  Martynia & Barleria — Luridae
XLI. Asperifoliae 21
XLII. Verticillatae 39
XLIII. Dumosae 19
XLIV. Sepiariae 9
XLV. Umbellatae 50
  Sium & Ninsi — Hederaceae
XLVI. Hederaceae 7
  Panax — Umbelliferae
XLVII. Stellatae 25
  Phyllis — Umbelliferae
XLVIII. Aggregatae 30
  Dipsacus & Globularia — Compositae
XLIX. Compositae 120
  Capitatae
  Sphaeranthus & Echinops — Aggregatae
L. Amentaceae 14
  Casuarina — Filices
  Betula — Coniferae
LI. Coniferae 7
  Cupressus — Amentaceae
LII. Coadunatae 8
LIII. Scabridae 12
LIV. Miscellaneae
LV. Filices 18
  Zamia & Cycas — Palmae
  Osmunda regalis — Musci
LVI. Musci 10
  Lycopodium clavatum — Filices
  Hypnum — Algae
LVII. Algae 12
  Jungermannia — Muscae
  Marchantia & Tremella — Fungi
LVIII. Fungi 11
  Agaricus & Peziza — Algae

7. C. Linnaeus, *The Elements of Botany...Being a Translation of the Philosophia Botanica and Other Treatises of the Celebrated Linnaeus*, Hugh Rose, tr. (London: 1775), pp. 51–52.

8. *Ibid.*, p. 232. See also *The 'Critica Botanica' of Linnaeus*, Sir Arthur Hort, tr. (London: 1938), pp. 196–97.

9. Linnaeus, *Elements of Botany*, p. 169.

10. C. Linnaeus, *A Dissertation on the Sexes of Plants*, J. E. Smith, tr. (Dublin: 1786), pp. 55–56.

11. C. Linnaeus, *A System of Vegetables, According to Their Classes, Orders, Genera, Species with Their Characters and Differences. Translated from the 13th Edition* (Lichfield, England: 1782), p. 26. Concerning the genera the passage cited reads: "That Nature then intermixed these generic plants by reciprocal marriages (which did not change the structure of the flower) and multiplied them into all existing species; excluding however from the

number of species, the mule-plants, produced from these marriages, as being barren. Each Genus therefore is natural, Nature assenting to it, if not making it."

12. Pulteney, *General View*, p. 556. In the *Genera plantarum* Linnaeus says that the crossing of the original species of each genus with the species of other genera may have taken place "either in the beginning or in the process of time." (Linnaeus, *The Families of Plants...Translated from the Last Edition...* (Lichfield, England: 1787), I, p.lxiv.) See also in this connection, Guyenot, *Les Sciences de la vie*, pp. 368 ff.; Zirkle, *Beginnings of Plant Hybridization.*

13. C. Linnaeus, *Species plantarum...*, (2nd ed.; Leyden: 1763), II, 1266. This example and others are given in Edward L. Greene, "Linnaeus as an Evolutionist," *Proc. Washington Acad. of Scis.* XI (1909), 21 ff. Greene summarizes his argument on p. 25: "My own impression is that few if any of the plants thought by Linnaeus to be hybrids are at all of that origin, according to the views of modern botanists...But what I have herein...clearly shown is, not only that Linnaeus accepted and admitted to his books, as species, forms he thought of as developed from other species, not by any crossing, but through mere environment — natural environment in some instances, artificial in others. And this bent of his mind was so strong that he could scarcely admit two members of a genus to be specifically distinct if found to occur always under the same physical conditions." Greene thinks that Linnaeus was an evolutionist at heart but feared to proclaim his views openly. However, the passage in Linnaeus' diary which deals with the origin of species stresses the accidents of interbreeding rather than the influence of the environment.

14. As quoted without citation by A. G. Nathorst, "Carl von Linné as a Geologist," *Annual Report of the Smithsonian Institution* (1908), p. 731.

15. *Ibid., passim.*

16. Linnaeus, *Reflections on the Study of Nature*, pp. 5–6. See also *General System of Nature*, VII, 3: "Genuine remains of the general deluge, as far as I have investigated, I have not found; much less the adamitic earth: but I have everywhere seen earths formed by the dereliction or deposition of waters, and in these the remains of a long and gradual lapse of ages."

17. D. H. Stoever, *The Life of Sir Charles Linnaeus* (London: 1794), xii. For an account of Linnaean anthropology, see Chapter 6, below.

18. Linnaeus, *Elements of Botany*, pp. 152–53.

19. *Ibid.*, p. 438. Sect. CCLXXXIII, pp. 331–32, sets forth Linnaeus' program of research and experimentation aimed at distinguishing the species from varieties: "Now that which promises certainty in distinguishing the species from the varieties, is to cultivate them in the most different and various soils; to examine attentively all the parts of a plant; to examine the fructification in all its parts...; to inspect the other species of the same genus; to attend to the constant laws of nature, which proceeds by slow degrees; to observe the remote modes of varieties; and, lastly, to place the species under the next different genus." In *The Families of Plants*, I, lxvii, Linnaeus declares: "I acknowledge no authority but inspection alone in Botany..."

20. Buffon, *Natural History, General and Particular*, William Smellie, tr., VI, 250. For a recent account of Buffon's life and work, see Leon Bertin and others, *Buffon (Les Grands Naturalistes français*, Roger Heim, ed., Paris: 1952). See also Louis Roule, *Buffon et la description de la nature* (Paris: 1924); P. Flourens, *Histoire des travaux et des idées de Buffon* (Paris: 1850). A comprehensive bibliography of works by and about Buffon is given in Jean Piveteau, ed., *Oeuvres philosophiques de Buffon (Corpus général des*

philosophes français, publié sous la direction de Raymond Bayer, Auteurs Modernes, XLI, 1, Paris: 1954) , 512–70.

21. *Loc. cit.*

22. Buffon, "De la Manière d'étudier et de traiter l'histoire naturelle," *Oeuvres complètes de Buffon...,* P. Flourens, ed., (Paris: 1853–1855) , I, 48. In this connection see Arthur O. Lovejoy, "Buffon and the Problem of Species," *Popular Science Monthly,* LXXIX (1911) , 464–73, 554–67.

23. Buffon, *Natural History,* III, "The Hog," 503. See also IV, "The Bat," 318.

24. *Ibid.,* V, "The Armadillo," 362–63. Believing the order of description to be quite arbitrary, Buffon proposed to describe domestic animals first, since they were better known and more useful to man than wild beasts. "Is it not better to follow the horse, which is soliped, with the dog, which is fissiped and which is accustomed to follow him in fact, than with a zebra, which is little known to us and which has perhaps no other affinity with the horse than to be soliped?" (*Oeuvres,* I, 53.)

25. *Ibid.,* II, "Examination of the Different Systems of Generation," 66. These passages contain an excellent statement of Buffon's metaphysical position. He was impressed with the high degree of abstraction involved in scientific reasoning and warned against the fallacy which A. N. Whitehead was later to call "the fallacy of misplaced concreteness."

26. *Ibid.,* II, "Of Reproduction in General," 28.

27. *Ibid.,* III, "Of the Nature of Animals," 287, 299; "The Hog," 505.

28. *Ibid.,* VIII, "Of Apes," 62–63; III, "Of the Nature of Animals," 216.

29. *Ibid.,* II, "Of Reproduction in General," 28.

30. *Ibid.,* II, 29–30; "Of Nutrition and Growth," 46–47.

31. *Ibid.,* IV, "Of Carnivorous Animals," 169.

32. *Ibid.,* II, "Of Reproduction in General," 34.

33. *Ibid.,* II, "Of Nutrition and Growth," 48. A theory of organic phenomena similar in some respects to Buffon's was propounded by Pierre Louis Moreau de Maupertuis. In his *Vénus physique* (1745) Maupertuis invoked the force of attraction to account for the formation of the foetus: "Suppose that there are in each of the seeds particles destined to form the heart, the head, the entrails, the arms, the legs, and that these particles have each a greater affinity for that one which must be its neighbor in the formation of the animal than for any other; the foetus will then form, and would form even if it were a thousand times more highly organized than it is." He showed how deviant forms might arise from the various combinations of these particles and how they might be perpetuated from generation to generation by human selection or, possibly, by environmental influence. In another work, published anonymously in 1751 and subsequently translated into French under the title *Système de la nature,* Maupertuis took the position that the attraction of the particles for each other in the formation of the foetus could only be explained by supposing that every bit of matter possesses in some degree properties analogous to intelligence, desire, aversion, and memory, so that each atom of the seminal fluid conserves a kind of memory of its former situation in the animal body even after the dissolution of that body. Having, as he thought, strengthened his theory of generation by this assumption, he proceeded to show how the theory would account for spontaneous generation, normal generation, the formation of monsters, the sterility of hybrids, and the proliferation of novel types. New species, he wrote, "would have owed their origin only to some fortuitous productions in which the elementary particles would not have retained the order which they had in

the mother and father animals; each degree of error would have made a new species, and, by repeated deviations, there would have arisen the infinite diversity of animals which we see today, a diversity which perhaps still increases with time, but to which the succession of ages makes but imperceptible additions." *(Système de la nature,* in *Oeuvres de Maupertuis* [Lyons: 1756], II, Sect. XLV, 148; *Vénus physique,* in *ibid.,* II, 89.) For a more extended discussion of Maupertuis' speculations, see Chapter 8 below; also Rostand, *Évolution des espèces,* pp. 31–32, Guyenot, *Les Sciences de la vie,* pp. 389–93, Lovejoy, "Some Eighteenth Century Evolutionists," *Popular Science Monthly,* LXV (1904), 238–51, A. C. Crombie, "P. J. Maupertuis, F.R.S. 1698–1759, précurseur du Transformisme," *Revue de Synthèse,* LXXVIII (1957), 35–56. The transformist implications of Maupertuis' argument were given wide publicity, not to say notoriety, in Diderot's *Pensées sur l'interprétation de la nature,* published in 1754.

34. *Ibid.,* VII, "Second View of Nature," 96–97.

35. *Ibid.,* VII, "Second View of Nature," 90. See also III, "The Ox," 423–24.

36. *Ibid.,* IV, "Of Wild Animals," 68.

37. *Ibid.,* II, "Of the Nature of Man," 356; III, "Homo Duplex," 299–300, VIII, "Of Apes," 66–67; V, "The Lion," 65.

38. *Ibid.,* VIII "Of Apes," 66; II, "Of the Nature of Man," 366–67.

39. *Ibid.,* III, "Of Domestic Animals," 302. Although Buffon ranked man first among living creatures, he insisted that all rankings of natural productions were relative to human modes of perception. See II, 2.

40. *Ibid.,* III, "The Ass," 405.

41. *Ibid.,* III, 406. Concerning plants Buffon wrote: "In plants, we have not the same advantage; for, though sexes have been attributed to them, and generic distinctions have been founded on the parts of the fructification; yet, as those characteristics are neither so certain nor so apparent as in animals; and, as the reproduction of plants can be accomplished by several methods which have no dependence on sexes, or the parts of fructification, this opinion has not been universally received; and it is only by the misapplication of analogy, that the sexual system has been pretended to be sufficient to enable us to distinguish the different species of the vegetable kingdom." *Ibid.,* II, 10. Although Buffon used the criterion of ability to produce fertile offspring as his test for distinguishing species, he recognized that it was not an absolutely certain test, nor a practical one in many cases.

42. *Ibid.,* IV, "The Hare," 141.

43. Unlike Linnaeus, Buffon believed in the possibility and actuality of spontaneous generation. From his own microscopic observations and those of J. T. Needham he thought to have demonstrated the existence of organic molecules; the origin of intestinal worms, vinegar eels, and the like might then be explained by supposing them to result from chance combinations of organic molecules. In general, Buffon regarded these creatures as incapable of reproduction in the ordinary manner, but he was forced to recognize some exceptions to this rule and thus to make room for the possibility that ordinary species had originated by spontaneous generation. This possibility, barely hinted at in his early discussion of generation, was to assume the dignity of a probability in his later writing. See his *Natural History,* II, 212–15, 252–53, 270, 347 ff. See also Jean Rostand, *La Formation de l'être: Histoire des idées sur la génération* (Paris: 1930), Chap. IX.

44. Buffon, *Natural History,* III, "The Ass," 402–3.

45. *Ibid.,* III, 411.

Chapter 5. FROM MONAD TO MAN

46. *Ibid.,* VI, "The Mouflon," 221.

47. *Ibid.,* IV, "Of Wild Animals," 66.

48. *Ibid.,* III, "The Goat," 487. Buffon believed, however, that domestically produced varieties reverted to their original form when returned to a state of nature. He suggested that experiments be undertaken to determine the time required for wheat to revert to its primitive type (IV, 12). He understood the process of artificial selection clearly: "The life of the dog is short; his prolific powers are great; and, as he is perpetually under the eye of man, whenever by any accident, which is not uncommon in Nature, some individuals, marked with singular characters, appeared, they would be perpetuated by preventing their intermixture with any other kinds, as is done at present when we want to procure new races of dogs, or of other animals" (IV, "The Dog," 10–11).

49. *Ibid.,* IV, "Of Wild Animals," 70–71; III, 344–45; VII, 397–98, 428 ff.; VI, 156.

50. *Ibid.,* VII, "Of the Degeneration of Animals," 399, 400–408; V, 364; VII, 98.

51. *Ibid.,* VII, 436–37, 414–15; V, 111 ff.

52. *Ibid.,* V, "Animals Common to Both Continents," 129, 132–39. Buffon offered still another explanation in his *Époques de la nature* (1778). See below, Note 59.

53. *Ibid.,* VII, "Of the Degeneration of Animals," 448–50, 451–52; V, 123 ff.

54. *Ibid.,* IX, 302–3.

55. *Ibid.,* V, "Animals Common to Both Continents," 150; IX, 45 ff.

56. *Ibid.,* V, 150.

57. *Ibid.,* VII, "Of the Degeneration of Animals," 420.

58. *Ibid.,* VII, "The Two-Toed and Three-Toed Sloths," 155.

59. Buffon, *Époques de la nature,* in *Histoire naturelle, générale et particulière, Supplément,* IX (Paris: 1778), 264–65. Buffon went on to suggest that animals of South America were smaller than those of the other continents because they had been generated at a later period in the earth's history.

60. Buffon, "'Addition à l'article des variétés dans la génération...& aux articles où il est question de la génération spontanée...,'" *Histoire naturelle, Supplément,* VIII, 59 ff. On p. 65: "This replacement of living Nature would be very incomplete at first, but in time all the creatures which were unable to reproduce would disappear; all the bodies imperfectly organized and all the defective species would vanish, and there would remain, as there remain today, only the most powerful and complete forms, whether among plants or animals, and these new beings would be, in general, similar to the old ones because, the brute matter and the living matter remaining always the same, the same general plan of organization and the same varieties within particular forms would result. On this hypothesis one must suppose, however, that this new nature would be shrunken, because the heat of the globe is a power which influences the extension of the molds, and this heat being weaker today than it was at the beginning of our nature, the largest species could not arise or could not arrive at their present dimensions." Here again Buffon brings in the "degenerate" animals of South America by way of illustration. Maupertuis, too, envisaged the possibility that the globe had formerly been denuded of its flora and fauna by some cosmic catastrophe: "But it would not be impossible that, if our earth should find itself again in some such state as we have spoken of...after such and such a deluge or conflagration, new unions of elements, new animals, and new plants, or rather entirely new things would be produced again." *Système de la nature,* Sect. XLIX, p. 153.

Chapter 5. FROM MONAD TO MAN

61. Buffon, *Natural History*, VIII, "Of Mules," 33–34.
62. *Ibid.*, VIII, "Of Mules," 34–35; III, "The Goat," 488; IX, 396, 409.
63. Jean Baptiste Pierre Antoine de Monet, Chevalier de Lamarck, *Histoire naturelle des animaux sans vertèbres...* (Paris: 1815–1822), I, 317. The interpretation of the genesis of Lamarck's evolutionary hypothesis presented in this chapter differs markedly from that recently advanced by Professor Charles Gillispie in two important articles: "The Formation of Lamarck's Evolutionary Theory," *Arch. Int. Hist. Sci.*, XXXV (1956), 323–38; "Lamarck and Darwin in the History of Science," *American Scientist*, XLVI (1958), 388–409. According to Professor Gillispie, "Lamarck's theory of evolution was the last attempt to make a science out of the instinct... that the world is flux and process, and that science is to study, not the configurations of matter, nor the categories of form, but the manifestations of that activity which is ontologically fundamental as bodies in motion and species of being are not.... Lamarck's philosophy, therefore, is no anticipation of Darwin but a medley of dying echoes; a striving toward perfection; an organic principle of order over against brute nature; a life process as the organism digesting its environment; a primacy of fire, seeking to return to its own, a world as flux and as becoming." The present book is not the place for an extended critique of this thesis. Suffice it to say that in my opinion, Professor Gillispie seizes upon one aspect of Lamarck's writings and builds it into a philosophy of nature which he attributes to Lamarck but which in many respects is plainly inconsistent with Lamarck's own statements. For a summary statement by Lamarck himself of his philosophy of nature, see Part VI, pp. 304–41, of his *History of Invertebrate Animals*, entitled: "Concerning Nature, or the Power, Mechanical so to Speak, Which Has Given Existence to Animals, and Which Has Necessarily Made Them What They Are."
64. *Ibid.*, I, 12, 121 ff., 53 ff.
65. Lamarck, *Philosophie zoologique...* (Paris: 1809), II, 172 ff.; Lamarck, *Histoire naturelle*, I, 253: "There are, then, degrees of intelligence, feeling, etc., because this is necessarily the case in all which nature does." Thus, Lamarck applied Buffon's axiom of nuance in nature to the phenomenon which Buffon excepted, namely, intelligence.
64. *Ibid.*, I, 12, 121 ff., 53 ff.
67. Lamarck, *Philosophie zoologique*, II, 174; also *Histoire naturelle*, I, 335 ff.
68. Lamarck, *Histoire naturelle*, I, 326–27.
69. Lamarck, *Système des animaux sans vertèbres...* (Paris: 1801), "Sur les fossiles," pp. 406 ff. This work contains Lamarck's earliest statement of his evolutionary view. The sequence of ideas is interesting. After stating that "extinct" forms are probably the progenitors of living forms, Lamarck proceeds: "Every observant and well-instructed man knows that nothing remains constantly in the same state on the surface of the earth... But, if, as I shall try to make clear elsewhere, the diversity of circumstances leads, in the case of living beings, to a diversity of life-habits, a different mode of existing, and consequently, to modifications or developments of their organs and of the form of their parts, it must appear that insensibly every living being whatsoever must vary in its organization and in its forms. It is further apparent that all the modifications which it undergoes in its organization and forms, in virtue of the circumstances which influence this being, will propagate themselves by generation, and that after a long series of ages, not only can new species, new genera and even new orders be formed, but each species will have varied necessarily in its organization

and forms." After attempting to explain the failure of some forms to change, Lamarck concludes: "...however, one may not conclude that any species has really been lost or annihilated. It is doubtless possible that among the largest animals there have been some species destroyed as a result of the multiplication of man in the places which they inhabit. But this conjecture cannot be founded solely upon the consideration of fossils; one cannot pronounce in this matter until every habitable part of the globe shall be perfectly known." *Système des animaux sans vertèbres,* pp. 409–11. Thus, Lamarck was, in his own way, averse to the notion of a real extinction of one of nature's productions. He felt that nature had provided for the survival of her productions by equipping them with the ability to change with changing conditions. See also *Philosophie zoologique,* I, 77 ff., 64 ff.

70. Lamarck, *Philosophie zoologique,* I, 3; *Histoire naturelle,* I, 257. On pp. 197–98 of the latter: "...I am firmly convinced that the races to which have been given the name *species* have only a limited or temporary constancy in their characters, and that there is no species which is absolutely constant. Doubtless they will subsist unchanged in the places which they inhabit so long as the circumstances which affect them do not change, do not force them to change their life-habits."

71. Lamarck, *Système des animaux sans vertèbres,* pp. 14–15.

72. Lamarck, *Histoire naturelle,* I, 181 ff., 185–86. In the lowest forms of life, Lamarck explained, new developments were the product of "mechanical causes," since these organisms did not possess the faculty of feeling and *le sentiment intérieur* which went with it. Thus, he made a clear distinction between purely mechanical causes and those involving a psychological factor.

73. Lamarck, *Système des animaux sans vertèbres,* p. 12; *Philosophie zoologique,* I, 238. On p. 236 of the latter work Lamarck states the issue flatly: "Naturalists having noticed that the forms of the parts of animals, compared to the uses of these parts, are always perfectly adjusted, have thought that the forms and the state of the parts had led to their use; but, this is an error, for it is easy to show by observation that it is, on the contrary, the needs and uses of the parts which have developed these same parts, which have even originated them when they did not exist, and which, consequently, have brought about the condition in which we observe them in each animal."

74. Lamarck, *Philosophie zoologique,* I, 5–6. On pp. 63–64, II, Lamarck declares that "nature, with the aid of heat and humidity, has created directly only those first beginnings of organization." Lamarck conceived the evolution of vegetable forms as parallel with that of animal forms, rather than continuous. He also believed that there was a great hiatus between living and nonliving bodies. *Histoire naturelle,* I, 108, 126.

75. Lamarck, *Histoire naturelle,* I, 382. In the *Philosophie zoologique,* II, 463, Lamarck set forth the first phylogeny, or genealogical tree, ever published, showing his idea of the *série rameuse* formed in the process of organic transformation. His classification in the *Philosophie zoologique* was based solely on anatomical and physiological characters, as contrasted with the physiological-psychological classification in the *Histoire naturelle.*

76. Lamarck, *Philosophie zoologique,* I, 357.

77. Lamarck, *Histoire naturelle,* I, 185 ff. Many, perhaps most, of the naturalists of that day shared Lamarck's belief in the transmissibility of acquired characters. See Zirkle, "The Knowledge of Heredity before 1900," p. 52.

Chapter 5. F R O M   M O N A D   T O   M A N

78. *Ibid.,* I, 348–49, 364. In the *Système des animaux sans vertèbres,* p. 18, Lamarck stated the principle which guided him in classification: "If there exists among living beings a graduated series at least in the principal divisions *(masses)*, relative to the complication or simplicity of organization, it is evident that in a truly natural distribution, whether of animals or of vegetables, one must necessarily place at the two extremities of the order of beings the most dissimilar types, those farthest apart in their affinities and, consequently, those which form the extreme limits which organization, whether animal or vegetable, can present." In the *Histoire naturelle* Lamarck called these relations of organization *rapports du rang.* He also recognized as bases of classification *rapports d'espèces, rapports de masses, principes qui concernent la comparaison de divers organes, considérés séparément,* and *rapports particuliers entre des parties modifiées.* These last, he declared, had resulted from accidental causes and hence were of inferior value for purposes of classification. *Histoire naturelle,* I, 356–62, 346 ff.

79. Lamarck, *Philosophie zoologique,* I, 101.

80. Lamarck, *Histoire naturelle,* I, 191–92.

81. Lamarck, *Philosophie zoologique,* II, 465; I, 113; *Histoire naturelle,* I, 311, 168, 323–24, 329–30.

82. In occasional passages, however, Lamarck recognized the possibility of accidental variations unrelated to the needs of the organism. Thus, *Histoire naturelle,* I, 198: ". . . sometimes one will even see varieties produced, not by the habits demanded by circumstances but by those which could have been contracted whether accidentally or otherwise. Thus man, being subject to the laws of nature with regard to his organization, exhibits himself some remarkable varieties in his species, and among these varieties there are some which appear to be due to the causes just cited." On p. 194 Lamarck lists competition among animals and the consequent need for protection as a cause "which has contributed to diversify animal structures and multiply races."

83. In his reluctance to admit the real extinction of species, in his notion of a "plan of operations" constituting the basis of a "natural method" of classification, in his emphasis on need rather than accidental fitness as the explanation of organic change, and in his conviction that nature was so ordered as to promote and preserve "progress" in perfecting organization Lamarck leaned away from a truly mechanistic intepretation of nature. He considered himself a bitter opponent of teleology in natural history, but he was never able to question the notion of a harmony of nature emerging from conflict. To this eighteenth century axiom he added the notion of "progress" in organization, involving the emergence of new powers in nature. Many historians of science have failed to notice the tension in Lamarck's thought between his radical positivistic materialism and his recognition of a semi-purposive, psychological factor in the evolution of organic forms. This tension is keenly appreciated and brilliantly discussed in Edward S. Russell, *Form and Function,* Chap. XIII.

84. Erasmus Darwin, *Zoonomia; Or the Laws of Organic Life,* (4th American ed.; Philadelphia: 1818). Section I, Vol. I, is entitled "Of Motion" and begins: "The whole of nature may be supposed to consist of two essences or substances; one of which may be termed spirit, and the other matter. The former of these possesses the power to commence or produce motion, and the latter to receive and communicate it. So that motion, considered as a cause, immediately precedes every effect; and considered as an effect, it immediately succeeds every cause. And the laws of motion therefore are

Chapter 5. F R O M   M O N A D   T O   M A N

the laws of nature." On p. 9, Darwin defines motion as variation of figure: "for the whole universe may be considered as one thing possessing a certain figure; the motions of any of its parts are a variation of this figure of the whole." As to the status of "spirit" in nature, Darwin is rather vague. The spirit of animation, which he posits as the immediate cause of the contraction of animal fibers, is sometimes described in terms suggesting a material ether. On p. 5, however, it is described as "that living principle . . . which resides throughout the body, without being cognizable to our senses, except by its effect." Lamarck was more consistently materialistic in referring the phenomena of life to certain states of matter capable of producing them, rather than to a nonmaterial principle.

85. *Ibid.,* I, 392.

86. *Ibid.,* I, 390. Rostand discusses the bearing of theories of generation on the development of evolutionary theory in his various books cited above; he notes that the pioneers of transformism were epigenesists rather than preformationists in their embryology. They believed that the new individual was not already present in the reproductive cell in minute form but gradually took shape in an evolutionary development leading from a relatively undifferentiated bit of organic matter to the completed individual. *Esquisse,* pp. 81 ff.

87. *Ibid.,* I, 397. Darwin believed that living matter had been endowed with its peculiar properties by divine agency, that it retained those properties upon the dissolution of the organism, and hence that it could give rise to microscopic forms of life by spontaneous generation. *Ibid.,* I, 435 ff.; E. Darwin, *The Temple of Nature; Or, the Origin of Society. A Poem, With Philosophical Notes* (Baltimore: 1804), pp. 3–4, 39 n. Darwin was closer to Buffon than to Lamarck in this conception, but like Lamarck he believed that the peculiar property of living matter, however it originated, was its capacity for development and transformation in interaction with the environment.

88. Darwin, *Zoonomia,* I, 399.

89. *Ibid.,* I, 395.

90. E. Darwin, *The Botanic Garden. A Poem in Two Parts...With Philosophical Notes,* (1st American ed.; New York: 1798), p. 105 n. Also, *Temple of Nature,* p. 68 n.

91. Darwin, *Zoonomia,* I, 400–401, 437.

92. The account of Cuvier's views set forth here is based on those of his works published before 1820, *i. e.,* the *Tableau élémentaire de l'histoire naturelle des animaux* (1797–1798), the *Leçons d'anatomie comparée* (1800–1805), the *Ossemens fossiles* (1812), and *Le Règne animal* (1817).

93. Georges Cuvier, *Lectures on Comparative Anatomy,* William Ross, tr. (London: 1802), I, 4–5. Cuvier rejected completely the notion of spontaneous generation.

94. Georges Cuvier, *The Animal Kingdom Arranged in Conformity With its Organization,* H. M'Murtrie, tr. and ed. (New York: 1831), I, 9.

95. *Ibid.,* I, 8–10. Cuvier defined a species as "the reunion of individuals descended from one another, or from common parents, or from such as resemble them as strongly as they resemble each other." See also *Tableau élémentaire,* 11–12.

96. Cuvier, *Animal Kingdom,* I, 11, 10; Cuvier, *Essay on the Theory of the Earth...,* Robert Jameson, ed. (New York: 1818), pp. 59–60, 118–28. Cuvier tried to keep the number of successive creations to a minimum. Thus, p.

128 of the *Essay*: "...I do not pretend that a new creation was required for calling our present races of animals into existence. I only urge that they did not anciently occupy the same places, and that they must have come from some other part of the globe." The phrase "calling into existence" betrays the general view of nature involved. Note also the language of the quotation in the text: "...assign to them the parts they are to play on the great stage of the universe."

97. Cuvier, *Lectures*, I, xxiii–xxiv, 47–48.

98. Cuvier, *Animal Kingdom*, I, 3–4.

99. *Ibid.*, I, 4–5.

100. *Ibid.*, I, 3–4.

101. *Ibid.*, I, 30; *Lectures*, I, 63.

102. Cuvier, *Lectures*, I, 60. Also *Animal Kingdom*, I, xvii, on the great chain of being. Cuvier was careful to distinguish man from the apes and monkeys anatomically. Thus, *Animal Kingdom*, I, 45: "The foot of Man is very different from that of the Monkey; it is large; the leg bears vertically upon it; the heel is expanded beneath....Man is the only true bimanous and biped animal." See also *Tableau*, p. 76.

103. Cuvier, *Lectures*, I, 58–59. Cuvier seems reluctant to confine nature to four combinations, for he repeatedly qualifies this position by using such words as "almost," etc.

104. Cuvier, *Essay on the Theory of the Earth,* pp. 118–19.

105. This is precisely the argument which Lamarck brought against the static view of creation, namely, that it denied nature, considered as a system of matter governed by law, any active role in producing the variety of the world. Cuvier was not nearly so explicit concerning his philosophy of nature as Buffon and Lamarck were. The "Introduction" to his *Animal Kingdom* contains some scattered reflections on nature but not a coherent philosophy.

106. It is interesting to note that botanical science failed to produce evolutionary speculations comparable to those which sprang up in the field of zoology. Eighteenth century botany was dominated by the search for the natural method of classification. Although the French botanist Michel Adanson went a long way toward recognizing the mutability of species in his *Familles des plantes* (1763), his primary interest was in classification. The "great work" to which he proposed to devote himself after the publication of his *Familles* was the discovery and accurate description and classification of the four or five families, the four to six thousand genera, and the eighteen to twenty-five thousand species which he believed still remained to be brought within the purview of scientific botany. *Familles des plantes* (Paris: 1763), p. cccxxiii. A. L. de Jussieu (1748–1836) carried the quest for the natural method of classification a step closer to realization by distinguishing essential from non-essential characters and founding the classes and orders on those characters deemed most essential because of their connection with reproduction and hence with survival. "It is only the parts of the fructification," he declared, "which can give the primitive characters of the natural order; thus, the calyx and the corolla must be put aside because these two envelopes of the flower can be lacking together or singly in a plant without its being less perfect or less capable of reproducing. This last property, which supposes all the others, constitutes the veritable perfection of a vegetable; it is inherent in the species formed by nature and resides in the sexual organs. If sometimes mutilation or abortion (*avortement*) has suppressed these organs in a particular individual it is a denatured being which falls short of its destination,

a monstrosity which constitutes an exception in the general order." Exposition d'un nouvel ordre de plantes adopté dans les démonstrations du Jardin Royal," *Hist. Acad. Roy. Scis.* (1774) , p. 179.

This idea of the subordination of characters proved a valuable guide to classification not only in botany but also in zoology, where it was applied by Cuvier, but in neither field did it upset the static view of nature. Just as Cuvier exorcised the specter of mutability in zoology, so Jussieu's disciple A. P. de Candolle (1778–1841) banished it from botany. In his *Théorie élémentaire de la botanique* (Paris: 1813) Candolle rejected the argument for the mutability of species root and branch. The advocates of mutability, he declared, based their arguments on a few ambiguous, rare, or tiny plants instead of on the generality of plants well known to botanists. They imagined the influence of life habits on organic forms. They talked of slow changes over vast periods of time, but failed to reckon with the fact that no important change had taken place in plant species during recorded history. Their theory, even if true, was impractical; it gave the botanist nothing to work on. *Théorie élémentaire,* pp. 160 ff.

Candolle agreed with Cuvier that the importance of organs and hence of characters was to be judged by the importance of their contribution to the performance of the vital functions. Like Cuvier he believed that comparative anatomy disclosed the existence of a relatively small number of primitive types from which organic creatures deviated in varying degree, each from its own prototype. In this conception Candolle encountered the same difficulty which beset Cuvier. If the immutability of species and genera was to be maintained, the "deviation" from the class or family prototype could only be a conceptual, not a temporal, deviation. But Candolle's discussion of variation in plant forms — arrested development, adherence of parts, unification of parts originally separate, etc. — showed clearly that he conceived important changes of form as resulting from material circumstances in the temporal world. He noted, for example, that adherence of parts may occur accidentally in a member of a given species, but it may also become normal for a species. Thus, if the two ovaries of a plant were close together, "it is clear that the occasions of adherence between them will be more frequent and that they can be so much so that we will never see the ovaries separate; this adherence is nothing but a *constant accident,* and although these two words seem contradictory, this kind of phenomenon is nonetheless very common in nature." *Ibid.,* pp. 112–13. Again, in discussing "the accidental causes which disturb the primitive symmetry of each system," Candolle said: "...I declare that the Personée are only alterations of the type of the Solanée because a Personée regularized in thought does not differ from a Solanée." *Ibid.,* p. 144. Finally, of the theory of abortions (*avortemens*) : "This theory explains the changes of form, and consequently the changes of use, which are so frequent in the organization [of a plant]." *Ibid.,* p. 110. But, like Cuvier, Candolle did not try to relate changes of organization and function to the external conditions of existence. For him the chief concern of botany was not the origin of species but "the study of the symmetry proper to each family and the relations of these families among themselves." *Ibid.,* p. 206. The notion of a symmetry proper to nature's productions was also developed by J. W. Goethe in his *Metamorphosis of Plants* (1790) and in his subsequent zoological writings, but without any clear evolutionary implication.

Notes to Chapter 6

1. Carolus Linnaeus, *Systema naturae*... (Leyden: 1735), as quoted with partial translation in Thomas Bendyshe, "On the Anthropology of Linnaeus, 1735-1776," *Memoirs Read Before the Anthropological Society of London*, I (1863-1864), 422. For a brief history of the growth of knowledge concerning the anthropoid apes, see Robert M. Yerkes and Ada W. Yerkes, *The Great Apes: A Study of Anthropoid Life* (New Haven and London: 1929), Part I, 1-46. This work contains excellent illustrations. See also Thomas Bendyshe, "The History of Anthropology," *Memoirs Read Before the Anthropological Society of London*, I (1863-64), 335-458; Thomas H. Huxley, *Man's Place in Nature and Other Anthropological Essays* (New York: 1898), p. 1-77.

2. Edward Tyson, *Orang-Outang: Or the Anatomy of a Pygmy Compared with That of a Monkey, an Ape, and a Man. With an Essay Concerning the Pygmies, Etc. of the Ancients*... (2nd ed.; London: 1751). The "Epistle Dedicatory" is not paged. A full account of Tyson's life and work may be found in M. F. Ashley Montagu, *Edward Tyson, M.D., F.R.S. 1650-1708* (*Memoirs of the American Philosophical Society*, XX, Philadelphia: 1943).

3. *Ibid.*, pp. 91, 3.

4. *Ibid.*, pp. 51-52, 55. Tyson is quoting here the words of the Parisian Academists, who dissected several species of monkeys. See [Claude Perrault], *Mémoires pour servir à l'histoire naturelle des animaux*, 2 vols. (Paris: 1671 and 1676). This work was translated into English by Alexander Pitfeild in 1688.

5. *Ibid.*, p. 55.

6. On page 82 Tyson says: "...we may safely conclude, that *Nature* intended it [the pygmy] to be a *Biped*, and hath not been wanting in anything, in forming the *Organs*, and all Parts accordingly; and if not altogether so exactly as in a *Man*, yet much more than in any other *Brute* besides. For I own it, as my constant Opinion, (notwithstanding the ill surmise and suggestion made by a forward Gentleman) that tho' our *Pygmie* has many advantages above the rest of it's Species [*sic*], yet I still think it but a sort of *Ape* and a meer *Brute*..."

7. For a good discussion of Tyson's treatment of previous accounts of apes and other anthropoid creatures, see Montagu, *op. cit.*, p. 244. Montagu concludes, p. 276: "In his survey of the literature, Tyson did not omit a single work of significance relating to the man-like apes." See also H. W. Janson, *Apes and Ape Lore in the Middle Ages and the Renaissance* (London: 1952); William C. McDermott, *The Ape in Antiquity* (The Johns Hopkins University Studies in Archaeology, No. 27, Baltimore: 1938).

8. Samuel Purchas, *Hakluytus Posthumus, or Purchas his Pilgrimes* (4th ed.; London: 1625), II, pp. 981-82, as quoted in Robert M. and Ada W. Yerkes. *The Great Apes: A Study of Anthropoid Life* (New Haven: 1929), p. 10. A less complete account of Battell's narrative may be found in the first edition of Purchas' work, published in 1613.

9. Buffon, *A Natural History. General and Particular;*...William Smellie, tr., (3rd ed.; London: 1791), VIII, 39, 43-44, 52-53.

10. *Ibid.*, VIII, 40.

11. *Ibid.*, VIII, 96, 85-86, 66-67.

12. *Ibid.*, VIII, 66, 76.

13. Letter from C. Linneaus to J. G. Gmelin (1747), as quoted in Edward L. Greene, "Linnaeus as an Evolutionist," *Proc. Washington Acad. of Scis.*, XI (March 31, 1909), 25-26.

14. Carolus Linnaeus, *Fauna suecica* (Leyden: 1746), as quoted in Bendyshe, "The Anthropology of Linnaeus," p. 445.

15. C. Linnaeus, *Systema naturae*... (10th ed.; Stockholm: 1758), as quoted in Bendyshe, "The Anthropology of Linnaeus," p. 425.

16. *Ibid.*, pp. 428–29.

17. C. E. Hoppius, "Anthropomorpha, Quae Praeside D. D. Car. Linnaeo...," *Amoenitates Academicae*..., VI (Stockholm: 1763), as translated in Bendyshe, "The Anthropology of Linnaeus," pp. 451–52. For Linnaeus' own remarks on *Homo caudatus vulgo dictus*, see *Systema naturae*, 10th ed., in Bendyshe, *op. cit.*, p. 454.

18. C. Linnaeus, *Systema naturae*..., (12th ed.; Stockholm: 1766), I, 34.

19. Arnout Vosmaer, "Naturlyke Historie van den Orang-outang, van Borneo," in *Description d'un recueil exquis d'animaux rares*... (Amsterdam: 1804). Each description is paged separately. See also in the same work: "Beschryving van de Oest-Indische Orang-outang." I am indebted to Mr. George Karreman for translating these works for me. Vosmaer measures by the Rhineland foot, which equals .314 meters as compared to .305 meters for the English foot.

20. Petrus Camper, "Account of the Organs of Speech of the Orang-outang. By Peter Camper, M. D., Late Professor of Anatomy, etc. in the University of Gröningen, and F.R.S. in a Letter to Sir John Pringle, F.R.S.," *Philos. Trans. Roy. Soc.*, LXIX (1779), 155–56. Camper was puzzled by the difference between the vocal organ he described and that described by Tyson. He suggested that Tyson might have observed carelessly.

21. Petrus Camper, *Peter Campers Naturgeschichte des Orang-outang und einiger andern Affenarten, des Africanischen Nashorns, und des Rennthiers. Ins Deutsche übersetzt, und mit dem neuesten Beobachtungen des Verfassers herausgegeben von J.F.M. Herbe* (Düsseldorf: 1791), pp. 195, 139 ff. The original edition, entitled *Natuurkundige Verhandelingen*, appeared in 1782.

22. Pierre [Petrus] Camper, *Dissertation sur les variétés naturelles qui caractérisent la physionomie des hommes des divers climats et des différens ages... Ouvrage posthume de M. Pierre Camper traduit du Hollandais par Henri J. Jansen* (Paris: 1792), pp. 12, 40.

23. *Ibid.*, pp. 33–34.

24. Buffon, "Additions aux singes de l'ancien continent," *Oeuvres complètes de Buffon...revue sur l'édition in-4° de l'Imprimerie Royale et annotées par M. Flourens* (Paris: 1853–1855), IV, 70 ff. These "Additions" formed part of Vol. 7 of *Suppléments* to the original edition.

25. *Ibid.*, IV, 81. Oddly enough, Buffon makes no mention of Camper's researches on the orang-outang, although Vosmaer states that Camper kept both him and Buffon informed of his findings.

26. Georges Cuvier, *Tableau élémentaire de l'histoire naturelle des animaux* (Paris: 1798). In the "Preface" Cuvier acknowledges the assistance he received from Geoffroy Saint-Hilaire in his classification of the mammals. Concerning his debt to Buffon, Flourens notes: "Since Buffon the number of species has increased greatly, and, in proportion as it has increased, all the original genera have had to be subdivided; but they have only been subdivided: at bottom the classification has been preserved." Buffon, *Oeuvres complètes*, IV, 8 n.

27. Georges Cuvier, *Leçons d'anatomie comparée...recueillies et publiées sous ses yeux par C. Dumeril...* (Paris: 1800–1805), I, 9–10.

Chapter 6. MAN'S PLACE IN NATURE

28. Cuvier, *Tableau*, pp. 95 ff. The "Wouwou" was later classified as a species of gibbon.

29. *Ibid.*, pp. 99–100.

30. F. Baron von Wurmb, "Beschryving van de groote Borneosche Orang Outang of de Oost-Indische Pongo," *Verhandelingen van het Bataaviaasch Genootschap van Kunsten en Wetenschappen.* Tweede Deel. Tweede Druk. (Batavia: 1823), II, 137–47. In English: "Description of the large Orang Outang of Borneo. By F. B. von Wurmb...," *The Philosophical Magazine*, I (August, 1798), 225–31. The French translation was by H. J. Jansen in the *Décade Philosophique, Littéraire, et Politique*, LXXIX (June, 1796), 1–8. See also J. C. Radermacher, "Beschryving van het Eiland Borneo, Voor Zoo Verne Hetzelva Tot Nu Toe Bekend Is...," *Verhandelingen van het Bataviaasch Genootschap van Kunsten en Wetenschappen.* Tweede Deel. Tweede Druk. (Batavia: 1823), II, 64–67. On the question whether Wurmb's ape ever reached Holland, see Thomas H. Huxley, "On the Natural History of the Man-Like Apes," in *Man's Place in Nature and Other Anthropological Essays* (New York: 1898), pp. 24 ff. Huxley quotes Camper's statement that he has seen the skeleton of a man-sized ape in the museum of the Prince of Orange but cites reasons for thinking that this was not the skeleton of Wurmb's ape. He conjectures that the skeleton which Camper saw was carried off to France by French soldiers "in accordance with the usual marauding habits of the Revolutionary armies" and so found its way into the Museum of Natural History in Paris. For an early nineteenth-century discussion and review of the evidence concerning anthropoid creatures see William Lawrence, *Lectures on Physiology, Zoology, and the Natural History of Man Delivered at the Royal College of Surgeons* (London: 1819), pp. 117 ff.

31. Étienne Geoffroy Saint-Hilaire, "Observations on the Account of the Supposed Orang Outang of the East Indies, published in the Transactions of the Batavian Society in the Island of Java...From the Journal de Physique, 1798," *The Philosophical Magazine*, I (1798), 324. See also the same author's "Tableau des quadrumanes...," *Ann. Mus. Natl. Hist. Nat. Paris*, XIX (1812), 89.

32. Georges Cuvier, *Le Règne animal, distribué d'après son organisation...* (Paris: 1817), I, 111–12; also, (2nd ed.; Paris: 1829–1830), I, 88–89. In the note on p. 88 of the second edition Cuvier explains: "The idea that it [Wurmb's ape] could be an adult orang came to me on seeing the head of an ordinary orang with a muzzle much more projecting than those of the very young individuals which have been described heretofore. I reported it in a memoir read before the Academy of Science in 1818. M. Tilesius and M. Rudolphi appear to have had the same idea. See the Memoirs of the Academy of Berlin for 1824, p. 131." The matter was placed beyond doubt by Richard Owen's memoir "On the Osteology of the Chimpanzee and Orang Utan," *Trans. Zool. Soc. London*, I (1835), 343–80, and by Conrad J. Temminck's *Monographies de mammologie*, 2 Vols. (Paris: 1829–1841).

33. Thomas S. Savage and Jeffries Wyman, "Notice of the External Characters and Habits of Troglodytes Gorilla, a New Species of Orang from the Gaboon River," *Boston Jour. Nat. Hist.*, V (1845–1847), 417–41.

34. Lamarck, *Philosophie zoologique*, I, 353 ff. There is evidence, however, that the transmutation hypothesis lurked in the background of scientific discussion of the anthropoid apes in the years preceding Darwin's *Origin of Species*. See, for example, the article by Richard Owen cited above, and also his *On the Classification and Geographical Distribution of the Mammalia...To Which Is Added an Appendix "On the Gorilla" and "On the Extinction and Transmutation of Species"* (London: 1859).

Notes to Chapter 7

1. James Burnet (Lord Monboddo), *Of the Origin and Progress of Language,* (2nd ed.; Edinburgh: 1774–1792), I, 24.

2. Buffon, *A Natural History. General and Particular;* ...William Smellie, tr., III, 172–73.

3. *Ibid.,* VIII, 64–71. Buffon distinguished between the education of the individual and the education of the species. The latter type of education is peculiar to the human species, he declared.

4. *Ibid.,* IX, 381. The passage continues with a brief sketch of the progress of civilization. See also III, 304–5; IV, 185–86.

5. *Ibid.,* IX, 383.

6. Jean Jacques Rousseau, *Discours sur l'origine et les fondements de l'inégalité parmi les hommes,* in *Oeuvres complètes de J. J. Rousseau avec des notes historiques* (Paris: 1835), I, "Preface," 531 ff. Also the "Notes," p. 567 ff. Rousseau cites evidence from comparative anatomy to show that man was probably originally fruit-eating and a biped. On the question of Scripture, see p. 535. The general interpretation of the views of Rousseau and Lord Monboddo presented in this chapter is similar in most respects to that expounded by Arthur O. Lovejoy in his two articles: "The Supposed Primitivism of Rousseau's Discourse on Inequality," *Modern Philology,* XXI (1923), 165–86, and "Monboddo and Rousseau," *Modern Philology,* XXX (1933), 275–96, both reprinted in *Essays in the History of Ideas* (Baltimore: 1948), pp. 14–37, 38–61. See also Felix Gunther, *Die Wissenschaft vom Menschen. Ein Beitrag zum Deutschen Geistesleben im Zeitalter des Rationalismus mit besonderer Rücksicht auf die Entwickelung der Deutschen Geschichtsphilosophie im 18 Jahrhundert* (Gotha: 1907); Hoxie N. Fairchild, *The Noble Savage: A Study in Romantic Naturalism* (New York: 1928); Gladys Bryson, *Man and Society: The Scottish Inquiry of the Eighteenth Century* (Princeton: 1945).

7. *Ibid.,* p. 573, Note 10. Also p. 538 ff.

8. *Ibid.,* p. 541–42.

9. J. J. Rousseau, *Essai sur l'origine des langues...,* in *Oeuvres,* III, 495–522. This essay was not published during Rousseau's lifetime. It appears to have been composed *circa* 1750–1755. See pp. 508–10.

10. Rousseau, *Oeuvres Complètes,* I, 540.

11. As quoted in James Boswell, *The Life of Samuel Johnson, LL.D...,* J. W. Croker, ed. (Boston: 1832), I, 358. See also William Knight, *Lord Monboddo and Some of His Contemporaries* (London: 1900).

12. James Burnet (Lord Monboddo), *Of the Origin and Progress of Language* (2nd ed.; Edinburgh: 1774–1792), I, i-ii. For his opinion of travel books see his *Antient Metaphysics* (Edinburgh: 1779–1799), III, 266.

13. Monboddo, *Antient Metaphysics,* III, 26–27. On p. 8 of this volume Monboddo says that "the mind also is degenerated in these later times." He appears here to refer to brain power, that is, the physiological basis of intellectual effort. This might presumably degenerate at the same time that human culture developed and expanded from primitive beginnings.

14. On nature's frugality, see *Antient Metaphysics,* III, 71–73; *Origin and Progress,* I, 150. On Scripture, see *Origin and Progress,* I, 368, 380.

15. Monboddo, *Origin and Progress,* I, 175. On the Scale of Being, *ibid.,* I, 182–83.

16. *Ibid.,* I, 24. On pp. 366–67: "This progress in civil society, and the many changes and revolutions it is subject to, plainly shew, that it is not from nature, but of human institution. For *nature* is permanent and unchangeable, like its *author:* And, accordingly, the wild animals, who are undoubtedly in

a state of nature, still preserve the same oeconomy and manner of life with no variation, except such as change of circumstances may make absolutely necessary for the preservation of the individual or the species; and the variation goes no farther than that necessity requires." Thus, Monboddo did recognize that change of circumstances might require organic variation if organisms were to remain adapted to their environment, but he was opposed to the idea that material circumstances could determine organic form. See pp. 215–16 for an account of his discussion of Buffon's environmentalism.

17. Horace, *Satires*, I, 3, lines 99–106, as translated in E. C. Wickham, *Horace for English Readers* (Oxford: 1903), 170. Monboddo quotes the Latin in his *Origin and Progress*, I, 410 n., and adds: "This system, I believe will, upon the strictest examination, be found the true system of human nature; and a history of man would be nothing else than a commentary upon these few lines."

18. Monboddo, *Antient Metaphysics*, III, 69–70, 103. See also *Origin and Progress*, I, 438–41, vii ff.

19. Monboddo, *Origin and Progress*, I, Chap. IV: "Of the Orang Outang — The Account Buffon and Linnaeus give of him examined."

20. *Ibid.*, I, 340–41.

21. Jean-Marc-Gaspard Itard, *The Wild Boy of Aveyron (Rapports et mémoires sur le sauvage de l'Aveyron)*, George and Muriel Humphrey, trs. (New York and London: 1932), pp. 49–50. The great French physician Pinel pronounced the boy an incurable idiot. Itard was able to develop greatly the child's powers of attention, perception, comparison, and the like, but he never succeeded in making him into a normal human being. For Monboddo's account of his experience with feral children and orang-outangs, see *Antient Metaphysics*, IV, 25–33.

22. "As to the humanity of the Orang-outangs, and the story of the men with tails, I think neither the one nor the other is necessarily connected with my system; and if I am in error, I have only followed Linnaeus, and I think I have given a better reason than he has done for the Orang-outang belonging to us, I mean, his use of a stick. From which, and many other circumstances, it appears to me evident that he is much above the Simian race, to which I think you very rightly disclaim the relation of brother, though I think that race is of kin to us, though not so nearly related.

"For the large monkeys, or baboons, appear to me to stand in the same relation to us, that the ass does to the horse, or our gold-finch to the canary bird. For it is certain, as you observe, that the baboon has a desire for our female, and — if we can believe the Swedish traveller, Roeping [Koeping] — they copulate together." Letter from Lord Monboddo to Sir John Pringle, June 16, 1773, quoted by William Knight, *op. cit.*, pp. 84–85. See also *Origin and Progress*, I, 311–12, on monkeys and apes. Lovejoy, in his essay on "Monboddo and Rousseau," quotes parts of this letter to Pringle as showing that Monboddo "accepted in principle the general possibility of the transformation of species and...definitely asserted, as a probable hypothesis, the community of descent of most or all of the Anthropoidea." (Lovejoy, *Essays*, pp. 52–53). This seems a rather sweeping conclusion to draw from one or two passages of uncertain meaning. The probable meaning of the letter to Pringle is indicated in the last sentence, not quoted by Lovejoy. Monboddo seems to be suggesting that there may exist hybrid creatures formed by unions between baboons and members of the human species. Conjectures of this sort were common in Monboddo's day, but they seldom implied an evolutionary conception, since hybrid species were notoriously infertile. On the other hand, the whole drift of Monboddo's philosophy of nature, as set forth in his *Antient*

Metaphysics, ran counter to the kind of evolutionism which was springing up in the intellectual climate of the late eighteenth century.

23. Monboddo, *Antient Metaphysics,* III, 246.

24. Monboddo, *Origin and Progress,* I, 259–60. Monboddo himself was Linnaeus' questioner. See also page 267, n.; *Antient Metaphysics,* IV, 48.

25. *Ibid.,* I, 359, 444, 209 ff. Like many of his contemporaries, Monboddo thought that the study of language would unravel the history of man. In a letter to Sir William Jones, June 20, 1789, he reports that he has been studying Sanscrit and finds that it has so great an affinity to Greek that "either the Greek is a dialect of the Sanscrit, or they are both dialects of the same parent language." As quoted in Knight, *Lord Monboddo,* p. 269.

26. *Ibid.,* I, 360–61; 403–7. *Antient Metaphysics,* III, 226.

27. Cuvier, *Tableau,* p. 77 ff.

28. Cuvier, *Le Règne animal,* I, 94.

29. Auguste Comte, *Early Essays on Social Philosophy,* Henry D. Hutton, tr. (London: 1911), pp. 237–38. For Comte's discussion of Lamarck's views, see his *Cours de philosophie positive* (Paris: 1908), III, 296 ff.; IV, 201. See also Herbert Spencer, *Social Statics: Or, the Conditions Essential to Human Happiness Specified, and the First of Them Developed* (New York: 1882), 80; reprinted from the first edition, 1850. According to Spencer: "Progress...is not an accident, but a necessity. Instead of civilization being artificial, it is a part of nature; all of a piece with the development of the embryo or the unfolding of a flower. The modifications mankind have undergone, and are still undergoing, result from a law underlying the whole organic creation; and provided the human race continues, and the constitution of things remains the same, those modifications must end in completeness." I discuss more fully the interaction of social and biological evolutionism in the nineteenth century in "Biology and Social Theory in the Nineteenth Century: Auguste Comte and Herbert Spencer," *Critical Problems in the History of Science,* Marshall Clagett, ed. (Madison, Wisconsin: 1959), 419–46.

Notes to Chapter 8

1. Linnaeus was troubled to find a generic character by which to differentiate man from some of the anthropoid creatures described in travel books, but he never doubted that the American, the European, the Asiatic, and the African belonged to one species. For Buffon and Kant the question of the biological unity of mankind was settled by the capacity of all races to interbreed successfully. Although Blumenbach rejected this criterion, he was convinced that the varieties of man could be explained by "known causes of degeneration." Cuvier and Camper were monogenists too. Representative selections from many of the naturalists discussed in this chapter may be found in Earl W. Count, ed., *This Is Race* (New York: 1950). See also Earl W. Count, "The Evolution of the Race Idea in Modern Western Culture During the Period of the Pre-Darwinian Nineteenth Century," *Trans. New York Acad. of Scis.,* Ser. 2, VIII, 139–65; Wilhelm E. Mühlmann, *Geschichte der Anthropologie* (Bonn: 1948); Alfred C. Haddon, *History of Anthropology* (London: 1934); Thomas Bendyshe, "The History of Anthropology," *Memoirs Read Before the Anthropological Society of London,* I, (1863–1864), 335–458; Karl Walter Scheidt, "Beiträge zur Geschichte der Anthropologie. Der Begriff der Rasse in der Anthropologie und die Einteilung der Menschenrassen von Linné bis

Chapter 8. THE ORIGIN OF HUMAN RACES

Deniker," *Archiv für Rassen-und Gesellschaftsbiologie,* XV (1924), 280–306, 383–97; XVI (1925), 178–202, 382–403 (abridged translation in Count, *This Is Race,* pp. 354–91) ; D. J. Cunningham, "Anthropology in the 18th Century," *Jour. Roy. Anthrop. Instit. Gr. Brit. and Ireland,* XXXVIII (1908), 14–23.

2. Among the polygenists with scientific training were Georg Forster (1754–1794) and Christoph Meiners (1747–1810) ; for an account of their controversy with Kant, see Mühlmann, *Geschichte der Anthropologie,* pp. 56–66. See also the widely read polygenist treatise by the Manchester physician Charles White: *An Account of the Regular Gradation in Man, and in Different Animals and Vegetables...* (London: 1799). Lord Kames (Henry Home) was a Scottish jurist and literary critic, whose *Sketches of the History of Man* first appeared in 1774; it was reissued in a larger, revised edition in 1788. Although the racial interpretation of history was by no means confined to the polygenist camp, it is perhaps significant that the qualities of purity, permanence, and divine contrivance which later came to be associated with the idea of a "pure race" were qualities which the eighteenth century attributed to species rather than to varieties.

3. Samuel Stanhope Smith, *An Essay on the Causes of the Variety of Complexion and Figure in the Human Species...,* (2nd ed.; New Brunswick, N. J.: 1810), p. 240, n.

4. Georges Louis Leclerc, Comte de Buffon, *A Natural History. General and Particular;...*W. Kenrick and J. Murdoch, trs. (London, 1775–1776), I, 270.

5. Thomas Bendyshe, ed. and tr., *The Anthropological Treatises of Johann Friedrich Blumenbach...With Memoirs of Him by Marx and Flourens, and an Account of His Anthropological Museum by Professor R. Wagner...* (London: 1865), pp. 347 ff.

6. J. F. Blumenbach, *On the Natural Variety of Mankind,* (3rd ed.; Göttingen: 1795), as translated in Bendyshe, *Anthropological Treatises,* pp. 236, 269.

7. Buffon, *Natural History,* I, 291–92.

8. Blumenbach, *On the Natural Variety of Mankind,* in Bendyshe, *Anthropological Treatises,* p. 204. According to Blumenbach, a group of animals constitute a single species "if they agree so well in form and constitution, that those things in which they do differ may have arisen from degeneration." *(Ibid.,* p. 188). For example: "I see...that the molar teeth of the African elephant differ most wonderfully in their conformation from those of the Asiatic. I do not know whether these elephants...have ever copulated together [Buffon's criterion]; nor do I know any more how constant this conformation of the teeth may be in each [Linnaeus]. But since so far in all the specimens I have seen, I have observed the same difference; and since I have never known any example of molar teeth so changed by mere *degeneration,* I conjecture from analogy that those elephants are not to be considered as mere varieties, but must be held to be different species." *(Ibid.,* p. 190). For Blumenbach's account of the *nisus formativus,* see *Ibid.,* pp. 194–95, 200–201.

9. J. F. Blumenbach, *Contributions to Natural History,* Part I, (2nd ed.; Göttingen: 1806), as translated in Bendyshe, *Anthropological Treatises,* p. 292.

10. Buffon, *Natural History,* I, 274–75. See also *Oeuvres complètes de Buffon...,* P. Flourens, ed. (Paris: 1853–1855), II, 233–34, 277–78; IX, 614.

11. Buffon, *Oeuvres complètes,* IX, 578–79.

12. Pierre Louis Moreau de Maupertuis, *Vénus physique* (1745), in *Oeuvres de Maupertuis,* (2nd ed.; Lyons: 1768), II, 119–24. Maupertuis' discussion of sexdigitarianism may be found in his "Lettres," *Oeuvres,* II, Lettre XIV, 275 ff. See also Pierre Brunet, *Maupertuis: Étude biographique* (Paris: 1929), pp. 289 ff; Yves Delage, *L'Hérédité et les grands problèmes de la biologie,* (2nd

Chapter 8. THE ORIGIN OF HUMAN RACES

ed.; Paris: 1903), 581–84; Bentley Glass, "Maupertuis and the Beginnings of Genetics," *Quart. Rev. Biol.,* XXII (1947), 196–210.

13. *Ibid.,* II, 130. Maupertuis also suggests that these races might have become established "par la convenance des climats." (p. 129).

14. See above, Chap. IV, n. 31.

15. Immanuel Kant, *Bestimmung des Begriffs einer Menschenrasse,* in *Gesammelte Schriften* (Berlin: 1912), VIII, 89–107. See also Kant's essay "On the Different Races of Man," translated in Count, ed., *This Is Race,* pp. 16–24. Kant made a distinction between the description of nature *(Naturbeschreibung)* and natural history *(Naturgeschichte).* "Academic taxonomy," he wrote, "deals with classes; it merely arranges according to similarities; while a natural taxonomy arranges according to kinships determined by generation. The former supplies a school system for the sake of memorizing; the latter a natural system for the comprehension; the former has for its purpose only to bring creatures under a system of labeling; but the latter seeks to bring them under a system of laws." *(This Is Race,* p. 16). He developed this distinction further in his essay of 1785: "The wolf, the fox, the jackal, the hyena, and the house dog are so many kinds of four-footed beasts. If one assumes that each of them has had to have a separate ancestry, then they are that many species, but if one concedes that they could all have descended from one stem, then they are only races thereof. Species and genus are not distinguished in natural history [*Naturgeschichte*] (which has to do only with ancestry and origin). Only in the description of nature [*Naturbeschreibung*], since it is a matter of comparing distinguishing marks, does this distinction come into play. What is species here must there often be called only race." (*Gesammelte Schriften,* VIII, 100 n.)

16. Kant, *Gesammelte Schriften,* VIII, 102.

17. Kant, "On the Different Races of Men," in *This Is Race,* p. 19. Kant's argument makes it plain that he would also reject the idea of random variation and natural selection as the key to the origin of races.

18. Kant, *Gesammelte Schriften,* VIII, 103. Kant thought that his theory of race formation explained why the four skin colors and these alone were invariably transmitted from generation to generation. "What else can be the origin of this than that they must have lain in the seeds of the original stem, to us unknown, of the human race, and that, as natural dispositions necessary for the perpetuation of the race, at least in the first epoch of its expansion, they must unfailingly appear in succeeding generations?" *(Ibid.,* p. 98). Skin color he regarded as the "outward sign" of an internal organization necessary for survival under certain conditions. But he did not explain why skin color should be the *only* remaining vestige of these early adaptive responses of the human stock to particular environments or how the persistence of these traits after they had ceased to be adaptive could be reconciled with the "wise foresight of Nature."

19. Georges Cuvier, *Essay on the Theory of the Earth...,* (1st American ed.; New York: 1818), p. 160. See also Cuvier's *Tableau élémentaire de l'histoire naturelle des animaux* (Paris: 1797–1798), p. 71 ff., and his *Le Règne animal distribué d'après son organisation...* (Paris: 1817), I, 94 ff.

20. John Frere, "Account of Flint Weapons Discovered at Hoxne in Suffolk...," *Archaeologia: Or Miscellaneous Tracts Relating to Antiquity,* (1st ed.; London: 1800), XIII, 204–5. In this connection see Glyn E. Daniel, *A Hundred Years of Archaeology* (London: 1950), pp. 25–28.

21. Cuvier, *Essay on the Theory of the Earth,* pp. 146 ff. This essay was first published in 1812 as a preliminary discourse to Cuvier's *Ossemens fossiles.*

Chapter 8. THE ORIGIN OF HUMAN RACES

22. James Cowles Prichard, *Researches into the Physical History of Man* (London: 1813). The present account of Prichard's views is based entirely on this first edition of his *Researches*. For the later development of his thought see the subsequent editions.
23. *Ibid.*, pp. 5–6.
24. *Ibid.*, pp. 7–8.
25. *Ibid.*, pp. 9, 16–17.
26. *Ibid.*, pp. 18, 46–53, 61, 84.
27. *Ibid.*, pp. 194–95.
28. *Ibid.*, p. 233.
29. *Ibid.*, p. 235.
30. *Ibid.*, p. 41.
31. William C. Wells, "An Account of a Female of the White Race of Mankind, Part of Whose Skin Resembles That of a Negro; With Some Observations on the Causes of the Differences in Colour and Form Between the White and Negro Races of Men," appended to W. C. Wells, *Two Essays: One upon Single Vision With Two Eyes; the Other on Dew...* (London: 1818), pp. 431–32. This volume also contains Wells' "Memoir" of his own life. See also Richard H. Shryock, "The Strange Case of Wells' Theory of Natural Selection (1813): Some Comments on the Dissemination of Scientific Ideas," in *Studies and Essays in the History of Science and Learning Offered in Homage to George Sarton...* (New York: 1946), 197–207; Conway Zirkle, "Natural Selection Before the 'Origin of Species'," *Proc. Amer. Philos. Soc.*, LXXXIV (1941), 71–123.
32. *Ibid.*, pp. 435–6.
33. *Ibid.*, p. 436.

Notes to Chapter 9

1. Charles Lyell, *Principles of Geology, Being an Attempt to Explain the Former Changes of the Earth's Surface, by Reference to Causes Now in Operation* (London: 1830–1833), II, 20–21.
2. *Ibid.*, II, 24–25.
3. *Ibid.*, II, 41, 44.
4. *Ibid.*, II, 125.
5. *Ibid.*, II, 174–75.
6. Letter from Charles Lyell to Sir John Herschel, London, June 1, 1836, quoted in *Life, Letters and Journals of Sir Charles Lyell, Bart.*, Katharine M. Lyell, ed. (London: 1881), I, 467–68. How near, yet how far Lyell was from Darwin's theory of natural selection is shown by the following passage in his letter to Herschel: "When I first came to the notion...of a succession of extinction of species, and creation of new ones, going on perpetually now, and throughout an indefinite period of the past, and to continue for ages to come, all in accommodation to the changes which must continue in the inanimate and habitable earth, the idea struck me as the grandest which I had ever conceived, so far as regards the attributes of the *Presiding Mind*. For one can in imagination summon before us a small past [part?] at least of the circumstances that must be contemplated and foreknown, before it can be decided what powers and qualities a new species must have in order to enable it to

endure for a given time, and to play its part in due relation to all other beings destined to coexist with it, before it dies out. . . .

"It may be seen that unless some slight additional precaution be taken, the species about to be born would at a certain era be reduced to too low a number. There may be a thousand modes of insuring its duration beyond that time; one, for example may be the rendering it more prolific, but this would perhaps make it press too hard upon other species at other times. Now if it be an insect it may be made in one of its transformations to resemble a dead stick, or a leaf, or a lichen, or a stone, so as to be somewhat less easily found by its enemies; or if this would make it too strong, an occasional variety of the species may have this advantage conferred upon it; or if this would still be too much, one sex of a certain variety. Probably there is scarcely a dash of colour on the wing or body of which the choice would be quite arbitrary, or which might not affect its duration for thousands of years. I have been told that the leaf-like expansions of the abdomen and thighs of a certain Brazilian Mantis turn from green to yellow as autumn advances, together with the leaves of the plants among which it seeks for its prey. Now if species come in in succession, such contrivances must sometimes be made, and such relations predetermined between species, as the Mantis, for example, and plants not then existing, but which it was foreseen would exist together with some particular climate at a given time. But I cannot do justice to this train of speculation in a letter, and will only say that it seems to me to offer a more beautiful subject for reasoning and reflecting on, than the notion of great batches of new species all coming in, and afterwards going out at once" (pp. 468–69). Thus, Lyell saw clearly that slight variations could give an organism a decisive advantage in the struggle for existence, but instead of conceiving these variations as arising "by chance" (i. e., without reference to their utility or disutility to the organism concerned), he thought of them as planned and predetermined in advance by the "Presiding Mind" in order that the organism might perpetuate its kind for a given period of time. This notion of divine direction of organic variation carried over into Lyell's criticism of Darwin's *Origin of Species*, as will be shown. Lyell's letters also make it clear that Sir John Herschel in his thinking about the "intermediate causes" which might give rise to new species was as far from the Darwinian conception as Lyell.

7. Darwin to W. D. Fox, Lima, July, 1835, quoted in *The Life and Letters of Charles Darwin Including an Autobiographical Chapter,* Francis Darwin, ed. (New York: 1898) , I, 234.

8. Letter from Darwin to Alfred Russel Wallace, Down, April 6, 1859, quoted in *More Letters of Charles Darwin. A Record of His Work in a Series of Hitherto Unpublished Letters,* Francis Darwin, ed. (London: 1903) , I, 118. See also *The Autobiography of Charles Darwin,* Nora Barlow, ed. (New York: 1959) , p. 118.

9. Quoted from Darwin's notebook of 1837 in *Life and Letters,* I, 370. Gertrude Himmelfarb examines the evidence bearing on the genesis of Darwin's theory of natural selection in her book *Darwin and the Darwinian Revolution* (New York: 1959) , Chap. 7. She concludes that Darwin was converted to evolutionism soon after his return to England but that he did not hit upon the idea of natural selection until he read Malthus. It seems strange, however, that Darwin should have called the general evolutionary hypothesis "my theory" in his notebook of 1837. This expression suggests that he already had the idea of natural selection in 1837.

10. See in this connection Loren C. Eiseley, "Charles Darwin, Edward Blyth, and the Theory of Natural Selection," *Proc. Amer. Philos. Soc.* CIII (February,

1959), pp. 94–158. Professor Eiseley presents a detailed argument designed to show that Darwin probably derived the idea of natural selection from two articles written by his acquaintance Edward Blyth and published in *The Magazine of Natural History* in 1835 and 1837. If these articles were in fact the source of Darwin's theory, Darwin was guilty of grave intellectual dishonesty. In the present writer's opinion, Professor Eiseley fails to establish his case beyond reasonable doubt, although the evidence he presents is sufficiently disturbing to merit further investigation aimed at establishing or disproving his thesis. The present chapter proceeds on the assumption that Darwin's statements about the genesis of his ideas, although they may contain errors due to faulty memory, were not motivated by a desire to conceal the truth. See also Loren C. Eiseley, "Charles Lyell," *Sci. Amer.*, CCI (1959), 98–106, No. 2.

11. Darwin to L. Jenyns (Rev. L. Blomefield), Down, 1845 (?), *Life and Letters*, I, 394–95.

12. *The Foundations of the Origin of Species; Two Essays Written in 1842 and 1844, by Charles Darwin*, Francis Darwin, ed. (Cambridge, England: 1909), pp. 57–58. The page numbers of subsequent quotations from the essay of 1844 are indicated in parentheses in the text.

13. Darwin to Joseph Dalton Hooker, Down, 1844, *Life and Letters*, I, 390. An excellent collection of essays on Darwin's precursors may be found in *Forerunners of Darwin: 1745–1859*, Bentley Glass, Owsei Temkin, and William L. Straus, Jr., eds. (Baltimore: 1959). See also Milton Millhauser, *Just Before Darwin: Robert Chambers and the* Vestiges (Middletown, Conn.: 1959).

14. Darwin to Charles Lyell, Down, March 12, 1863, *Life and Letters*, II, 198–99.

15. Thomas Huxley to George Romanes, London, May 9, 1882, quoted in *Life and Letters of Thomas Henry Huxley*, Leonard Huxley, ed. (London: 1913), II, 317.

16. Letter from Charles Darwin to the *Athenaeum*, Down, May 5, 1863, *Life and Letters*, II, 207. Printed in the *Athenaeum* May 9, 1863.

17. Darwin to J. D. Hooker, Down, September 25, 1853, *Life and Letters*, I, 400.

18. Darwin to Charles Lyell, Down, September 30, 1859, *Life and Letters*, I, 524–25.

19. Darwin to J. D. Hooker, Down, 1855, *Life and Letters*, I, 415–16.

20. Biographical notes written by Joseph Dalton Hooker, quoted in *Life and Letters*, I, 387–88.

21. Darwin to Asa Gray, Down, July 20, 1856, *Life and Letters*, I, 437.

22. Darwin to Asa Gray, Down, September 5, 1857, quoted in *Darwin, Wallace and the Theory of Natural Selection Including the Linnean Society Papers*, Bert J. Loewenberg, ed. (New Haven: 1957), pp. 60–61. Darwin's relation to Wallace is discussed in Professor Loewenberg's introduction to the Darwin-Wallace papers.

23. Joseph Dalton Hooker to Charles Darwin, 1859 or 1860, quoted in *More Letters of Charles Darwin*, I, 135. For an account of the public reaction to Darwin's writings see Alvar Ellegard, *Darwin and the General Reader: The Reception of Darwin's Theory of Evolution in the British Periodical Press, 1859–1872* (Gothenburg Studies in English, VIII; Frank Behre, ed., Göteborg: 1958).

24. Darwin to the Editor of *Nature*, Down, November 5, 1880, *More Letters*, I, 389.

25. Darwin to J. D. Hooker, Down, March 18, 1862, *More Letters*, I, 198.

Chapter 9. TRIUMPH OF CHANCE AND CHANGE

26. Darwin to George Bentham, Down, May 22, 1863, *Life and Letters*, II, 210.
27. Charles Darwin, *The Origin of Species by Charles Darwin: A Variorum Text*, Morse Peckham, ed. (Philadelphia: 1959), pp. 168–69.
28. Darwin to J. D. Hooker, Down, 1854, *More Letters*, I, 76.
29. Darwin to J. D. Hooker, Down, December 30, 1858, *More Letters*, I, 114–15.
30. Darwin to Charles Lyell, Ilkley, Yorkshire, October 25, 1859, *Life and Letters*, I, 531.
31. Alfred Russel Wallace to Charles Darwin, Hurstpierpoint, Sussex, July 2, 1866, *More Letters*, I, 267–68.
32. Sir John F. W. Herschel, *Physical Geography of the Globe* (Edinburgh: 1861), p. 12, n., quoted in *More Letters*, I, 191, n. The discussion of religious and philosophical issues in this chapter and the next draws to a considerable extent on a paper entitled "Darwin and Religion" which I read before the American Philosophical Society, April 25, 1959. See the *Proceedings* of the Society, CIII (October, 1959) pp. 716 ff. See also the discussion in Himmelfarb, *Darwin and the Darwinian Revolution*, Chaps. 16–20.
33. Darwin to Charles Lyell, Torquay, August 21, 1861, *More Letters*, I, 194.
34. Darwin to J. D. Hooker, Down, February 8, 1867, *Life and Letters*, II, 245.
35. Darwin to Asa Gray, Down, May 22, 1860, *Life and Letters*, II, 105.
36. Thomas Huxley, "On the Reception of the 'Origin of Species,'" *Life and Letters*, I, 554–55.
37. Charles Peirce, MSS., IB 3a "Folder of Late Fragments" ("Why should the Doctrine of Chances raise Science to a Higher Plane?") (Jan. 25, 1909), p. 15, quoted in Philip P. Wiener, *Evolution and the Founders of Pragmatism* (Cambridge, Mass.: 1949), p. 81. This is an excellent study of the reception of Darwin's writings by the American pragmatists. According to Wiener (p. 83), "Peirce in his statistical conception of law went much farther than Darwin in biology, Maxwell in physics, or Quetelet in sociology. The latter three scientists, in keeping with an established philosophy of mechanical determinism, regarded chance as no explanation at all but a makeshift concept to patch up our ignorance of more fundamental mechanical or dynamical laws governing every individual event, thing, or character with strict necessity. It is in its moral opposition to this mechanistic or necessitarian assumption in both scientists and philosophers from Ockham's time to the 19th century (Hegel and Spencer) that Peirce's philosophy of evolution may be historically understood." [Above quotation and quotations pp. 305–6, reprinted by permission of the publishers, Harvard University Press.] See also *Evolutionary Thought in America*, Stow Persons, ed. (New Haven: 1950).
38. Charles Peirce, *Collected Papers of Charles Sanders Peirce*, Charles Hartshorne and Paul Weiss, eds. (Cambridge, Mass.: 1931–1935), VI, paragraph 33, quoted in Wiener, *Evolution and the Founders of Pragmatism*, p. 84.
39. Alfred North Whitehead, *Science and the Modern World* (New York: 1947), p. 149. See also the same author's *Nature and Life* (Chicago: 1934).

Notes to Chapter 10

1. Darwin to Alfred Russel Wallace, Down, December 22, 1857, *The Life and Letters of Charles Darwin*, I, 467.
2. Herbert Spencer, *The Principles of Psychology* (London: 1855), p. 529.
3. Thomas Huxley, *Man's Place in Nature and Other Anthropological Essays* (New York: 1898), p. 151. The first three essays were originally published in 1863.
4. *Ibid.*, p. 208.

376

Chapter 10. DARWIN AND ADAM

5. Sir Charles Lyell, *The Geological Evidences of the Antiquity of Man with Remarks on Theories of the Origin of Species by Variation* (London: 1863), p. 412.
6. *Ibid.*, p. 469.
7. John Bird Sumner, *A Treatise on the Records of the Creation* ... (London: 1816), quoted in Lyell, *Geological Evidences of the Antiquity of Man*, p. 497.
8. Alfred Russel Wallace, "The Origin of Human Races and the Antiquity of Man Deduced from the Theory of 'Natural Selection,'" *Journal of the Anthropological Society of London* II (1864), clviii.
9. *Ibid.*, pp. clxv–clxvi.
10. *Ibid.*, pp. clxiv–clxv.
11. *Ibid.*, p. clxviii.
12. *Ibid.*, pp. clxix–clxx.
13. Herbert Spencer, *Social Statics: Or, the Conditions Essential to Human Happiness Specified, and the First of Them Developed* (London: 1850). Wallace acknowledges his debt to Spencer in a note on the last page of his article on the origin of human races. Wallace developed his ideas further in a review of the 10th edition of Lyell's *Principles of Geology*, published in *The Quarterly Review* CXXVI (1869), 359–94.
14. Darwin to Alphonse de Candolle, Down, July 6, 1868, *Life and Letters*, II, 280.
15. Charles Darwin, *The Descent of Man, and Selection in Relation to Sex*, 2nd ed. (New York: 1886), pp. 160–61.
16. *Ibid.*, pp. 46–47.
17. *Ibid.*, pp. 594–95.
18. *Ibid.*, pp. 563–64.
19. *Ibid.*, p. 595.
20. *Ibid.*, p. 145.
21. Herbert Spencer, "Progress: Its Law and Cause," in *Illustrations of Universal Progress; A Series of Discussions* (New York: 1878), p. 58. This essay was originally published in 1857.
22. Darwin to Charles Lyell, Ilkley, Yorkshire, October 11, 1859, *Life and Letters*, II, 7. See also *More Letters*, II, 30.
23. Darwin, *Descent of Man*, p. 129.
24. *Ibid.*, pp. 99–100.
25. *Ibid.*, p. 132.
26. *Charles Darwin's Diary of the Voyage of H. M. S. Beagle*, Nora Barlow, ed. (Cambridge, England: 1933), entry dated New South Wales, January 12, 1836, p. 378.
27. Darwin, *Descent of Man*, pp. 191–92.
28. *Ibid.*, pp. 133–34.
29. *Ibid.*, p. 618.
30. *Ibid.*, p. 125.
31. Thomas Huxley, *Evolution and Ethics and Other Essays* (New York: 1896), p. 83.
32. Thomas Huxley, *Science and Hebrew Tradition: Essays* (New York: 1910), pp. 160–61, 362.
33. Darwin to William Graham, Down, July 3, 1881, *Life and Letters*, I, 285.
34. Darwin to J. D. Hooker, Down, February 9, 1865, *More Letters*, I, 260–61.
35. Theodosius Dobzhansky, *The Biological Basis of Human Freedom* (New York: 1956), pp. 87–88. See also Sir Julian Huxley, *Evolution in Action* (New York: 1957), p. 116.

ACKNOWLEDGMENTS

The author wishes to thank the publishers below for permission to quote from the following books: pp. 262-84 *passim:* Cambridge University Press, Cambridge (*The Foundations of the Origin of Species; Two Essays Written in 1842 and 1844, by Charles Darwin,* Francis Darwin, ed.) ; p. 31: Cambridge University Press, Cambridge, Copyright 1933, (*The Herschel Chronicle: The Life Story of William and His Sister Caroline Herschel,* Constance A. Lubbock, ed.) ; p. 287: St. Martin's Press, New York, Copyright 1913, (*Life and Letters of Thomas Henry Huxley,* Leonard Huxley, ed.) ; p. 305: Harvard University Press, Cambridge, Mass., Copyright 1949, by The President and Fellows of Harvard College (*Evolution and the Founders of Pragmatism,* Philip P. Wiener) ; p. 338: Columbia University Press, New York, Copyright 1956 (*The Biological Basis of Human Freedom,* Theodosius Dobzhansky; pp. 234, 372n15: Abelard-Schuman, Limited, New York, Copyright 1950 (*This is Race,* Earl W. Count, ed.) .

INDEX

385

Transitional forms (*Continued*)
  between man and apes, 144, 179, 181, 192, 311–13, 315, 323
Tulp, Nicolaas, 179, 181, 184, 186, 190, 192, 199
Turner, George, 105, 351n*21*
Tuscany, geological history of, 49–51
Tyson, Edward
  Anatomy of a Pygmy, 176 ff.
  concept of nature and natural history, 177
  on scale of nature, 177
  on structure and function, 177–78

UNIFORMITARIANISM. *See* Geology.
Uranus, discovery of, 31

VARIATION
  causes of, 45, 99, 119, 130, 133, 148–50, 216, 224, 241, 245, 262–63, 275–76, 293, 296, 324, 356n*33*, 361n*82*, 374n*6* (*See also* Environment)
  use and disuse, principle of, 159, 162, 204, 251, 252, 262, 269, 280, 295, 324, 326, 360n*73*
  and definition of species, 129–30, 226, 239, 263 ff., 288, 293, 311, 355, 371n*8*
  descent, with modification, 45, 130, 146–47, 148–51, 158–60, 221–47, 271 ff., 278, 282–84, 287, 321–22, 327, 369n*22*
  under domestication, 130, 133, 147, 148, 150, 210, 228, 241–42, 246, 253, 257–58, 261 ff., 267, 285, 286, 323, 358n*48*
  limits of, 119, 150 ff., 170, 172, 216, 221–22, 228–29, 240, 252, 270, 356–57n*33*, 364n*106*, 369n*16*
  random, and natural selection, 235, 244 ff., 264, 266, 296, 302, 323–24, 372n*17*, 374n*6*
Volcanoes. *See* Geology.

Voltaire, 201, 221
Vosmaer, Arnout, 187–88, 196, 366n*19*
Vulcanists. *See* Geology.

WALLACE, Alfred R.
  antiquity of man, 316, 318
  limits of natural selection, 320
  monogenist-polygenist controversy, 316, 318
  objects to term "natural selection," 300, 375
  origin of human races, 315–19
  perfectibility of man, 319
  and Spencer, 319–20, 377n*13*
Wells, William
  natural selection, 245–46
  origin of Negro race, 245
Werner, Abraham G.
  attack on Vulcanism, 70
  classification of rocks, 70–71
  theory of the earth, 70–72, 347n*36*
Whiston, William
  Buffon's reaction to, 24–25
  interpretation of Scripture, 18–19
  and Newton, 342n2, 343n*10*
  scholarly career, 342n2
  theory of the earth, 17 ff.
Whitehead, Alfred N., 306–7, 356n*25*
Wild men. *See* Man, "wild."
Wistar, Caspar, 103–4, 109, 111, 350–51n*20*, 352n*32*
Woman, 325–26. *See also* Selection, sexual.
Woodward, John, 52–54, 55, 59, 91
Wright, Thomas (of Durham)
  career of, 343n*14*
  on dissolution of worlds, 343–44n*16*
  and J. H. Lambert, 343n*16*
  posthumous reputation of, 343n*14*
  theory of the universe, 26, 27–28
Wurmb, F., Baron von, 194, 195, 196
Wyman, Jeffries, 367n*33*